Inorganic Structural Chemistry

Second Edition

Inorganic Chemistry

A Wiley Series of Advanced Textbooks

Previously Published Books In this Series

Lanthanide and Actinide Chemistry
Author: S. A. Cotton

Mass Spectrometry of Inorganic and Organometallic Compounds: Tools–Techniques–Tips
Authors: W. Henderson & J. J. McIndoe

Main Group Chemistry, Second Edition
Author: A. G. Massey

Synthesis of Organometallic Compounds: A Practical Guide
Edited by: S. Komiya

Chemical Bonds: A Dialog
Author: J. K. Burdett

Molecular Chemistry of the Transition Elements: An Introductory Course
Authors: F. Mathey & A. Sevin

Stereochemistry of Coordination Compounds
Author: A. Von Zelewsky

Bioinorganic Chemistry: Inorganic Elements in the Chemistry of Life – An Introduction and Guide
Author: W. Kaim

Inorganic Structural Chemistry

Second Edition

Ulrich Müller
Philipps-Universität Marburg, Germany

BICENTENNIAL 1807 WILEY 2007 BICENTENNIAL

John Wiley & Sons, Ltd

Telephone (+44) 1243 779777

Email (for orders and customer service enquiries): cs-books@wiley.co.uk
Visit our Home Page on www.wiley.com

German version
Anorganische Strukturchemie
First edition 1991
Second edition 1992
Third edition 1996
Fourth edition 2004
Fifth edition 2006

The author was awarded the Prize for Chemical Literature for this book by the *Verband der Chemischen Industrie* (German Federation of Chemical Industries) in 1992

Other Wiley Editorial Offices

John Wiley & Sons Inc., 111 River Street, Hoboken, NJ 07030, USA

Jossey-Bass, 989 Market Street, San Francisco, CA 94103-1741, USA

Wiley-VCH Verlag GmbH, Boschstr. 12, D-69469 Weinheim, Germany

John Wiley & Sons Australia Ltd, 42 McDougall Street, Milton, Queensland 4064, Australia

John Wiley & Sons (Asia) Pte Ltd, 2 Clementi Loop #02-01, Jin Xing Distripark, Singapore 129809

John Wiley & Sons Canada Ltd, 6045 Freemont Blvd, Mississauga, ONT, L5R 4J3, Canada

Wiley also publishes its books in a variety of electronic formats. Some content that appears
in print may not be available in electronic books.

British Library Cataloguing in Publication Data

A catalogue record for this book is available from the British Library

ISBN-13: 978-0-470-01864-4 (HBK) 978-0-470-01865-1 (PBK)
ISBN-10: 0-470-01864-X (HBK) 0-470-018650-8 (PBK)

Produced from LaTeX files supplied by the author
Many figures were drawn with the programs ATOMS by E. Dowty and DIAMOND by K. Brandenburg
Printed and bound in Great Britain by Antony Rowe Ltd, Chippenham, England
This book is printed on acid-free paper responsibly manufactured from sustainable forestry
in which at least two trees are planted for each one used for paper production.

Ulrich Müller

Born in 1940 in Bogotá, Colombia. School attendance in Bogotá, then in Elizabeth, New Jersey, and finally in Fellbach, Germany. Studied chemistry at the Technische Hochschule in Stuttgart, Germany, obtaining the degree of Diplom-Chemiker in 1963. Work on the doctoral thesis in inorganic chemistry was performed in Stuttgart and at Purdue University in West Lafayette, Indiana, in the research groups of K. Dehnicke and K. S. Vorres, respectively. The doctor's degree in natural sciences (Dr. rer. nat.) was awarded by the Technische Hochschule Stuttgart in 1966. Subsequent post-doctoral work in crystallography and crystal structure determinations was performed in the research group of H. Bärnighausen at the Universität Karlsruhe, Germany. Appointed in 1972 as professor of inorganic chemistry at the Philipps-Universität Marburg, Germany, then from 1992 to 1999 at the Universität Kassel, Germany, and since 1999 again in Marburg. Helped installing a graduate school of chemistry as visiting professor at the Universidad de Costa Rica from 1975 to 1977. Courses in spectroscopic methods were repeatedly given at different universities in Costa Rica, Brazil and Chile. Main areas of scientific interest: synthetic inorganic chemistry, crystallography and crystal structure systematics, crystallographic group theory. Co-author of *Chemie*, a textbook for beginners, *Schwingungsspektroskopie*, a textbook about the application of vibrational spectroscopy, and of *Schwingungsfrequenzen I* and *II* (tables of characteristic molecular vibrational frequencies); co-author and co-editor of *International Tables for Crystallography*, Vol. *A*1.

Contents

Preface

Given the increasing quantity of knowledge in all areas of science, the imparting of this knowledge must necessarily concentrate on general principles and laws while details must be restricted to important examples. A textbook should be reasonably small, but essential aspects of the subject may not be neglected, traditional foundations must be considered, and modern developments should be included. This introductory text is an attempt to present inorganic structural chemistry in this way. Compromises cannot be avoided; some sections may be shorter, while others may be longer than some experts in this area may deem appropriate.

Chemists predominantly think in illustrative models: they like to "see" structures and bonds. Modern bond theory has won its place in chemistry, and is given proper attention in Chapter 10. However, with its extensive calculations it corresponds more to the way of thinking of physicists. Furthermore, albeit the computational results have become quite reliable, it often remains difficult to understand structural details. For everyday use, simple models such as those treated in Chapters 8, 9 and 13 are usually more useful to a chemist: "The peasant who wants to harvest in his lifetime cannot wait for the *ab initio* theory of weather. Chemists, like peasants, believe in rules, but cunningly manage to interpret them as occasion demands" (H.G. VON SCHNERING [112]).

This book is mainly addressed to advanced students of chemistry. Basic chemical knowledge concerning atomic structure, chemical bond theory and structural aspects is required. Parts of the text are based on a course on inorganic crystal chemistry by Prof. H. Bärnighausen at the University of Karlsruhe. I am grateful to him for permission to use the manuscript of his course, for numerous suggestions, and for his encouragement. For discussions and suggestions I also thank Prof. D. Babel, Prof. K. Dehnicke, Prof. C. Elschenbroich, Prof. D. Reinen and Prof. G. Weiser. I thank Prof. T. Fässler for supplying figures of the electron localization function and for reviewing the corresponding section. I thank Prof. S. Schlecht for providing figures and for reviewing the chapter on nanostructures. I thank Ms. J. Gregory and Mr. P. C. Weston for reviewing and correcting the English version of the manuscript.

In this second edition the text has been revised and new scientific findings have been taken into consideration. For example, many recently discovered modifications of the elements have been included, most of which occur at high pressures. The treatment of symmetry has been shifted to the third chapter and the aspect of symmetry is given more attention in the following chapters. New sections deal with quasicrystals and other not strictly crystalline solids, with phase transitions and with the electron localization function. There is a new chapter on nanostructures. Nearly all figures have been redrawn.

Ulrich Müller — Marburg, Germany, April 2006

1 Introduction

Structural chemistry or *stereochemistry* is the science of the structures of chemical compounds, the latter term being used mainly when the structures of molecules are concerned. Structural chemistry deals with the elucidation and description of the spatial order of atoms in a compound, with the explanation of the reasons that lead to this order, and with the properties resulting therefrom. It also includes the systematic ordering of the recognized structure types and the disclosure of relationships among them.

Structural chemistry is an essential part of modern chemistry in theory and practice. To understand the processes taking place during a chemical reaction and to render it possible to design experiments for the synthesis of new compounds, a knowledge of the structures of the compounds involved is essential. Chemical and physical properties of a substance can only be understood when its structure is known. The enormous influence that the structure of a material has on its properties can be seen by the comparison of graphite and diamond: both consist only of carbon, and yet they differ widely in their physical and chemical properties.

The most important experimental task in structural chemistry is the *structure determination.* It is mainly performed by X-ray diffraction from single crystals; further methods include X-ray diffraction from crystalline powders and neutron diffraction from single crystals and powders. Structure determination is the analytical aspect of structural chemistry; the usual result is a static model. The elucidation of the spatial rearrangements of atoms during a chemical reaction is much less accessible experimentally. *Reaction mechanisms* deal with this aspect of structural chemistry in the chemistry of molecules. *Topotaxy* is concerned with chemical processes in solids, in which structural relations exist between the orientation of educts and products. Neither dynamic aspects of this kind are subjects of this book, nor the experimental methods for the preparation of solids, to grow crystals or to determine structures.

Crystals are distinguished by the regular, periodic order of their components. In the following we will focus much attention on this order. However, this should not lead to the impression of a perfect order. Real crystals contain numerous faults, their number increasing with temperature. Atoms can be missing or misplaced, and dislocations and other imperfections can occur. These faults can have an enormous influence on the properties of a material.

Inorganic Structural Chemistry, Second Edition Ulrich Müller

2 Description of Chemical Structures

In order to specify the structure of a chemical compound, we have to describe the spatial distribution of the atoms in an adequate manner. This can be done with the aid of chemical nomenclature, which is well developed, at least for small molecules. However, for solid-state structures, there exists no systematic nomenclature which allows us to specify structural facts. One manages with the specification of *structure types* in the following manner: 'magnesium fluoride crystallizes in the rutile type', which expresses for MgF_2 a distribution of Mg and F atoms corresponding to that of Ti and O atoms in rutile. Every structure type is designated by an arbitrarily chosen representative. How structural information can be expressed in formulas is treated in Section 2.1.

Graphic representations are useful. One of these is the much used valence-bond formula, which allows a succinct representation of essential structural aspects of a molecule. More exact and more illustrative are perspective, true-to-scale figures, in which the atoms are drawn as balls or — if the always present thermal vibrations are to be expressed — as ellipsoids. To achieve a better view, the balls or ellipsoids are plotted on a smaller scale than that corresponding to the effective atomic sizes. Covalent bonds are represented as sticks. The size of a thermal ellipsoid is chosen to represent the probability of finding the atom averaged over time (usually 50 % probability of finding the center of the atom within the ellipsoid; *cf.* Fig. 2.1 **b**). For more complicated structures the perspective image can be made clearer with the aid of a stereoscopic view (*cf.* Fig. 7.5, p. 56). Different types of drawings can be used to stress different aspects of a structure (Fig. 2.1).

Quantitative specifications are made with numeric values for interatomic distances and angles. The interatomic distance is defined as the distance between the nuclei of two atoms

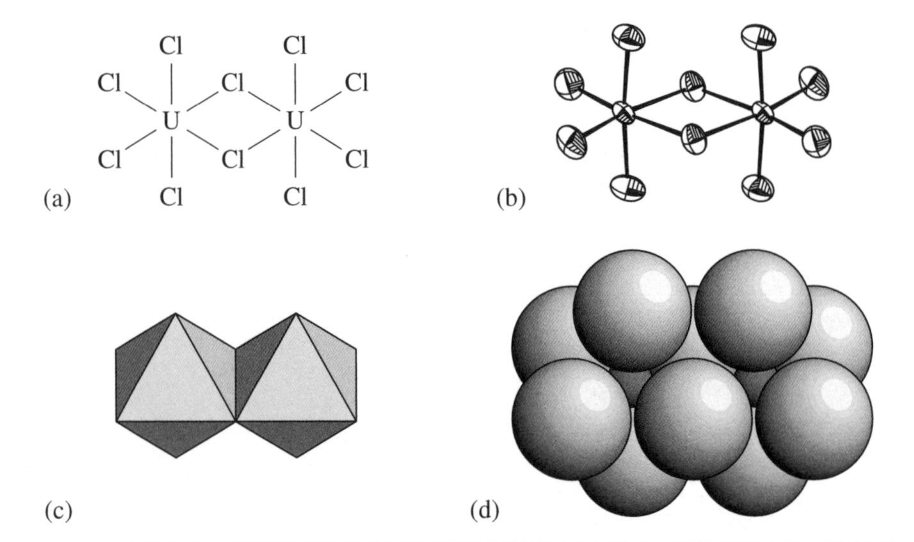

Fig. 2.1 Graphic representations for a molecule of $(UCl_5)_2$, all drawn to the same scale. (**a**) Valence-bond formula. (**b**) Perspective view with ellipsoids of thermal motion. (**c**) Coordination polyhedra. (**d**) Emphasis of the space requirements of the chlorine atoms

Inorganic Structural Chemistry, Second Edition Ulrich Müller

in their mean positions (mean positions of the thermal vibration). The most common method to determine interatomic distances experimentally is X-ray diffraction from single crystals. Other methods include neutron diffraction from crystals and, for small molecules, electron diffraction and microwave spectroscopy with gaseous samples. X-ray diffraction determines not the positions of the atomic nuclei but the positions of the centers of the negative charges of the atomic electron shells, because X-rays are diffracted by the electrons of the atoms. However, the negative charge centers coincide almost exactly with the positions of the atomic nuclei, except for covalently bonded hydrogen atoms. To locate hydrogen atoms exactly, neutron diffraction is also more appropriate than X-ray diffraction for another reason: X-rays are diffracted by the large number of electrons of heavy atoms to a much larger extent, so that the position of H atoms in the presence of heavy atoms can be determined only with low reliability. This is not the case for neutrons, as they interact with the atomic nuclei. (Because neutrons suffer incoherent scattering from H atom nuclei to a larger extent than from D atom nuclei, neutron scattering is performed with deuterated compounds.)

2.1 Coordination Numbers and Coordination Polyhedra

The coordination number (c.n.) and the coordination polyhedron serve to characterize the immediate surroundings of an atom. The *coordination number* specifies the number of coordinated atoms; these are the closest neighboring atoms. For many compounds there are no difficulties in stating the coordination numbers for all atoms. However, it is not always clear up to what limit a neighboring atom is to be counted as a closest neighbor. For instance, in metallic antimony every Sb atom has three neighboring atoms at distances of 291 pm and three others at distances of 336 pm, which is only 15 % more. In this case it helps to specify the coordination number by 3+3, the first number referring to the number of neighboring atoms at the shorter distance.

Stating the coordination of an atom as a single number is not very informative in more complicated cases. However, specifications of the following kind can be made: in white tin an atom has four neighboring atoms at a distance of 302 pm, two at 318 pm and four at 377 pm. Several propositions have been made to calculate a mean or 'effective' coordination number (e.c.n. or ECoN) by adding all surrounding atoms with a weighting scheme, in that the atoms are not counted as full atoms, but as fractional atoms with a number between 0 and 1; this number is closer to zero when the atom is further away. Frequently a gap can be found in the distribution of the interatomic distances of the neighboring atoms: if the shortest distance to a neighboring atom is set equal to 1, then often further atoms are found at distances between 1 and 1.3, and after them follows a gap in which no atoms are found. According to a proposition of G. BRUNNER and D. SCHWARZENBACH an atom at the distance of 1 obtains the weight 1, the first atom beyond the gap obtains zero weight, and all intermediate atoms are included with weights that are calculated from their distances by linear interpolation:

$$\text{e.c.n.} = \Sigma_i (d_g - d_i)/(d_g - d_1)$$

d_1 = distance to the closest atom
d_g = distance to the first atom beyond the gap
d_i = distance to the i-th atom in the region between d_1 and d_g

For example for antimony: taking $3 \times d_1 = 291$, $3 \times d_i = 336$ and $d_g = 391$ pm one obtains e.c.n. = 4.65. The method is however of no help when no clear gap can be discerned.

A mathematically unique method of calculation considers the *domain of influence* (also called *Wirkungsbereich*, VORONOI polyhedron, WIGNER-SEITZ cell, or DIRICHLET domain). The domain is constructed by connecting the atom in question with all surrounding atoms; the set of planes perpendicular to the connecting lines and passing through their midpoints forms the domain of influence, which is a convex polyhedron. In this way, a polyhedron face can be assigned to every neighboring atom, the area of the face serving as measure for the weighting. A value of 1 is assigned to the largest face. Other formulas have also been derived, for example,

$$\text{ECoN} = \sum_i \exp[1 - (d_i/d_1)^n]$$

$n = 5$ or 6
d_i = distance to the i-th atom
d_1 = shortest distance or d_1 = assumed standard distance

With this formula we obtain ECoN = 6.5 for white tin and ECoN = 4.7 for antimony.

The kind of bond between neighboring atoms also has to be considered. For instance, the coordination number for a chlorine atom in the CCl_4 molecule is 1 when only the covalently bonded C atom is counted, but it is 4 (1 C + 3 Cl) when all atoms 'in contact' are counted. In the case of molecules one will tend to count only covalently bonded atoms as coordinated atoms. In the case of crystals consisting of monoatomic ions usually only the anions immediately adjacent to a cation and the cations immediately adjacent to an anion are considered, even when there are contacts between anions and anions or between cations and cations. In this way, an I^- ion in LiI (NaCl type) is assigned the coordination number 6, whereas it is 18 when the 12 I^- ions with which it is also in contact are included. In case of doubt, one should always specify exactly what is to be included in the coordination sphere.

The *coordination polyhedron* results when the centers of mutually adjacent coordinated atoms are connected with one another. For every coordination number typical coordination polyhedra exist (Fig. 2.2). In some cases, several coordination polyhedra for a given coordination number differ only slightly, even though this may not be obvious at first glance; by minor displacements of atoms one polyhedron may be converted into another. For example, a trigonal bipyramid can be converted into a tetragonal pyramid by displacements of four of the coordinated atoms (Fig. 8.2, p. 71).

Larger structural units can be described by connected polyhedra. Two polyhedra can be joined by a common vertex, a common edge, or a common face (Fig. 2.3). The common atoms of two connected polyhedra are called bridging atoms. In face-sharing polyhedra the central atoms are closest to one another and in vertex-sharing polyhedra they are furthest apart. Further details concerning the connection of polyhedra are discussed in chapter 16.

The coordination conditions can be expressed in a chemical formula using a notation suggested by F. MACHATSCHKI (and extended by several other authors; for recommendations see [35]). The coordination number and polyhedron of an atom are given in brackets in a right superscript next to the element symbol. The polyhedron is designated with a symbol as listed in Fig. 2.2. Short forms can be used for the symbols, namely the coordination number alone or, for simple polyhedra, the letter alone, *e.g.* t for tetrahedron, and in this case the brackets can also be dropped. For example:

$Na^{[6o]}Cl^{[6o]}$ or $Na^{[6]}Cl^{[6]}$ or $Na^{o}Cl^{o}$
$Ca^{[8cb]}F_2^{[4t]}$ or $Ca^{[8]}F_2^{[4]}$ or $Ca^{cb}F_2^{t}$

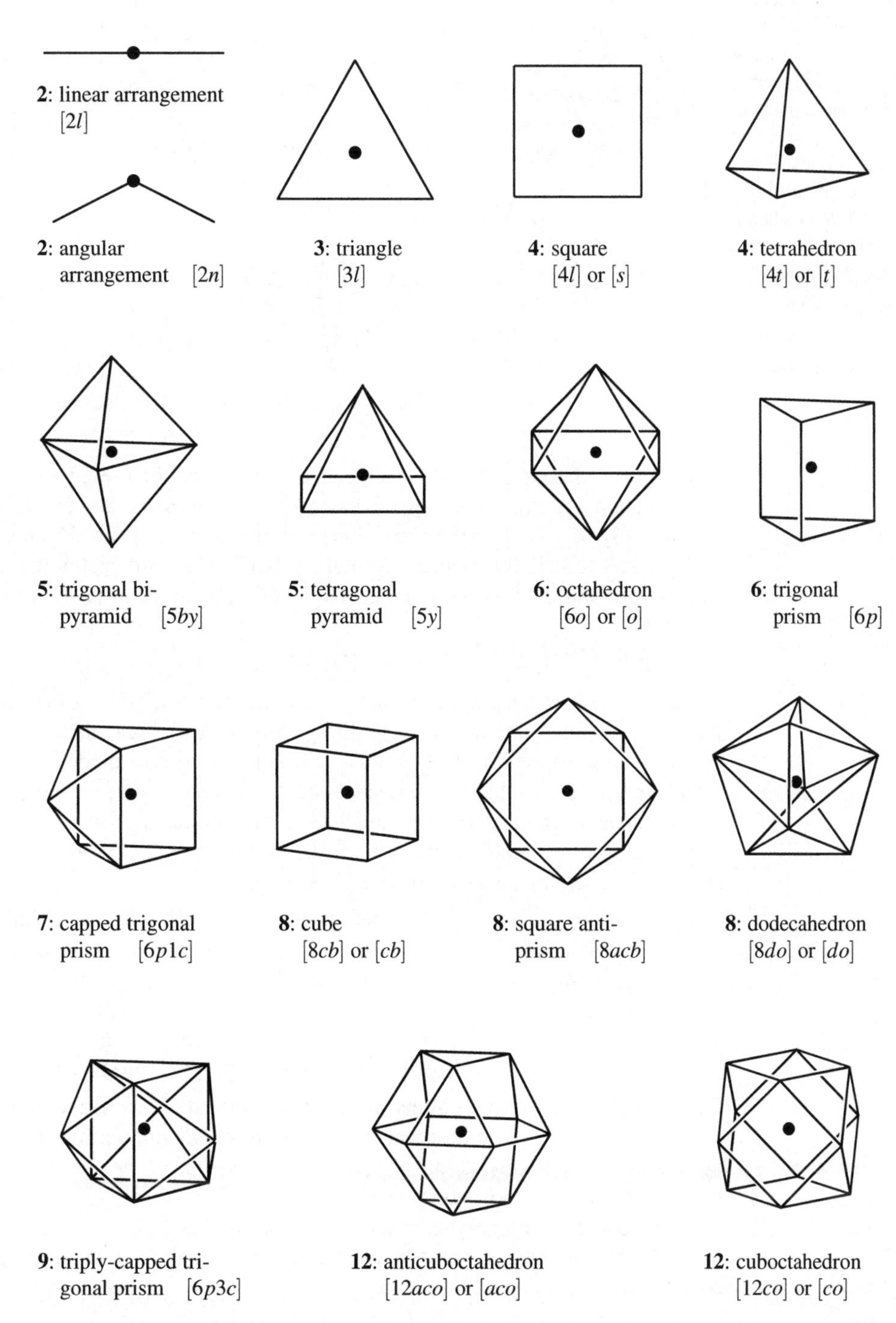

Fig. 2.2
The most important coordination polyhedra and their symbols; for explanation of the symbols see page 6

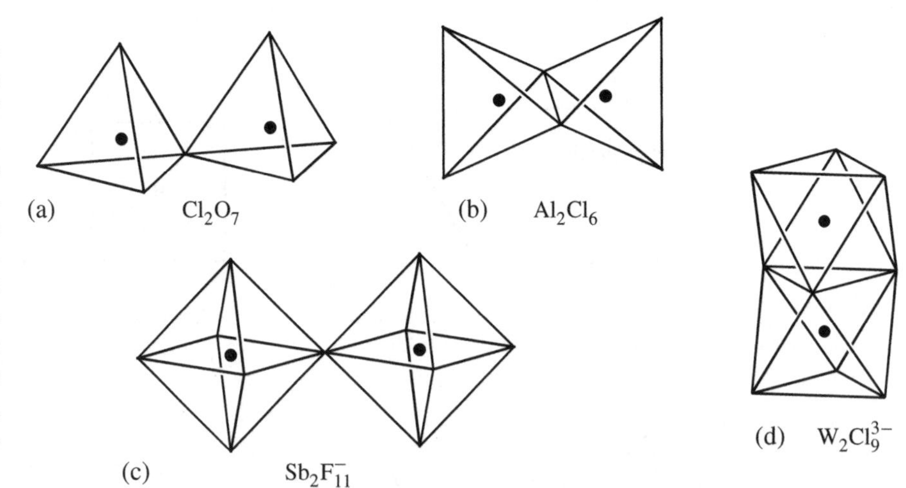

Fig. 2.3 Examples for the connection of polyhedra. (a) Two tetrahedra sharing a vertex. (b) Two tetrahedra sharing an edge. (c) Two octahedra sharing a vertex. (d) Two octahedra sharing a face. For two octahedra sharing an edge see Fig. 1

For more complicated cases an extended notation can be used, in which the coordination of an atom is expressed in the manner $A^{[m,n;p]}$. For m, n and p the polyhedra symbols are taken. Symbols before the semicolon refer to polyhedra spanned by the atoms B, C... , in the sequence as in the chemical formula $A_aB_bC_c$. The symbol after the semicolon refers to the coordination of the atom in question with atoms of the same kind. For example perovskite:

$Ca^{[,12co]}Ti^{[,6o]}O_3^{[4l,2l;8p]}$ (*cf.* Fig. 17.10, p. 203)

Since Ca is not directly surrounded by Ti atoms, the first polyhedron symbol is dropped; however, the first comma cannot be dropped to make it clear that the 12*co* refers to a cuboctahedron formed by 12 O atoms. Ti is not directly surrounded by Ca, but by six O atoms forming an octahedron. O is surrounded in planar (square) coordination by four Ca, by two linearly arranged Ti and by eight O atoms forming a prism.

In addition to the polyhedra symbols listed in Fig. 2.2, further symbols can be constructed. The letters have the following meanings:

l	collinear	*t*	tetrahedral	*do*	dodecahedral
	or coplanar	*s*	square	*co*	cuboctahedral
n	not collinear	*o*	octahedral	*i*	icosahedral
	or coplanar	*p*	prismatic	*c*	capped
y	pyramidal	*cb*	cubic	*a*	anti-
by	bipyramidal	*FK*	Frank–Kasper polyhedron (Fig. 15.5)		

For example: [3*n*] = three atoms not coplanar with the central atom as in NH_3; [12*p*] = hexagonal prism. When lone electron pairs in polyhedra vertices are also counted, a symbolism in the following manner can be used: $[\psi - 4t]$ (same meaning as [3*n*]), $[\psi - 6o]$ (same as [5*y*]), $[2\psi - 6o]$ (same as [4*l*]).

When coordination polyhedra are connected to chains, layers or a three-dimensional network, this can be expressed by the preceding symbols $^1_\infty$, $^2_\infty$ or $^3_\infty$, respectively. Examples:

$^3_\infty Na^{[6]}Cl^{[6]}$ $^3_\infty Ti^{[o]}O_2^{[3l]}$ $^2_\infty C^{[3l]}$ (graphite)

To state the existence of individual, finite atom groups, $^0_\infty$ can be set in front of the symbol. For their further specification, the following less popular symbols may be used:

chain fragment	$\{f\}$ or $\wedge$
ring	$\{r\}$ or ○
cage	$\{k\}$ or Ⓥ

For example: $Na_2\wedge S_3$; $\{k\}P_4$; Na_3○$[P_3O_9]$.

The packing of the atoms can be specified by inserting the corresponding part of the formula between square brackets and adding a label between angular brackets < >, for example $Ti^o[CaO_3]$<c>. The c means that the combined Ca and O atoms form a cubic closest-packing of spheres (packings of spheres are treated in Chapters 14 and 17). Some symbols of this kind are:

Tc or c	cubic closest-packing of spheres
Th or h	hexagonal closest-packing of spheres
Ts	stacking sequence $AA\ldots$ of hexagonal layers
Qs	stacking sequence $AA\ldots$ of square layers
Qf	stacking sequence $AB\ldots$ of square layers

For additional symbols of further packings *cf.* [38, 156]. T (triangular) refers to hexagonal layers, Q to layers with a periodic pattern of squares. The packing Qs yields a primitive cubic lattice (Fig. 2.4), Qf a body-centered cubic lattice (*cf.* Fig. 14.3, p. 153). Sometimes the symbols are set as superscripts without the angular brackets, for example $Ti[CaO_3]^c$.

Another type of notation, introduced by P. NIGGLI, uses fractional numbers in the chemical formula. The formula $TiO_{6/3}$ for instance means that every titanium atom is surrounded by 6 O atoms, each of which is coordinated to 3 Ti atoms. Another example is: $NbOCl_3 = NbO_{2/2}Cl_{2/2}Cl_{2/1}$ which has coordination number 6 for the niobium atom ($= 2+2+2 =$ sum of the numerators), coordination number 2 for the O atom and coordination numbers 2 and 1 for the two different kinds of Cl atoms (*cf.* Fig. 16.11, p. 176).

2.2 Description of Crystal Structures

In a crystal atoms are joined to form a larger network with a periodical order in three dimensions. The spatial order of the atoms is called the *crystal structure.* When we connect the periodically repeated atoms of one kind in three space directions to a three-dimensional grid, we obtain the *crystal lattice*. The crystal lattice represents a three-dimensional order of points; all points of the lattice are completely equivalent and have the same surroundings. We can think of the crystal lattice as generated by periodically repeating a small parallelepiped in three dimensions without gaps (Fig. 2.4; parallelepiped = body limited by six faces that are parallel in pairs). The parallelepiped is called the *unit cell.*

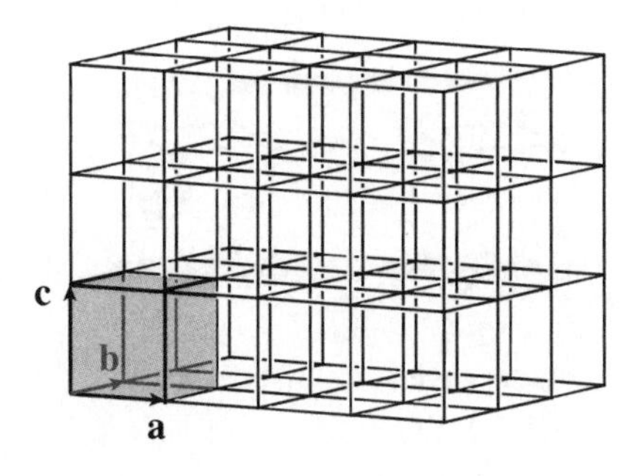

Fig. 2.4 Primitive cubic crystal lattice. One unit cell is marked

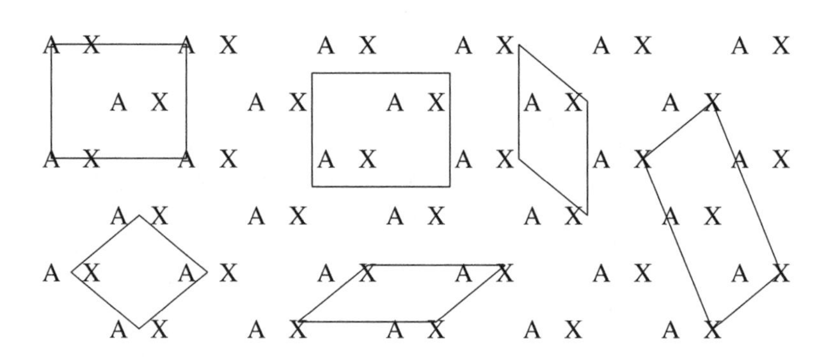

Fig. 2.5 Periodical, two-dimensional arrangement of A and X atoms. The whole pattern can be generated by repeating any one of the plotted unit cells.

The unit cell can be defined by three *basis vectors* labeled **a**, **b** and **c**. By definition, the crystal lattice is the complete set of all linear combinations $\mathbf{t} = u\mathbf{a} + v\mathbf{b} + w\mathbf{c}$, u, v, w comprising all positive and negative integers. Therefore, the crystal lattice is an abstract geometric construction, and the terms 'crystal lattice' and 'crystal structure' should not be confounded. The lengths a, b, and c of the basis vectors and the angles α, β, and γ between them are the *lattice parameters* (or lattice constants; α betweeen **b** and **c** etc.). There is no unique way to choose the unit cell for a given crystal structure, as is illustrated for a two-dimensional example in Fig. 2.5. To achieve standardization in the description of crystal structures, certain conventions for the selection of the unit cell have been settled upon in crystallography:

1. The unit cell is to show the symmetry of the crystal, *i.e.* the basis vectors are to be chosen parallel to symmetry axes or perpendicular to symmetry planes.
2. For the origin of the unit cell a geometrically unique point is selected, with priority given to an inversion center.
3. The basis vectors should be as short as possible. This also means that the cell volume should be as small as possible, and the angles between them should be as close as possible to 90°.
4. If the angles between the basis vectors deviate from 90°, they are either chosen to be all larger or all smaller than 90° (preferably > 90°).

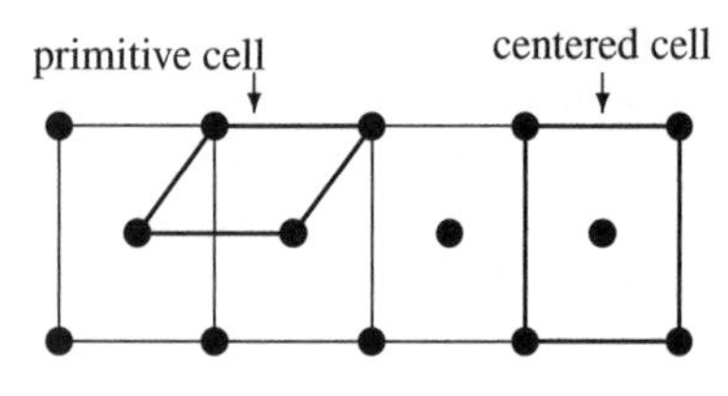

A unit cell having the smallest possible volume is called a *primitive* cell. For reasons of symmetry according to rule 1 and contrary to rule 3, a primitive cell is not always chosen, but instead a *centered* cell, which is *double, triple* or *fourfold primitive*, *i.e.* its volume is larger by a corresponding factor. The centered cells to be considered are shown in Fig. 2.6.

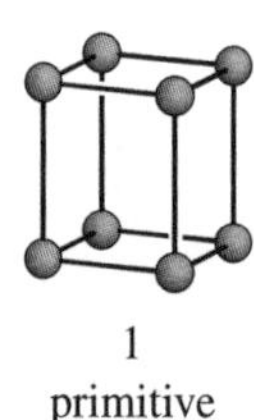

1 primitive *P*

2 base centered *C* (or *A*, *B*)

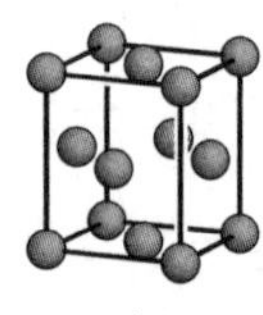

4 face centered *F*

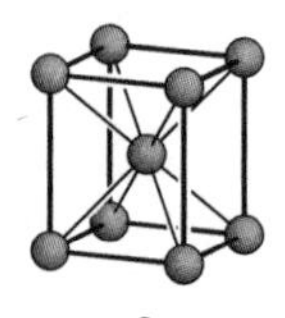

2 body centered *I*

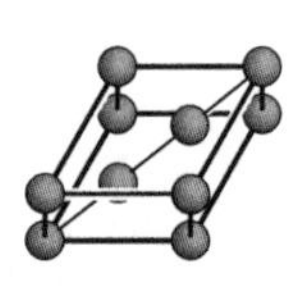

3 rhombohedral *R*

Fig. 2.6 Centered unit cells and their symbols. The numbers specify how manifold primitive the respective cell is

Aside from the conventions mentioned for the cell choice, further rules have been developed to achieve standardized descriptions of crystal structures [36]. They should be followed to assure a systematic and comparable documentation of the data and to facilitate the inclusion in databases. However, contraventions of the standards are rather frequent, not only from negligence or ignorance of the rules, but often for compelling reasons, for example when the relationships between different structures are to be pointed out.

Specification of the lattice parameters and the positions of all atoms contained in the unit cell is sufficient to characterize all essential aspects of a crystal structure. A unit cell can only contain an integral number of atoms. When stating the contents of the cell one refers to the chemical formula, *i.e.* the number of 'formula units' per unit cell is given; this number is usually termed Z. How the atoms are to be counted is shown in Fig. 2.7.

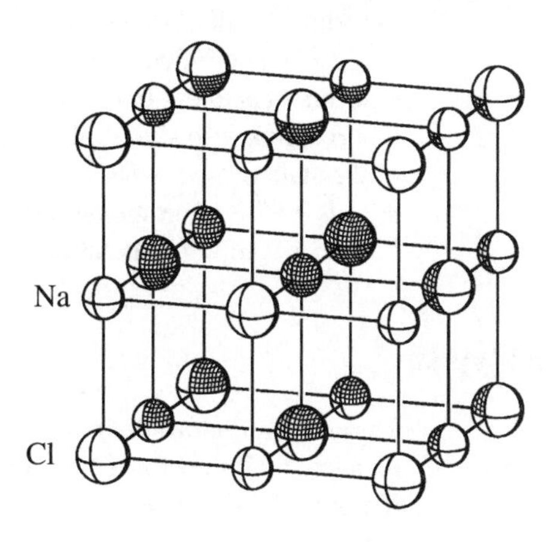

Fig. 2.7
The way to count the contents of a unit cell for the example of the face-centered unit cell of NaCl: 8 Cl^- ions in 8 vertices, each of which belongs to 8 adjacent cells makes $8/8 = 1$; 6 Cl^- ions in the centers of 6 faces belonging to two adjacent cells each makes $6/2 = 3$. 12 Na^+ ions in the centers of 12 edges belonging to 4 cells each makes $12/4 = 3$; 1 Na^+ ion in the cube center, belonging only to this cell. Total: 4 Na^+ and 4 Cl^- ions or four formula units of NaCl ($Z = 4$).

2.3 Atomic Coordinates

The position of an atom in the unit cell is specified by a set of *atomic coordinates, i.e.* by three coordinates x, y and z. These refer to a coordinate system that is defined by the basis vectors of the unit cell. The unit length taken along each of the coordinate axes corresponds to the length of the respective basis vector. The coordinates x, y and z for every atom within the unit cell thus have values between 0.0 and <1.0. The coordinate system is *not* a Cartesian one; the coordinate axes can be inclined to one another and the unit lengths on the axes may differ from each other. Addition or subtraction of an *integral* number to a coordinate value generates the coordinates of an equivalent atom in a different unit cell. For example, the coordinate triplet $x = 1.27, y = 0.52$ and $z = -0.10$ specifies the position of an atom in a cell neighboring the origin cell, namely in the direction $+\mathbf{a}$ and $-\mathbf{c}$; this atom is equivalent to the atom at $x = 0.27, y = 0.52$ and $z = 0.90$ in the origin cell.

Commonly, only the atomic coordinates for the atoms in one *asymmetric unit* are listed. Atoms that can be 'generated' from these by symmetry operations are not listed. Which symmetry operations are to be applied is revealed by stating the *space group* (*cf.* Section 3.3). When the lattice parameters, the space group, and the atomic coordinates are known, all structural details can be deduced. In particular, all interatomic distances and angles can be calculated.

The following formula can be used to calculate the distance d between two atoms from the lattice parameters and atomic coordinates:

$$d = \sqrt{(a\Delta x)^2 + (b\Delta y)^2 + (c\Delta z)^2 + 2bc\Delta y\Delta z\cos\alpha + 2ac\Delta x\Delta z\cos\beta + 2ab\Delta x\Delta y\cos\gamma}$$

$\Delta x = x_2 - x_1$, $\Delta y = y_2 - y_1$ and $\Delta z = z_2 - z_1$ are the differences between the coordinates of the two atoms. The angle ω at atom 2 in a group of three atoms 1, 2 and 3 can be calculated from the three distances d_{12}, d_{23} and d_{13} between them according to the cosine formula:

$$\cos\omega = -\sqrt{\frac{d_{13}^2 - d_{12}^2 - d_{23}^2}{2d_{12}d_{23}}}$$

When specifying atomic coordinates, interatomic distances etc., the corresponding standard deviations should also be given, which serve to express the precision of their experimental determination. The commonly used notation, such as '$d = 235.1(4)$ pm' states a standard deviation of 4 units for the last digit, *i.e.* the standard deviation in this case amounts to 0.4 pm. Standard deviation is a term in statistics. When a standard deviation σ is linked to some value, the probability of the true value being within the limits $\pm\sigma$ of the stated value is 68.3 %. The probability of being within $\pm 2\sigma$ is 95.4 %, and within $\pm 3\sigma$ is 99.7 %. The standard deviation gives no reliable information about the trueness of a value, because it only takes into account statistical errors, and not systematic errors.

2.4 Isotypism

The crystal structures of two compounds are *isotypic* if their atoms are distributed in a like manner and if they have the same symmetry. One of them can be generated from the other if atoms of an element are substituted by atoms of another element without changing their positions in the crystal structure. The absolute values of the lattice dimensions and the interatomic distances may differ, and *small* variations are permitted for the atomic coordinates. The angles between the crystallographic axes and the relative lattice dimensions (axes ratios) must be similar. Two isotypic structures exhibit a one-to-one relation for all atomic positions and have coincident geometric conditions. If, in addition, the chemical bonding conditions are also similar, then the structures also are *crystal-chemical isotypic*.

The ability of two compounds which have isotypic structures to form mixed crystals, *i.e.* when the exchange process of the atoms can actually be performed continuously, has been termed *isomorphism*. However, because this term is also used for some other phenomena, it has been recommended that its use be discontinued in this context.

Two structures are *homeotypic* if they are similar, but fail to fulfill the aforementioned conditions for isotypism because of different symmetry, because corresponding atomic positions are occupied by several different kinds of atoms (substitution derivatives) or because the geometric conditions differ (different axes ratios, angles, or atomic coordinates). An example of substitution derivatives is: C (diamond)–ZnS (zinc blende)–Cu_3SbS_4 (famatinite). The most appropriate method to work out the relations between homeotypic structures takes advantage of their symmetry relations (*cf.* Chapter 18).

If two ionic compounds have the same structure type, but in such a way that the cationic positions of one compound are taken by the anions of the other and vice versa ('exchange of cations and anions'), then they sometimes are called 'antitypes'. For example: in Li_2O the Li^+ ions occupy the same positions as the F^- ions in CaF_2, while the O^{2-} ions take the same positions as the Ca^{2+} ions; Li_2O crystallizes in the 'anti-CaF_2 type'.

2.5 Problems

2.1 Calculate effective coordination numbers (e.c.n.) with the formula given on page 3 for:
(a) Tellurium, $4 \times d_1 = 283$ pm, $2 \times d_2 = 349$ pm, $d_g = 444$ pm;
(b) Gallium, $1 \times d_1 = 247$ pm, $2 \times d_2 = 270$ pm, $2 \times d_3 = 274$ pm, $2 \times d_4 = 279$ pm, $d_g = 398$ pm;
(c) Tungsten, $8 \times d_1 = 274.1$ pm, $6 \times d_2 = 316.5$ pm, $d_g = 447.6$ pm.
2.2 Include specifications of the coordination of the atoms in the following formulas:
(a) $FeTiO_3$, Fe and Ti octahedral, O coordinated by 2 Fe and by 2 Ti in a nonlinear arrangement;
(b) $CdCl_2$, Cd octahedral, Cl trigonal-nonplanar;
(c) MoS_2, Mo trigonal-prismatic, S trigonal-nonplanar;
(d) Cu_2O, Cu linear, O tetrahedral;
(e) PtS, Pt square, S tetrahedral;
(f) $MgCu_2$, Mg FRANK-KASPER polyhedron with c.n. 16, Cu icosahedral;
(g) $Al_2Mg_3Si_3O_{12}$, Al octahedral, Mg dodecahedral, Si tetrahedral;
(h) UCl_3, U tricapped trigonal-prismatic, Cl 3-nonplanar.
2.3 Give the symbols stating the kind of centering of the unit cells of CaC_2 (Fig. 7.6, heavily outlined cell), K_2PtCl_6 (Fig. 7.7), cristobalite (Fig. 12.9), $AuCu_3$ (Fig. 15.1), K_2NiF_4 (Fig. 16.4), perovskite (Fig. 17.10).
2.4 Give the number of formula units per unit cell for:
CsCl (Fig. 7.1), ZnS (Fig. 7.1), TiO_2 (rutile, Fig. 7.4), $ThSi_2$ (Fig. 13.1), ReO_3 (Fig. 16.5), α-$ZnCl_2$ (Fig. 17.14).
2.5 What is the I–I bond length in solid iodine? Unit cell parameters: $a = 714$, $b = 469$, $c = 978$ pm, $\alpha = \beta = \gamma = 90°$. Atomic coordinates: $x = 0.0$, $y = 0.1543$, $z = 0.1174$; A symmetrically equivalent atom is at $-x, -y, -z$.
2.6 Calculate the bond lengths and the bond angle at the central atom of the I_3^- ion in RbI_3. Unit cell parameters: $a = 1091$, $b = 1060$, $c = 665.5$ pm, $\alpha = \beta = \gamma = 90°$. Atomic coordinates: I(1), $x = 0.1581$, $y = \frac{1}{4}$, $z = 0.3509$; I(2), $x = 0.3772$, $y = \frac{1}{4}$, $z = 0.5461$; I(3), $x = 0.5753$, $y = \frac{1}{4}$, $z = 0.7348$.

In the following problems the positions of symmetrically equivalent atoms (due to space group symmetry) may have to be considered; they are given as coordinate triplets to be calculated from the generating position x, y, z. To obtain positions of adjacent (bonded) atoms, some atomic positions may have to be shifted to a neighboring unit cell.
2.7 MnF_2 crystallizes in the rutile type with $a = b = 487.3$ pm and $c = 331.0$ pm. Atomic coordinates: Mn at $x = y = z = 0$; F at $x = y = 0.3050$, $z = 0.0$. Symmetrically equivalent positions: $-x, -x, 0$; $0.5-x, 0.5+x, 0.5$; $0.5+x, 0.5-x, 0.5$. Calculate the two different Mn–F bond lengths (< 250 pm) and the F–Mn–F bond angle referring to two F atoms having the same x and y coordinates and z differing by 1.0.
2.8 $WOBr_4$ is tetragonal, $a = b = 900.2$ pm, $c = 393.5$ pm, $\alpha = \beta = \gamma = 90°$. Calculate the W–Br, W=O and W···O bond lengths and the O=W–Br bond angle. Make a true-to-scale drawing (1 or 2 cm per 100 pm) of projections on to the *ab* and the *ac* plane, including atoms up to a distance of 300 pm from the z axis and covering $z = -0.5$ to $z = 1.6$. Draw atoms as circles and bonds (atomic contacts shorter than 300 pm) as heavy lines. What is the coordination polyhedron of the W atom?

Atomic coordinates:

	x	y	z	
W	0.0	0.0	0.0779	
O	0.0	0.0	0.529	Symmetrically equivalent positions:
Br	0.2603	0.0693	0.0	$-x, -y, z$; $-y, x, z$; $y, -x, z$

2.9 Calculate the Zr–O bond lengths in baddeleyite (ZrO_2), considering only interatomic distances shorter than 300 pm. What is the coordination number of Zr?
Lattice parameters: $a = 514.5$, $b = 520.7$, $c = 531.1$ pm, $\beta = 99.23°$, $\alpha = \gamma = 90°$.

Atomic coordinates:

	x	y	z	
Zr	0.2758	0.0411	0.2082	Symmetrically equivalent positions:
O(1)	0.0703	0.3359	0.3406	$-x, -y, -z$; $x, 0.5-y, 0.5+z$;
O(2)	0.5577	0.2549	0.0211	$-x, 0.5+y, 0.5-z$

3 Symmetry

The most characteristic feature of any crystal is its symmetry. It not only serves to describe important aspects of a structure, but is also related to essential properties of a solid. For example, quartz crystals could not exhibit the piezoelectric effect if quartz did not have the appropriate symmetry; this effect is the basis for the application of quartz in watches and electronic devices. Knowledge of the crystal symmetry is also of fundamental importance in crystal structure analysis.

In order to designate symmetry in a compact form, symmetry symbols have been developed. Two kinds of symbols are used: the *Schoenflies symbols* and the *Hermann–Mauguin symbols*, which are also called *international symbols*. Historically, Schoenflies symbols were developed first; they continue to be used in spectroscopy and to designate the symmetry of molecules. However, since they are less appropriate for describing the symmetry in crystals, they are now scarcely used in crystallography. We therefore discuss primarily the Hermann–Mauguin symbols. In addition, there are graphical symbols which are used in figures.

3.1 Symmetry Operations and Symmetry Elements

A symmetry operation transfers an object into a new spatial position that cannot be distinguished from its original position. In terms of mathematics, this is a mapping of an object onto itself that causes no distortions. A *mapping* is an instruction by which each point in space obtains a uniquely assigned point, the image point. 'Mapping onto itself' does not mean that each point is mapped exactly onto itself, but that after having performed the mapping, an observer cannot decide whether the object as a whole has been mapped or not.

After selecting a coordinate system, a mapping can be expressed by the following set of equations:

$$\left.\begin{array}{rcl} \tilde{x} & = & W_{11}x + W_{12}y + W_{13}z + w_1 \\ \tilde{y} & = & W_{21}x + W_{22}y + W_{23}z + w_2 \\ \tilde{z} & = & W_{31}x + W_{32}y + W_{33}z + w_3 \end{array}\right\} \qquad (3.1)$$

(x, y, z coordinates of the original point; $\tilde{x}$, $\tilde{y}$, $\tilde{z}$ coordinates of the image point)

A symmetry operation can be repeated infinitely many times. The *symmetry element* is a point, a straight line or a plane that preserves its position during execution of the symmetry operation. The symmetry operations are the following:

1. Translation (more exactly: symmetry-translation). Shift in a specified direction by a specified length. A *translation vector* corresponds to every translation. For example:

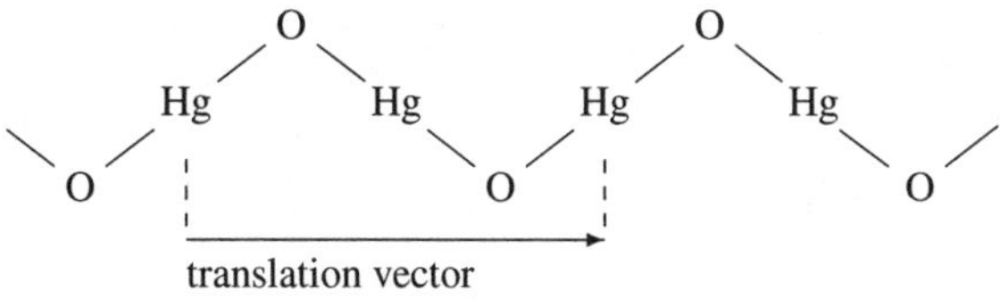

O Hg O Hg O Hg O Hg O (endless chain)

translation vector

Strictly speaking, a symmetry-translation is only possible for an infinitely extended object. An ideal crystal is infinitely large and has translational symmetry in three dimensions. To characterize its translational symmetry, three non-coplanar translation vectors **a**, **b** and **c** are required. A real crystal can be regarded as a finite section of an ideal crystal; this is an excellent way to describe the actual conditions.

As vectors **a**, **b** and **c** we choose the three basis vectors that also serve to define the unit cell (Section 2.2). Any translation vector **t** in the crystal can be expressed as the vectorial sum of three basis vectors, $\mathbf{t} = u\mathbf{a} + v\mathbf{b} + w\mathbf{c}$, where u, v and w are positive or negative integers.

Translational symmetry is the most important symmetry property of a crystal. In the Hermann–Mauguin symbols the three-dimensional translational symmetry is expressed by a capital letter which also allows the distinction of primitive and centered crystal lattices (*cf.* Fig. 2.6, p. 8):

P = primitive
A, B or C = base-centered in the **bc**-, **ac** or **ab** plane, respectively
F = face-centered (all faces)
I = body-centered (from *innenzentriert* in German)
R = rhombohedral

2. Rotation about some axis by an angle of $360/N$ degrees. The symmetry element is an N-fold *rotation axis*. The multiplicity N is an integer. After having performed the rotation N times the object has returned to its original position. Every object has infinitely many axes with $N = 1$, since an arbitrary rotation by 360° returns the object into its original position. The symbol for the onefold rotation is used for objects that have no symmetry other than translational symmetry. The Hermann–Mauguin symbol for an N-fold rotation is the number N; the Schoenflies symbol is C_N (*cf.* Fig. 3.1):

	Hermann–Mauguin symbol	Schoenflies symbol	graphical symbol	
onefold rotation axis	1	C_1	none	
twofold rotation axis	2	C_2	⬮	axis perpendicular to the plane of the paper
			← →	axis parallel to the plane of the paper
threefold rotation axis	3	C_3	▲	
fourfold rotation axis	4	C_4	◆	
sixfold rotation axis	6	C_6	⬢	

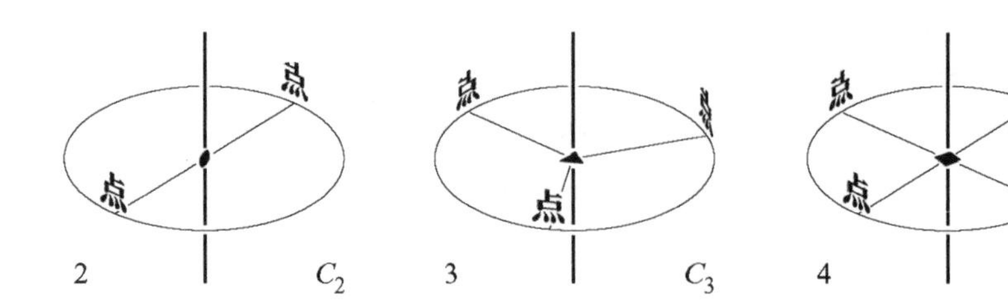

Fig. 3.1
Examples of rotation axes. In each case the Hermann–Mauguin symbol is given on the left side, and the Schoenflies symbol on the right side. 点 means point, pronounced dyǎn in Chinese, hoshi in Japanese

3. Reflection. The symmetry element is a *reflection plane* (Fig. 3.2).

Hermann–Mauguin symbol: m. Schoenflies symbol: σ (used only for a detached plane). Graphical symbols:

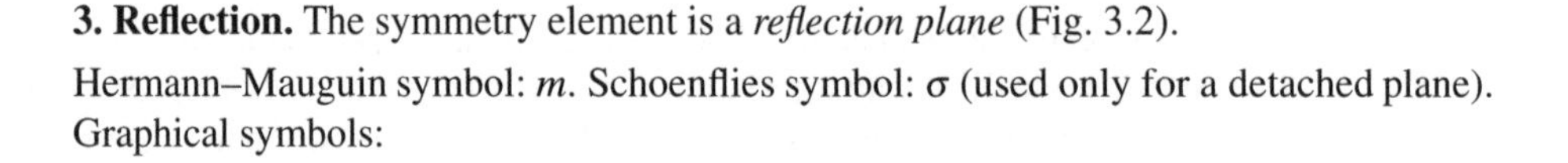

reflection plane perpendicular to the plane of the paper

reflection plane parallel to the plane of the paper

4. Inversion. 'Reflection' through a point (Fig. 3.2). This point is the symmetry element and is called *inversion center* or *center of symmetry*.

Hermann–Mauguin symbol: $\bar{1}$ ('one bar'). Schoenflies symbol: i. Graphical symbol: ∘

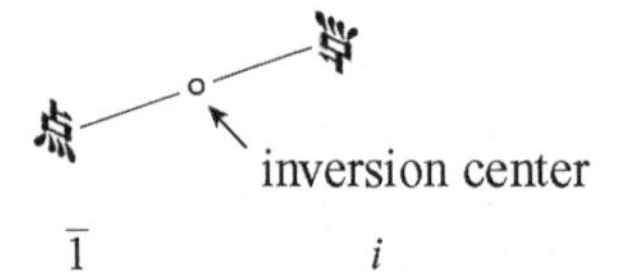

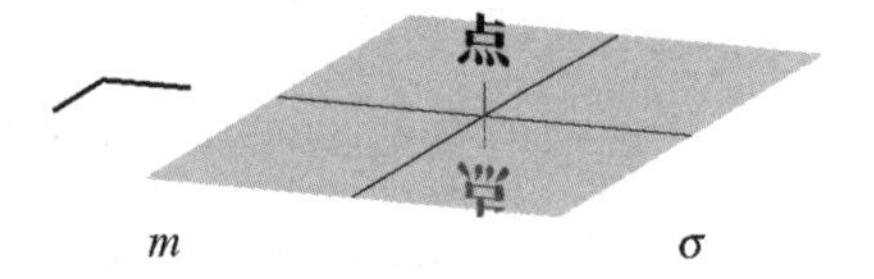

Fig. 3.2
Examples of an inversion and a reflection

5. Rotoinversion. The symmetry element is a *rotoinversion axis* or, for short, an *inversion axis*. This refers to a *coupled* symmetry operation which involves two motions: take a rotation through an angle of $360/N$ degrees immediately followed by an inversion at a point located on the axis (Fig. 3.3):

Hermann–Mauguin symbol	graphical symbol	
$\bar{1}$	∘	identical with an inversion center
$\bar{2} = m$		identical with a reflection plane perpendicular to the axis
$\bar{3}$	▲	
$\bar{4}$	◆	
$\bar{5}$	⬟	
$\bar{6}$	⬢	

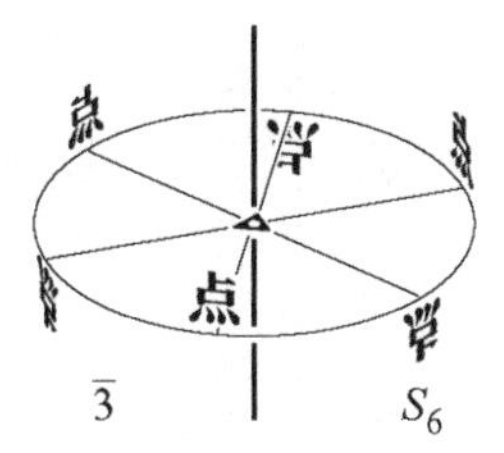

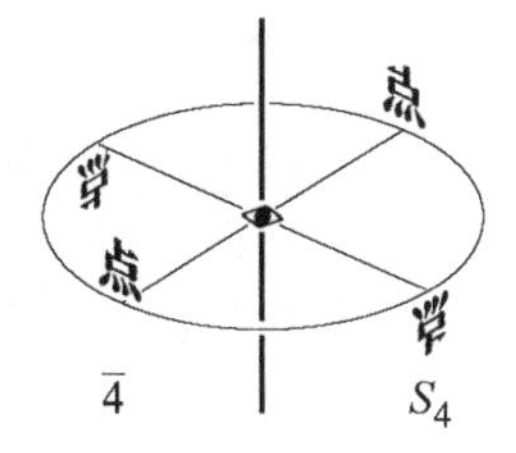

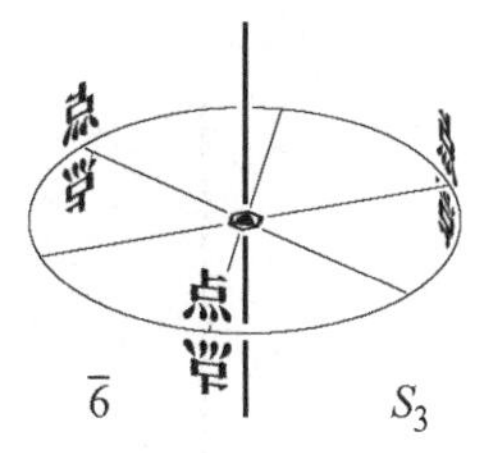

Fig. 3.3
Examples of inversion axes. If they are considered to be rotoreflection axes, they have the multiplicities expressed by the Schoenflies symbols S_N

If N is an even number, the inversion axis automatically contains a rotation axis with half the multiplicity. If N is an odd number, automatically an inversion center is present. This is expressed by the graphical symbols. If N is even but not divisible by 4, automatically a reflection plane perpendicular to the axis is present.

A **rotoreflection** is a coupled symmetry operation of a rotation and a reflection at a plane perpendicular to the axis. Rotoreflection axes are identical with inversion axes, but the multiplicities do not coincide if they are not divisible by 4 (Fig. 3.3). In the Hermann–Mauguin notation only inversion axes are used, and in the Schoenflies notation only rotoreflection axes are used, the symbol for the latter being S_N.

6. Screw rotation. The symmetry element is a *screw axis*. It can only occur if there is translational symmetry in the direction of the axis. The screw rotation results when a rotation of $360/N$ degrees is coupled with a displacement parallel to the axis. The Hermann–Mauguin symbol is N_M ('N sub M'); N expresses the rotational component and the fraction M/N is the displacement component as a fraction of the translation vector. Some screw axes are right or left-handed. Screw axes that can occur in crystals are shown in Fig. 3.4. Single polymer molecules can also have non-crystallographic screw axes, *e.g.* 10_3 in polymeric sulfur.

7. Glide reflection. The symmetry element is a *glide plane*. It can only occur if translational symmetry is present parallel to the plane. At the plane, reflections are performed, but every reflection is coupled with an immediate displacement parallel to the plane. The Hermann–Mauguin symbol is *a*, *b*, *c*, *n*, *d* or *e*, the letter designating the direction of the glide referred to the unit cell. *a*, *b* and *c* refer to displacements parallel to the basis vectors **a**, **b** and **c**, the displacements amounting to $\frac{1}{2}a$, $\frac{1}{2}b$ and $\frac{1}{2}c$, respectively. The glide planes *n* and *d* involve displacements in a diagonal direction by amounts of $\frac{1}{2}$ and $\frac{1}{4}$ of the translation vector in this direction, respectively. *e* designates two glide planes in one another with two mutually perpendicular glide directions (Fig. 3.5).

3.2 Point Groups

A geometric object can have several symmetry elements simultaneously. However, symmetry elements cannot be combined arbitrarily. For example, if there is only one reflection plane, it cannot be inclined to a symmetry axis (the axis has to be in the plane or perpendicular to it). Possible combinations of symmetry operations *excluding translations* are called *point groups*. This term expresses the fact that any allowed combination has one unique

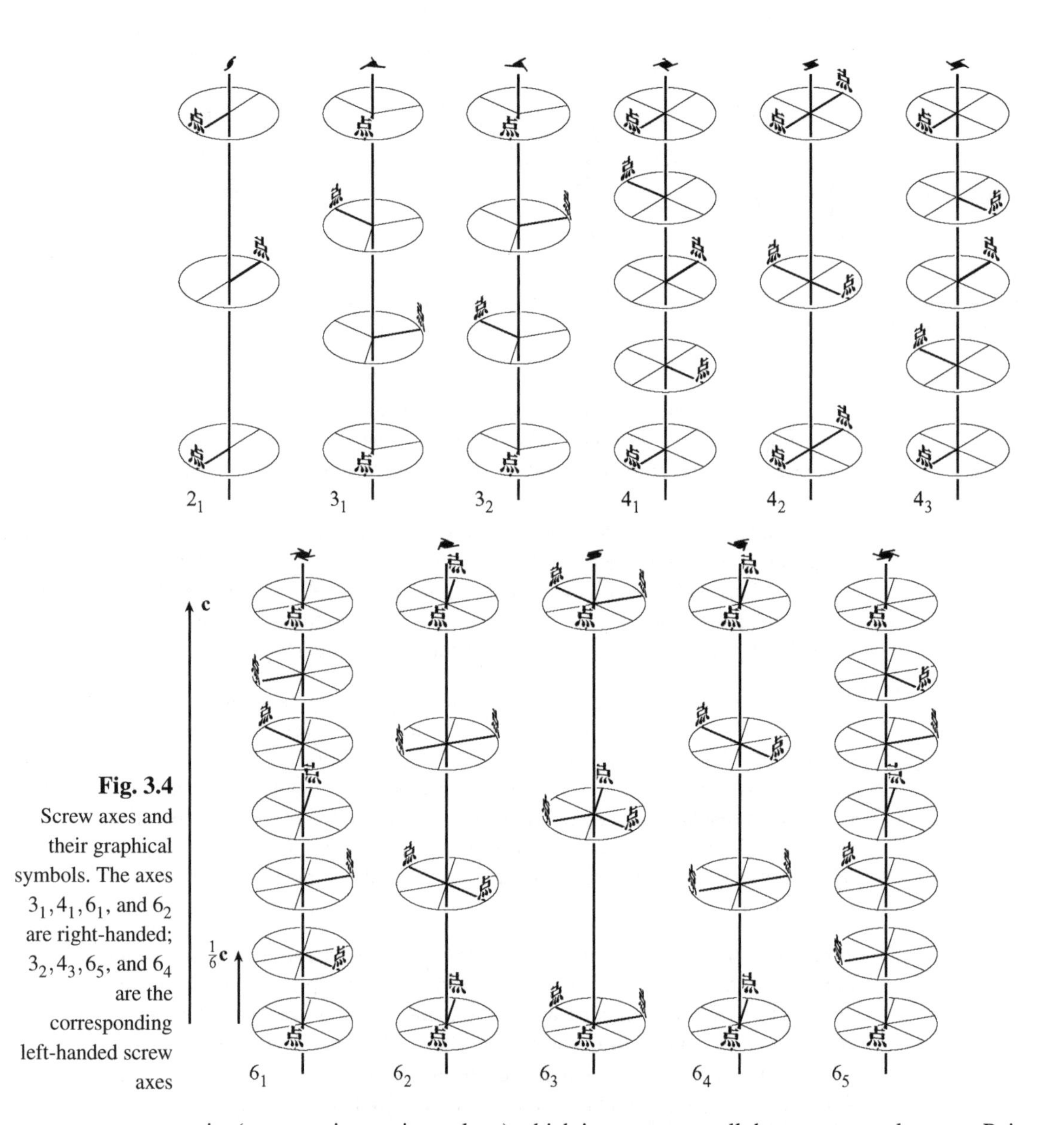

Fig. 3.4 Screw axes and their graphical symbols. The axes $3_1, 4_1, 6_1$, and 6_2 are right-handed; $3_2, 4_3, 6_5$, and 6_4 are the corresponding left-handed screw axes

point (or one unique axis or plane) which is common to all the symmetry elements. Point groups strictly fulfill the conditions set by group theory in mathematics. The symmetry operations are the elements that make up the group.

When two symmetry operations are combined, a third symmetry operation can result automatically. For example, the combination of a twofold rotation with a reflection at a plane perpendicular to the rotation axis automatically results in an inversion center at the site where the axis crosses the plane. It makes no difference which two of the three symmetry operations are combined (2, m or $\bar{1}$), the third one always results (Fig. 3.6).

Hermann–Mauguin Point-group Symbols

A Hermann–Mauguin point-group symbol consists of a listing of the symmetry elements that are present according to certain rules in such a way that their relative orientations can

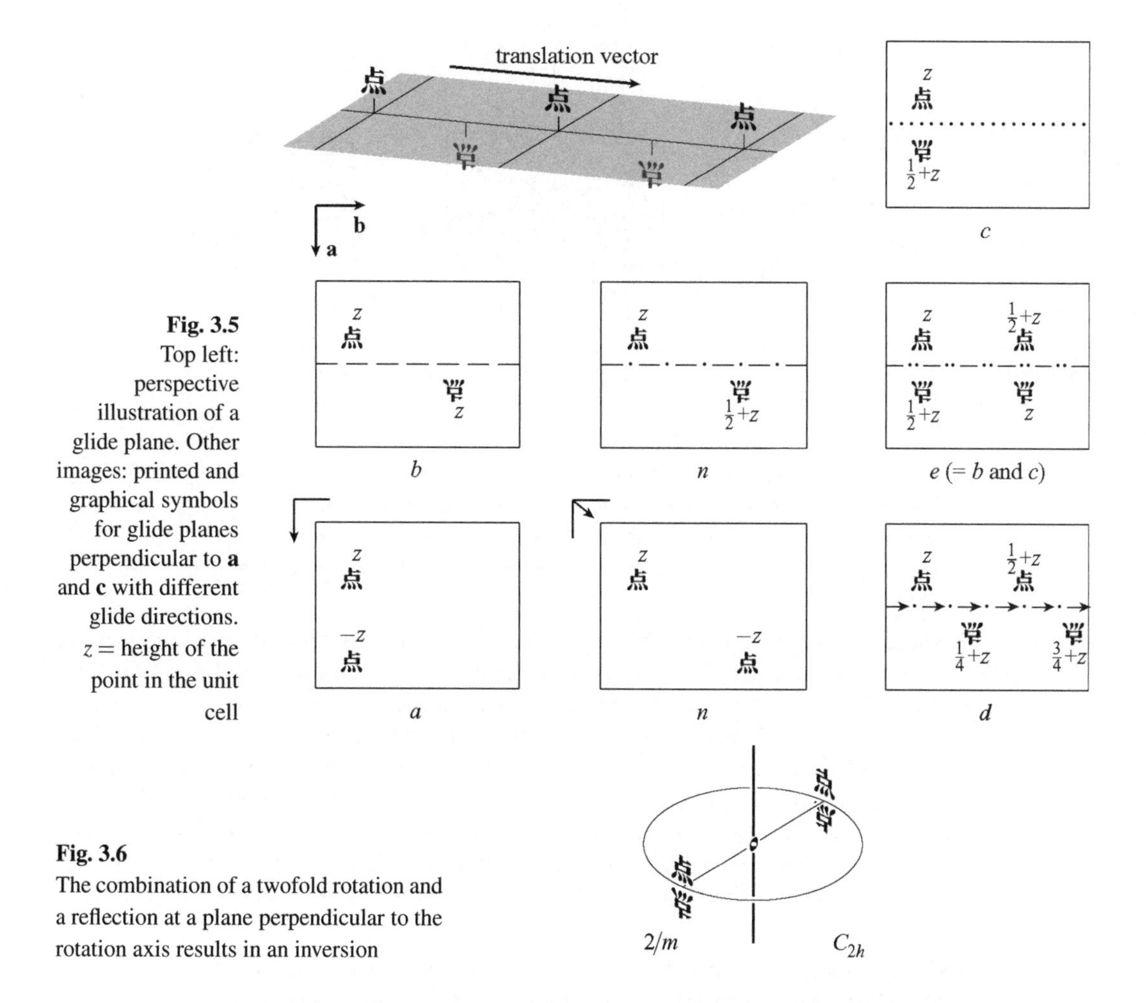

Fig. 3.5 Top left: perspective illustration of a glide plane. Other images: printed and graphical symbols for glide planes perpendicular to **a** and **c** with different glide directions. z = height of the point in the unit cell

Fig. 3.6 The combination of a twofold rotation and a reflection at a plane perpendicular to the rotation axis results in an inversion

also be recognized. In the *full Hermann–Mauguin symbol* all symmetry elements, with few exceptions, are listed. However, because they are more compact, usually only the *short Hermann–Mauguin symbols* are cited; in these, symmetry axes that result automatically from mentioned symmetry planes are not expressed; symmetry planes which are present are not omitted.

The following **rules** apply:

1. The orientation of symmetry elements is referred to a coordinate system xyz. If one symmetry axis is distinguished from the others by a higher multiplicity ('principal axis') or when there is only one symmetry axis, it is set as the z axis.
2. An inversion center is mentioned only if it is the only symmetry element present. The symbol then is $\overline{1}$. In other cases the presence or absence of an inversion center can be recognized as follows: it is present and only present if there is either an inversion axis with odd multiplicity ($\overline{N}$, with N odd) or a rotation axis with even multiplicity and a reflection plane perpendicular to it (N/m, with N even).
3. A symmetry element occurring repeatedly because it is multiplied by another symmetry operation is mentioned only once.

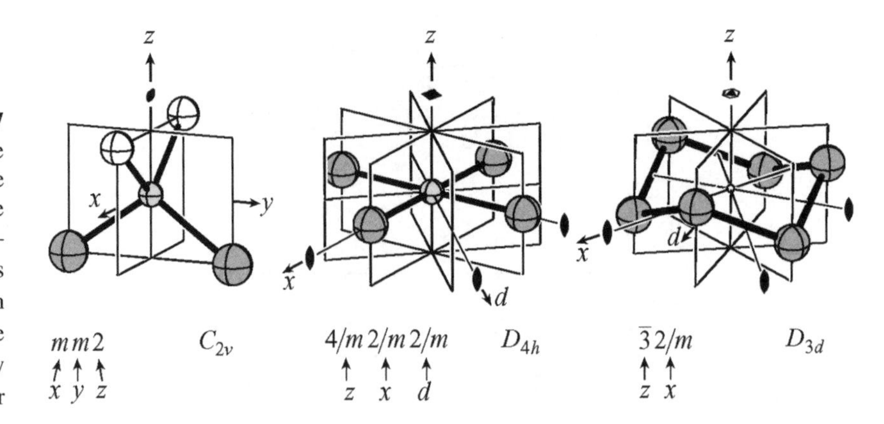

Fig. 3.7 Examples of three point groups. The letters under the Hermann–Mauguin symbols indicate to which directions the symmetry elements refer

4. A reflection plane that is perpendicular to a symmetry axis is designated by a slash, *e.g.* $2/m$ ('two over m') = reflection plane perpendicular to a twofold rotation axis. However, reflection planes perpendicular to rotation axes with odd multiplicities are not usually designated in the form $3/m$, but as inversion axes like $\bar{6}$; $3/m$ and $\bar{6}$ express identical facts.
5. The mutual orientation of different symmetry elements is expressed by the sequence in which they are listed. The orientation refers to the coordinate system. If the symmetry axis of highest multiplicity is twofold, the sequence is x–y–z, *i.e.* the symmetry element in the x direction is mentioned first etc.; the direction of reference for a reflection plane is *normal* to the plane. If there is an axis with a higher multiplicity, it is mentioned first; since it coincides by convention with the z axis, the sequence is different, namely z–x–d. The symmetry element oriented in the x direction occurs repeatedly because it is being multiplied by the higher multiplicity of the z axis; the bisecting direction between x and its next symmetry-equivalent direction is the direction indicated by d. See the examples in Fig. 3.7.
6. Cubic point groups have four threefold axes (3 or $\bar{3}$) that mutually intersect at angles of 109.47°. They correspond to the four body diagonals of a cube (directions **x+y+z**, **–x+y–z**, **–x–y+z** and **x–y–z**, added vectorially). In the directions **x**, **y**, and **z** there are axes 4, $\bar{4}$ or 2, and there can be reflection planes perpendicular to them. In the six directions **x+y**, **x–y**, **x+z**, ... twofold axes and reflection planes may be present. The sequence of the reference directions in the Hermann–Mauguin symbols is **z**, **x+y+z**, **x+y**. The occurrence of a 3 in the *second position* of the symbol (direction **x+y+z**) gives evidence of a cubic point group. See Fig. 3.8.

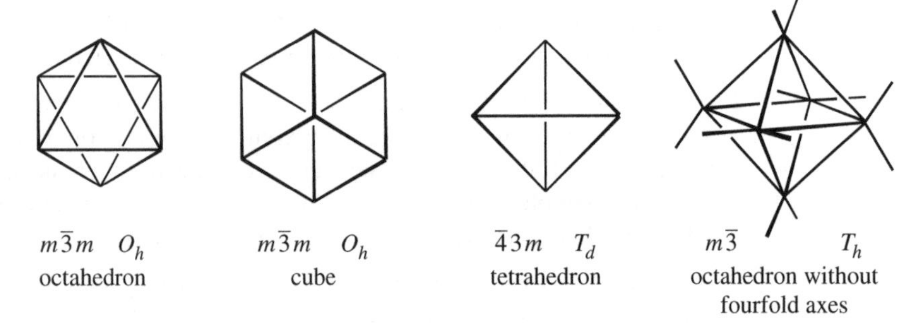

Fig. 3.8 Examples of three cubic point groups

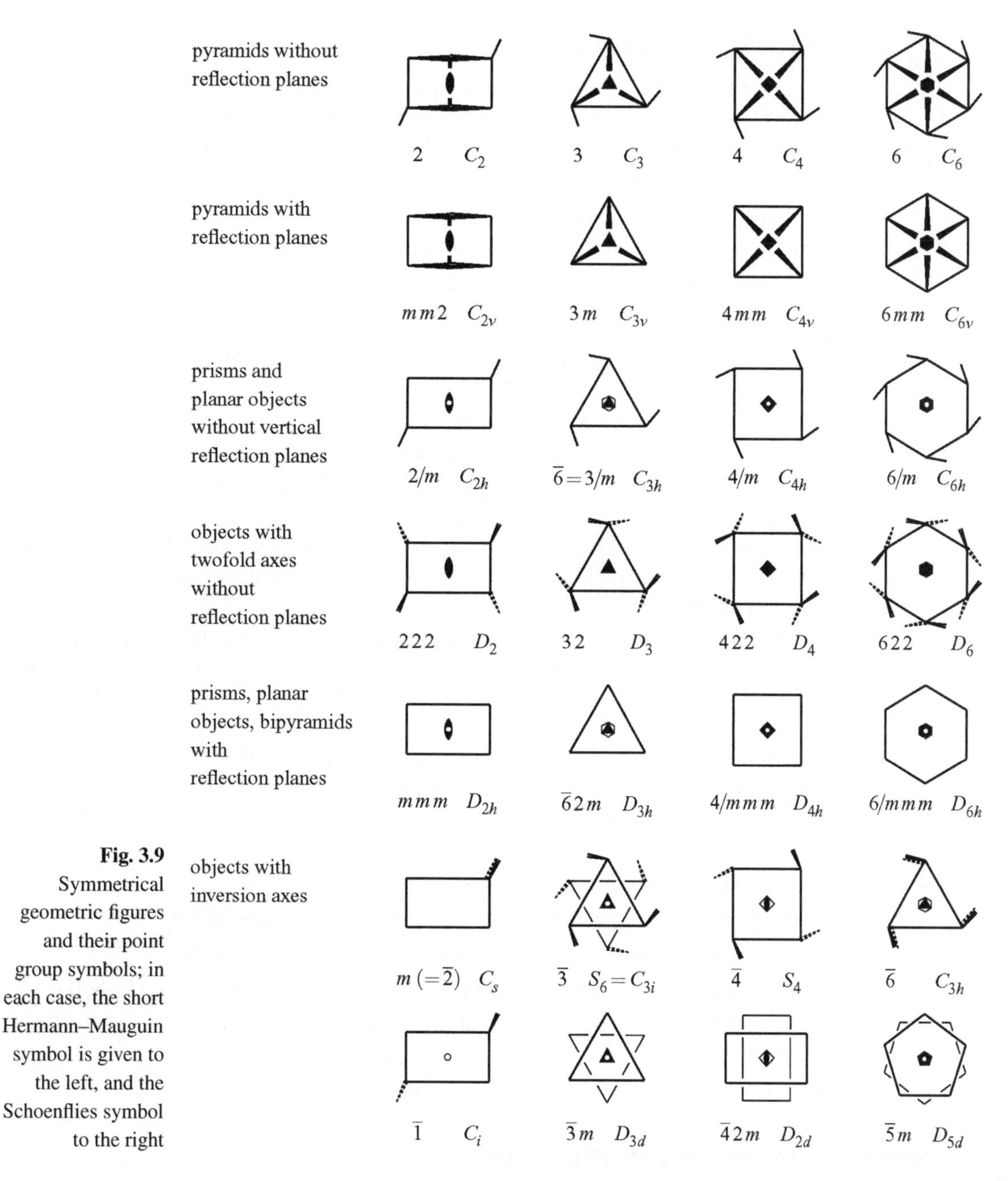

Fig. 3.9 Symmetrical geometric figures and their point group symbols; in each case, the short Hermann–Mauguin symbol is given to the left, and the Schoenflies symbol to the right

Figures 3.8 and 3.9 list point group symbols and illustrate them by geometric figures. In addition to the short Hermann–Mauguin symbols the Schoenflies symbols are also listed. Full Hermann–Mauguin symbols for some point groups are:

short	full	short	full
mmm	$2/m\,2/m\,2/m$	$\bar{3}m$	$\bar{3}\,2/m$
$4/mmm$	$4/m\,2/m\,2/m$	$m\bar{3}m$	$4/m\,\bar{3}\,2/m$
$6/mmm$	$6/m\,2/m\,2/m$	$m\bar{3}$	$2/m\,\bar{3}$

Schoenflies Point-group Symbols

The coordinate system of reference is taken with the vertical principal axis (z axis). Schoenflies symbols are rather compact — they designate only a minimum of the symmetry elements present in the following way (the corresponding Hermann–Mauguin symbols are given in brackets):

C_i = an inversion center is the only symmetry element $[\bar{1}]$.

C_s = a reflection plane is the only symmetry element $[m]$.

C_N = an N-fold rotation axis is the only symmetry element $[N]$.

C_{Ni} (N odd) = an N-fold rotation axis and an inversion center $[\bar{N}]$.

D_N = perpendicular to an N-fold rotation axis there are N twofold rotation axes $[N2$ if the value of N is odd; $N22$ if N is even$]$.

C_{Nh} = there is one N-fold (vertical) rotation axis and one horizontal reflection plane $[N/m]$.

C_{Nv} = an N-fold (vertical) rotation axis is at the intersection line of N vertical reflection planes $[Nm$ if the value of N is odd; Nmm if N is even$]$. $C_{\infty v}$ = symmetry of a cone $[\infty m]$.

D_{Nh} = in addition to an N-fold (vertical) rotation axis there are N horizontal twofold axes, N vertical reflection planes and one horizontal reflection plane $[\bar{N}2/m$ if N is odd; $N/m\,2/m\,2/m$, for short N/mmm, if N is even$]$. $D_{\infty h}$ = symmetry of a cylinder $[\infty/m\,2/m$, for short ∞/mm or $\bar{\infty}m]$.

D_{Nd} = the N-fold vertical rotation axis contains a $2N$-fold rotoreflection axis, N horizontal twofold rotation axes are situated at bisecting angles between N vertical reflection planes $[\bar{M}2m$ with $M = 2\times N]$. S_{Mv} has the same meaning as D_{Nd} and can be used instead, but it has gone out of use.

S_N = there is only an N-fold (vertical) rotoreflection axis (*cf.* Fig. 3.3). The symbol S_N is needed only if N is divisible by 4. If N is even but not divisible by 4, $C_{\frac{N}{2}i}$ can be used instead, *e.g.* $C_{5i} = S_{10}$. If N is odd, the symbol C_{Nh} is commonly used instead of S_N, *e.g.* $C_{3h} = S_3$.

T_d = symmetry of a tetrahedron $[\bar{4}3m]$.

O_h = symmetry of an octahedron and of a cube $[4/m\,\bar{3}\,2/m$, short $m\bar{3}m]$.

T_h = symmetry of an octahedron without fourfold axes $[2/m\bar{3}$, short $m\bar{3}]$.

I_h = symmetry of an icosahedron and of a pentagonal dodecahedron $[2/m\,\bar{3}\,\bar{5}$, short $m\bar{3}\bar{5}]$.

O, T and I = as O_h, T_h and I_h, but with no reflection planes $[432, 23$ and 235, respectively$]$.

K_h = symmetry of a sphere $[2/m\,\bar{\infty}$, short $m\bar{\infty}]$.

3.3 Space Groups and Space-group Types

Symmetry axes can only have the multiplicities 1, 2, 3, 4 or 6 when translational symmetry is present in three dimensions. If, for example, fivefold axes were present in one direction, the unit cell would have to be a pentagonal prism; space cannot be filled, free of voids, with prisms of this kind. Due to the restriction to certain multiplicities, symmetry operations can only be combined in a finite number of ways in the presence of three-dimensional translational symmetry. The 230 possibilities are called *space-group types* (often, not quite correctly, called the 230 space groups).

The Hermann–Mauguin symbol for a space-group type begins with a capital letter *P, A, B, C, F, I* or *R* which expresses the presence of translational symmetry in three dimensions and the kind of centering. The letter is followed by a listing of the other symmetry elements according to the same rules as for point groups; the basis vectors **a**, **b** and **c** define the coordinate system (Fig. 3.11). If several kinds of symmetry elements exist in one direction (*e.g.* parallel 2 and 2_1 axes), then only one is mentioned; as a rule, the denomination of mirror planes has priority over glide planes and rotation axes over screw axes.

The 230 space-group types are listed in full in *International Tables for Crystallography*, Volume A [48]. Whenever crystal symmetry is to be considered, this fundamental tabular work should be consulted. It includes figures that show the relative positions of the symmetry elements as well as details concerning all possible sites in the unit cell (*cf.* next section).

In some circumstances the magnitudes of the translation vectors must be taken into account. Let us demonstrate this with the example of the trirutile structure. If we triplicate the unit cell of rutile in the **c** direction, we can occupy the metal atom positions with two kinds of metals in a ratio of 1 : 2, such as is shown in Fig. 3.10. This structure type is known for several oxides and fluorides, *e.g.* $ZnSb_2O_6$. Both the rutile and the trirutile structure belong to the same space-group **type** $P4_2/mnm$. Due to the triplicated translation vector in the **c** direction, the density of the symmetry elements in trirutile is less than in rutile. The total number of symmetry operations (including the translations) is reduced to $\frac{1}{3}$. In other words, trirutile has a symmetry that is reduced by a factor of three. A structure with a specific symmetry *including* the translational symmetry has a specific *space group*; the space-group type, however, is independent of the special magnitudes of the translation vectors. Therefore, rutile and trirutile do *not* have the same space group. Although space group and space-group type have to be distinguished, the same symbols are used for both. However, this does not cause any problems since the specification of a space group is only used to designate the symmetry of a specific structure or a specific structure type, and this always involves a crystal lattice with definite translation vectors.

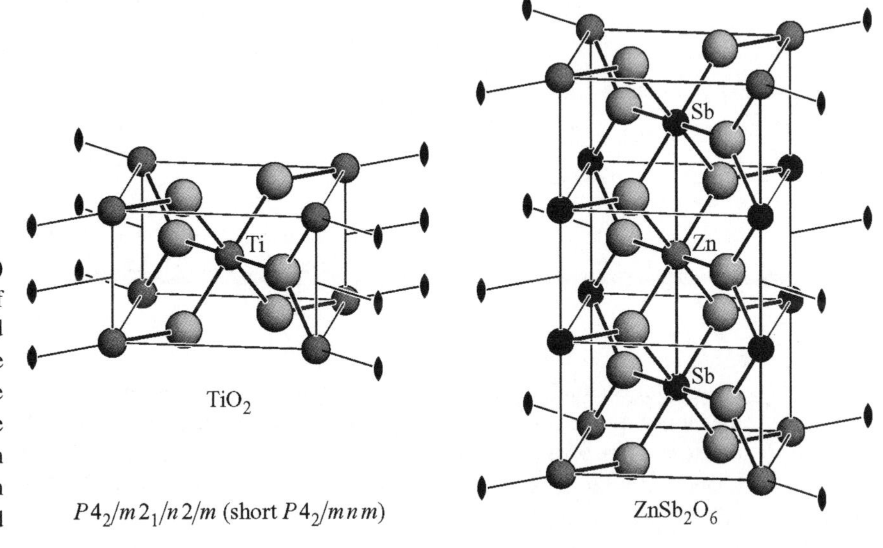

Fig. 3.10 Unit cells of the rutile and the trirutile structures. The positions of the twofold rotation axes have been included

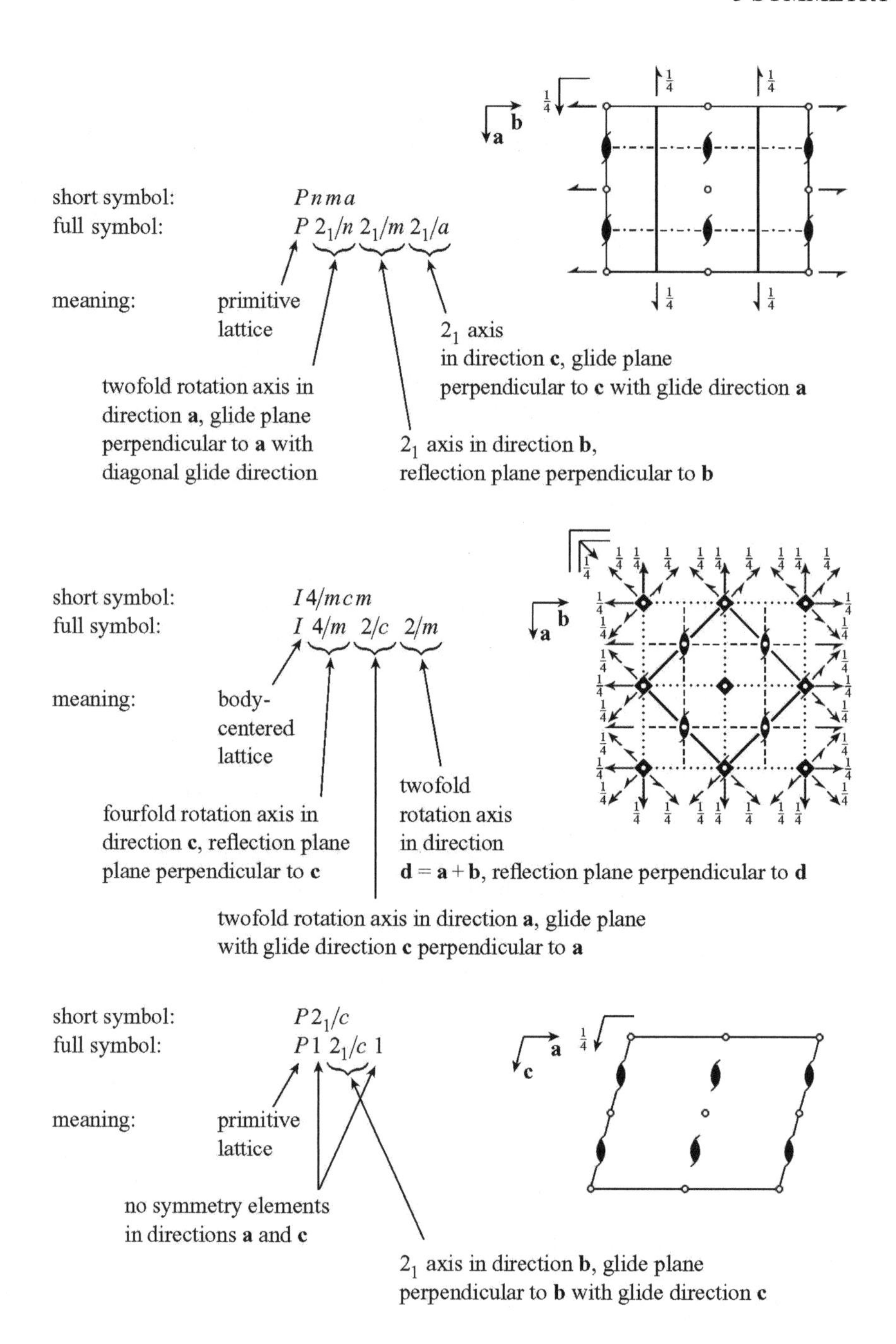

Fig. 3.11 Examples of space-group type symbols and their meanings

3.4 Positions

If an atom is situated on a center of symmetry, on a rotation axis or on a reflection plane, then it occupies a *special position*. On execution of the corresponding symmetry operation, the atom is mapped onto itself. Any other site is a *general position*. A special position is connected with a specific *site symmetry* which is higher than 1. The site symmetry at a general position is always 1.

Molecules or ions in crystals often occupy special positions. In that case the site symmetry may not be higher than the symmetry of the free molecule or ion. For example, an octahedral ion like $SbCl_6^-$ can be placed on a site with symmetry 4 if its Sb atom and two *trans* Cl atoms are located on the fourfold axis; a water molecule, however, cannot be placed on a fourfold axis.

The different sets of positions in crystals are called *Wyckoff positions*. They are listed for every space-group type in *International Tables for Crystallography*, Volume A, in the following way (example space-group type Nr. 87, $I4/m$):

Multiplicity, Wyckoff letter,	Site symmetry	Coordinates $(0,0,0)+$ $(\frac{1}{2},\frac{1}{2},\frac{1}{2})+$			
16 *i*	1	(1) x,y,z	(2) $\bar{x},\bar{y},z$	(3) $\bar{y},x,z$	(4) $y,\bar{x},z$
		(5) $\bar{x},\bar{y},\bar{z}$	(6) $x,y,\bar{z}$	(7) $y,\bar{x},\bar{z}$	(8) $\bar{y},x,\bar{z}$
8 *h*	m	$x,y,0$	$\bar{x},\bar{y},0$	$\bar{y},x,0$	$y,\bar{x},0$
8 *g*	2	$0,\frac{1}{2},z$	$\frac{1}{2},0,z$	$0,\frac{1}{2},\bar{z}$	$\frac{1}{2},0,\bar{z}$
8 *f*	$\bar{1}$	$\frac{1}{4},\frac{1}{4},\frac{1}{4}$	$\frac{3}{4},\frac{3}{4},\frac{1}{4}$	$\frac{3}{4},\frac{1}{4},\frac{1}{4}$	$\frac{1}{4},\frac{3}{4},\frac{1}{4}$
4 *e*	4	$0,0,z$	$0,0,\bar{z}$		
4 *d*	$\bar{4}$	$0,\frac{1}{2},\frac{1}{4}$	$\frac{1}{2},0,\frac{1}{4}$		
4 *c*	$2/m$	$0,\frac{1}{2},0$	$\frac{1}{2},0,0$		
2 *b*	$4/m$	$0,0,\frac{1}{2}$			
2 *a*	$4/m$	$0,0,0$			

In crystallography it is a common practice to write minus signs on top of the symbols; $\bar{x}$ means $-x$. The meaning of the coordinate triplets is: to a point with the coordinates x, y, z the following points are symmetry-equivalent:

$-x, -y, z;\ -y, x, z;\ y, -x, z$ etc.;
in addition, all points with $+(\frac{1}{2},\frac{1}{2},\frac{1}{2})$, *i.e.* $\frac{1}{2}+x, \frac{1}{2}+y, \frac{1}{2}+z;\ \frac{1}{2}-x, \frac{1}{2}-y, \frac{1}{2}+z;$
$\frac{1}{2}-y, \frac{1}{2}+x, \frac{1}{2}+z$ etc.
The coordinate triplets are just a shorthand notation for mappings according to equation (3.1).

The *Wyckoff symbol* is a short designation; it consists of a numeral followed by a letter, for example $8f$. The cipher 8 states the multiplicity, that is, the number of symmetry-equivalent points in the unit cell. The f is an alphabetical label $(a, b, c, \ldots)$ according to the sequence of the listing of the positions; a is always the position with the highest site symmetry.

A *(crystallographic) orbit* is the set of all points that are symmetry equivalent to a point. An orbit can be designated by the coordinate triplet of any of its points. If the coordinates of a point are fixed by symmetry, for example $0, \frac{1}{2}, \frac{1}{4}$, then the orbit and the Wyckoff position are identical. However, if there is a free variable, for example z in $0, \frac{1}{2}, z$, the Wyckoff position comprises an infinity of orbits. Take the points $0, \frac{1}{2}, 0.2478$ and $0, \frac{1}{2}, 0.3629$: they designate two *different* orbits; both of them belong to the *same* Wyckoff position $8g$ of the space group $I4/m$. Each of these points belongs to an orbit consisting of an infinity of points (don't get irritated by the singular form of the words 'Wyckoff position' and 'orbit').

3.5 Crystal Classes and Crystal Systems

A well-grown crystal exhibits a macroscopic symmetry which is apparent from its faces; this symmetry is intimately related to the pertinent space group. Due to its finite size, a macroscopic crystal can have no translational symmetry. In addition, due to the conditions of crystal growth, it hardly ever exhibits a perfect symmetry. However, the ideal symmetry of the crystal follows from the symmetry of the bundle of normals perpendicular to its faces. This symmetry is that of the point group resulting from the corresponding space group if translational symmetry is removed, screw axes are replaced by rotation axes, and glide planes are replaced by reflection planes. In this way the 230 space-group types can be correlated with 32 point groups which are called *crystal classes*. Examples of some space-group types and the crystal classes to which they belong are:

space-group type		crystal class	
full symbol	short symbol	full symbol	short symbol
$P12_1/c1$	$P2_1/c$	$12/m1$	$2/m$
$C2/m2/c2_1/m$	$Cmcm$	$2/m2/m2/m$	mmm
$P6_3/m2/m2/c$	$P6_3/mmc$	$6/m2/m2/m$	$6/mmm$

In general: the *P, A, B, C, F, I* or *R* of the space group symbol is removed, the subscript numbers are removed, and *a, b, c, d, e* or *n* are replaced by *m*

A special coordinate system defined by the basis vectors **a**, **b** and **c** belongs to each space group. Depending on the space group, certain relations hold among the basis vectors; they serve to classify seven *crystal systems*. Every crystal class can be assigned to one of these crystal systems, as listed in Table 3.1. The existence of the corresponding symmetry elements is relevant for assigning a crystal to a specific crystal system. The metric parameters of the unit cell alone are not sufficient (*e.g.* a crystal can be monoclinic even if $\alpha = \beta = \gamma = 90°$).

Table 3.1: The 32 crystal classes and the corresponding crystal systems

crystal system (abbreviation)	crystal classes	metric parameters of the unit cell
triclinic (*a*)	1; $\bar{1}$	$a \neq b \neq c$; $\alpha \neq \beta \neq \gamma \neq 90°$
monoclinic (*m*)	2; m; $2/m$	$a \neq b \neq c$; $\alpha = \gamma = 90°$, $\beta \neq 90°$ (or $\alpha = \beta = 90°$, $\gamma \neq 90°$)
orthorhombic (*o*)	222; $mm2$; mmm	$a \neq b \neq c$; $\alpha = \beta = \gamma = 90°$
tetragonal (*t*)	4; $\bar{4}$; $4/m$; 422; $4mm$; $\bar{4}2m$; $4/mmm$	$a = b \neq c$; $\alpha = \beta = \gamma = 90°$
trigonal (*h*)	3; $\bar{3}$; 32; $3m$; $\bar{3}m$	$a = b \neq c$; $\alpha = \beta = 90°$, $\gamma = 120°$
hexagonal (*h*)	6; $\bar{6}$; $6/m$; 622; $6mm$; $\bar{6}2m$; $6/mmm$	$a = b \neq c$; $\alpha = \beta = 90°$, $\gamma = 120°$
cubic (*c*)	23; $m\bar{3}$; 432; $\bar{4}3m$; $m\bar{3}m$	$a = b = c$; $\alpha = \beta = \gamma = 90°$

3.6 Aperiodic Crystals

Normally, solids are crystalline, *i.e.* they have a three-dimensional periodic order with three-dimensional translational symmetry. However, this is not always so. Aperiodic crystals do have a long-distance order, but no three-dimensional translational symmetry. In a formal (mathematical) way, they can be treated with lattices having translational symmetry in four- or five-dimensional 'space', the so-called superspace; their symmetry corresponds to a four- or five-dimensional *superspace group*. The additional dimensions are not dimensions in real space, but have to be taken in a similar way to the fourth dimension in space-time. In space-time the position of an object is specified by its spatial coordinates x, y, z; the coordinate t of the fourth dimension is the time at which the object is located at the site x, y, z.

We distinguish three kinds of aperiodic crystals:

1. Incommensurately modulated structures;
2. Incommensurate composite crystals;
3. Quasicrystals.

Incommensurately modulated structures can be described with a three-dimensional periodic average structure called *approximant*. However, the true atomic positions are shifted from the positions of the approximant. The shifts follow one or several modulation functions. An example is the modification of iodine that occurs at pressures between 23 and 28 GPa (iodine-V). The three-dimensional approximant is a face-centered structure in the orthorhombic space group $Fmmm$ (*cf.* Fig. 11.1, upper right, p. 104). In the incommensurately modulated structure the atoms are shifted parallel to **b** and follow a sine wave along **c** (Fig. 11.1, lower right). Its wave length is *incommensurate* with c, *i.e.* there is no rational numeric ratio with the lattice parameter c. The wave length depends on pressure; at 24.6 GPa it is $3.89c$. In this case the description is made with the three-dimensional space group $Fmmm$ with an added fourth dimension; the translation period of the axis in the fourth dimension is $3.89c$. The corresponding four-dimensional superspace group obtains the symbol $Fmmm(00q_3)0s0$ with $q_3 = 0.257 = 1/3.89$.

One of the structures of this kind that has been known for a long time is that of γ-Na_2CO_3. At high temperatures sodium carbonate is hexagonal (α-Na_2CO_3). It contains carbonate ions that are oriented perpendicular to the hexagonal c axis. Upon cooling below 481 °C, the c axis becomes slightly tilted to the ab plane, and the hexagonal symmetry is lost; the symmetry now is monoclinic (β-Na_2CO_3, space group $C2/m$). γ-Na_2CO_3 appears at temperatures between 332 °C and −103 °C. In the mean, it still has the structure of β-Na_2CO_3. However, the atoms are no longer arranged in a straight line along c, but follow a sine wave. In this case, the symbol of the superspace group is $C2/m(q_1 0 q_3)0s$. q_1 and q_3 are the reciprocal values of the components of the wave length of the modulation wave given as multiples of the lattice parameters a and c; they depend on pressure and temperature. Below −103 °C the modulation wave becomes commensurate with a wave length of $6\mathbf{a} + 3\mathbf{c}$. This structure can be described with a normal three-dimensional space group and a correspondingly enlarged unit cell.

In X-ray diffraction, modulated structures reveal themselves by the appearance of satellite reflections. In between the intense main reflections which correspond to the structure of the approximant, weaker reflections appear; they do not fit into the regular pattern of the main reflections.

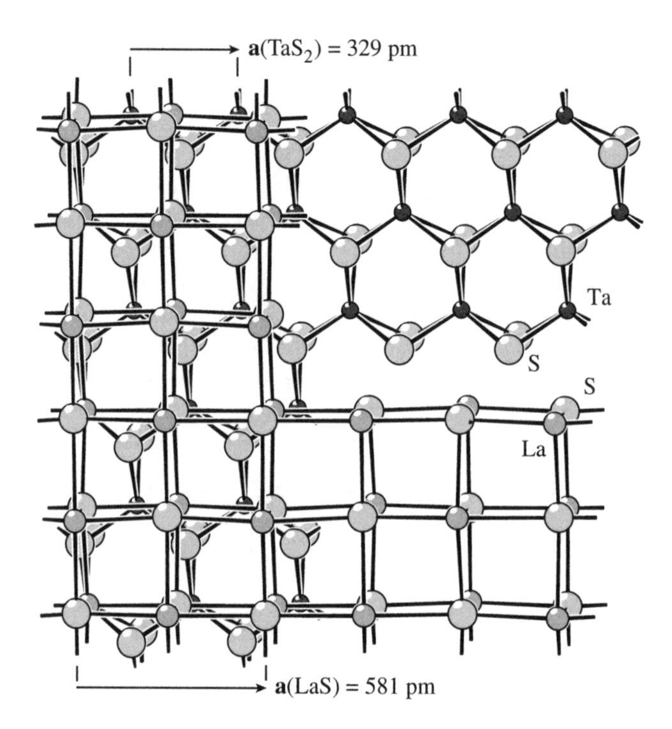

Fig. 3.12 $(LaS)_{1.14}TaS_2$: layers of the compositions LaS and TaS_2 are alternately stacked in the direction of view; to the right only one layer of each is shown

An **incommensurate composite crystal** can be regarded as the intergrowth of two periodic structures whose periodicities do not match with one another. The compound $(LaS)_{1.14}TaS_2$ offers an example. It consists of alternating stacked layers of the compositions LaS and TaS_2. Periodical order is present in the stacking direction. Parallel to the layers, their translational periods match in one direction, but not in the other direction. The translation vectors **a** are 581 pm in the LaS layer and 329 pm in the TaS_2 layer (Fig. 3.12). The chemical composition results from the numerical ratio $581/329 = 1.766$: $(LaS)_{2/1.766}TaS_2$ (the number of La atoms in a layer fraction of length 581 pm is twice that of the Ta atoms in 329 pm).

Quasicrystals exhibit the peculiarity of noncrystallographic symmetry operations. Most frequent are axial quasicrystals with a tenfold rotation axis. In addition, axial quasicrystals with five-, eight- and twelvefold rotation axes and quasicrystals with icosahedral symmetry have been observed. Axial quasicrystals have periodic order in the direction of the axis and can be described with the aid of five-dimensional superspace groups. Thus far, all observed quasicrystals are alloys. Generally, they have a complicated composition comprising one to three transition metals and mostly an additional main group element (mainly Mg, Al, Si or Te). In three-dimensional space, their structures can be described as nonperiodic tilings. At least two kinds of tiles are needed to attain a voidless filling of space. A well-known tiling is the PENROSE tiling. It has fivefold rotation symmetry and consists of two kinds of rhomboid tiles with rhombus angles of 72°/108° and 36°/144° (Fig. 3.13).

The X-ray diffraction pattern of a quasicrystal exhibits noncrystallographic symmetry. In addition, the number of observable reflections increases more and more the more intense the X-ray radiation is or the longer the exposure time is (in a similar way to the number of stars visible in the sky with a more potent telescope).

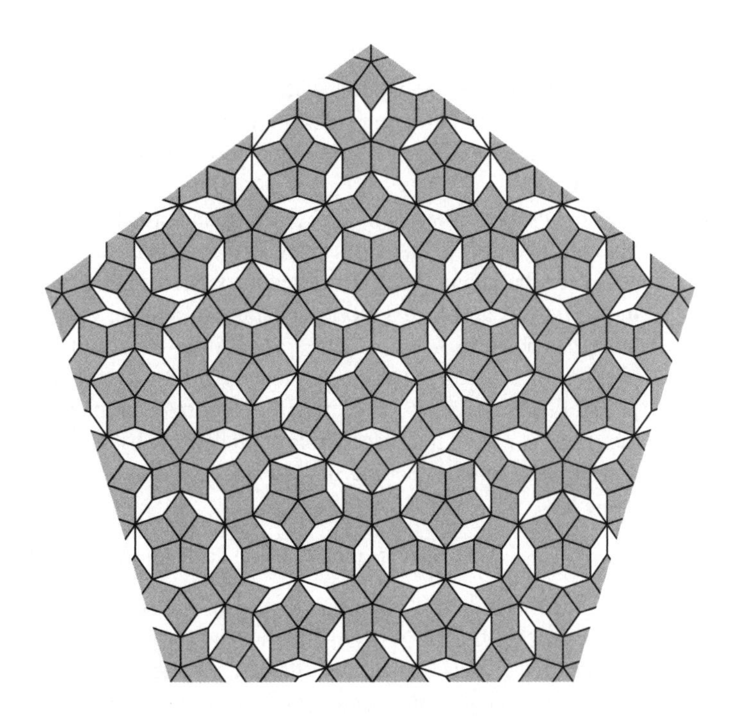

Fig. 3.13 PENROSE tiling with a fivefold symmetry axis consisting of two kinds of rhomboid tiles

3.7 Disordered Crystals

Several kinds of intermediate states exist between the state of highest order in a crystal having translational symmetry in three dimensions and the disordered distribution of particles in a liquid. *Liquid crystals* are closest to the liquid state. They behave macroscopically like liquids, their molecules are in constant motion, but to a certain degree there exists a crystal-like order.

In *plastic crystals* all or a part of the molecules rotate about their centers of gravity. Typically, plastic crystals are formed by nearly spherical molecules, for example hexafluorides like SF_6 or MoF_6 or white phosphorus in a temperature range immediately below the melting point. Such crystals often are soft and can be easily deformed.

The term plastic crystal is not used if the rotation of the particles is hindered, *i.e.* if the molecules or ions perform rotational vibrations (librations) about their centers of gravity with large amplitudes; this may include the occurrence of several preferred orientations. Instead, such crystals are said to have orientational disorder. Such crystals are annoying during crystal structure analysis by X-ray diffraction because the atoms can hardly be located. This situation is frequent among ions like BF_4^-, PF_6^- or $N(CH_3)_4^+$. To circumvent difficulties during structure determination, experienced chemists avoid such ions and prefer heavier, less symmetrical or more bulky ions.

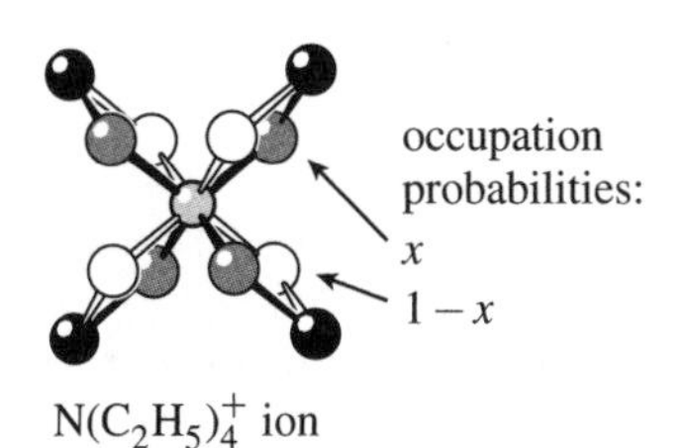

$N(C_2H_5)_4^+$ ion

Orientational disorder is also present if a molecule or part of a molecule occupies two or more different orientations in the crystal, even without performing unusual vibrations. For example, tetraethylammonium ions often occupy two orientations that are mutually rotated by 90°, in such a way that the positions of the C atoms of the methyl groups coincide, but the C atoms of the CH_2 groups occupy the vertices of a cube around the N atom, with two occupation probabilities.

Plastic crystals and crystals with orientational disorder still fulfill the three-dimensional translational symmetry, provided a mean partial occupation is assumed for the atomic positions of the molecules whose orientations differ from unit cell to unit cell ('split positions').

Crystals having *stacking faults* lack translational symmetry in one direction. Such crystals consist of layers that are stacked without a periodic order. Usually, only a few positions occur for the layers, but their sequence is statistical. For example, if we have two layers in positions A and B in a closest-packing of spheres and the following (third) layer again takes the position A with 100 % probability, the result is an ordered hexagonal closest-packing of spheres (stacking sequence $ABAB\ldots$, *cf.* Fig. 14.1, p. 151). However, if the probability is only 90 % and the third layer may adopt a position C with 10 % probability, the result is a stacking disorder. Essentially, *i.e.* by 90 %, the packing still corresponds to a hexagonal closest-packing of spheres, but on an average there is a stacking fault every ten layers:

$$\ldots {}_{A}B{}_{A}B{}_{A}B{}_{A}B{}_{A}B^{C}B^{C}B^{C}B{}_{A}B{}_{A}B{}_{A}B{}_{A}B{}_{A}B{}_{A}B{}_{A}B^{C}B^{C}B^{C}B^{C}B\ldots$$

Metallic cobalt exhibits this phenomenon, and so do layered silicates and layered halides like CdI_2 or BiI_3. In X-ray diffraction, stacking faults cause the appearance of diffuse streaks (continuous lines in the diffraction pattern).

If the stacking faults occur only rarely (say, every 10^5 layers on average), the result is a polysynthetic twinned crystal (*cf.* Fig. 18.8, p. 223). Depending on the frequency of the stacking faults, there is a smooth transition between crystals with stacking faults and polysynthetic twinning.

Among crystals with stacking faults the lack of a periodic order is restricted to one dimension; this is called a *one-dimensional disorder*. If only a few layer positions occur and all of them are projected into one layer, we obtain an *averaged structure*. Its symmetry can be described with a space group, albeit with partially occupied atomic positions. The real symmetry is restricted to the symmetry of an individual layer. The layer is a three-dimensional object, but it only has translational symmetry in two dimensions. Its symmetry is that of a *layer group*; there exist 80 layer-group types.

A two-dimensional disorder results when rod-like polymeric molecules are mutually shifted with statistical frequency. Translational symmetry then only exists in the direction of the molecules, and not in the transverse directions. The rod is a three-dimensional object with one-dimensional translational symmetry. Its symmetry is that of a *rod group*. Layer groups and rod groups are *subperiodic groups*. They are listed in detail in *International Tables for Crystallography*, Volume E.

Structures with one- or two-dimensional disorder are also called order–disorder structures (OD structures).

A solid that has no translational symmetry at all is said to be *amorphous*. Glasses are amorphous solids that behave like liquids with very high viscosity. The viscosity decreases with increasing temperature, *i.e.* the material softens, but it has no melting point.

3.8 Problems

3.1 Give the Hermann–Mauguin symbols for the following molecules or ions:
H_2O, $HCCl_3$, BF_3 (triangular planar), XeF_4 (square planar), $ClSF_5$, SF_6, *cis*-$SbF_4Cl_2^-$, *trans*-N_2F_2, $B(OH)_3$ (planar), $Co(NO_2)_6^{3-}$.

3.2 Plots of the following molecules or ions can be found on pp. 132, 133 and 146. State their Hermann–Mauguin symbols.
Si_4^{6-}, As_4S_4, P_4S_3, Sn_5^{2-}, As_4^{6-}, As_4^{4-}, P_6^{6-}, As_7^{3-}, P_{11}^{3-}, Sn_9^{2-}, Bi_8^{2+}.

3.3 What Hermann–Mauguin symbols correspond to the linked polyhedra shown in Fig. 16.1 (p. 166)?

3.4 What symmetry elements are present in the HgO chain shown on page 13? Does its symmetry correspond to a point group, rod group, layer group or space group?

3.5 Find out which symmetry elements are present in the structures of the following compounds. Derive the Hermann–Mauguin symbol of the corresponding space group (it may be helpful to consult *International Tables for Crystallography,* Vol. A).
Tungsten bronzes M_xWO_3 (Fig. 16.6, p. 172); CaC_2 (Fig. 7.6, heavily outlined cell, p. 57); CaB_6 (Fig. 13.13, p. 145).

3.6 State the crystal classes and crystal systems to which the following space groups belong:
(a) $P2_1/b2_1/c2_1/a$; (b) $I4_1/amd$; (c) $R\bar{3}2/m$; (d) $C2/c$; (e) $P6_3/m$; (f) $P6_322$; (g) $P2_12_12_1$; (h) $Fdd2$; (i) $Fm\bar{3}m$.

3.7 Rutile (TiO_2, Fig. 3.10) crystallizes in the space group $P4_2/mnm$ (Nr. 136 in *International Tables for Crystallography,* Vol. A). The atomic coordinates are: Ti 0, 0, 0; O 0.303, 0.303, 0. Which Wyckoff positions are occupied by the atoms? How many atoms are in one unit cell? What are the site symmetries of the atoms?

3.8 What is the point group of the PENROSE tiling (Fig. 3.13) if it consists of one layer of tiles? What is the point group if two layers are stacked with their midpoints one on top of the other, the second layer being rotated by 180°?

4 Polymorphism and Phase Transitions

4.1 Thermodynamic Stability

When the free enthalpy of reaction ΔG for the transformation of the structure of a compound to any other structure is positive, then this structure is *thermodynamically stable*. Since ΔG depends on the transition enthalpy ΔH and the transition entropy ΔS, and ΔH and ΔS in turn depend on pressure and temperature, a structure can be stable only within a certain range of pressures and temperatures. By variation of the pressure and/or the temperature, ΔG will eventually become negative relative to some other structure and a phase transition will occur. This may be a phase transition from a solid to another solid modification, or it may be a transition to another aggregate state.

According to the thermodynamic relations

$$\Delta G = \Delta H - T\Delta S \quad \text{and} \quad \Delta H = \Delta U + p\Delta V \tag{4.1}$$

the following rules can be given for the temperature and pressure dependence of thermodynamically stable structures:

1. With increasing temperature T structures with a low degree of order will be favored. Their formation involves a positive transition entropy ΔS and the value of ΔG then depends mainly on the term $T\Delta S$. For instance, among hexahalides such as MoF_6 two modifications are known in the solid state, one having molecules with well-defined orientations and the other having molecules rotating about their centers of gravity within the crystal. Since the order is lower for the latter modification, it is the thermodynamically stable one at higher temperatures. In the liquid state, the order is even lower and it is the lowest in the gaseous state. Raising the temperature will thus lead to melting and finally to evaporation of the substance.

2. Higher pressures p favor structures that occupy a lower volume, *i.e.* that have a higher density. As their formation involves a decrease in the volume (negative ΔV), ΔH will attain a negative value. For instance, diamond (density 3.51 g cm^{-3}) is more stable than graphite (density 2.26 g cm^{-3}) at very high pressures.

4.2 Kinetic Stability

A thermodynamically unstable structure can exist when its conversion to some other structure proceeds at a negligible rate. In this case we call the structure *metastable*, *inert* or *kinetically stable*. Since the rate constant k depends on the activation energy E_a and the temperature according to the ARRHENIUS equation,

$$k = k_0 \, \mathrm{e}^{-E_a/RT}$$

we have kinetic stability whenever a negligibly low k results from a large ratio E_a/RT. At sufficiently low temperatures any structure can be stabilized kinetically. Kinetic stability is

Inorganic Structural Chemistry, Second Edition Ulrich Müller

not a well-defined term because the limit below which a conversion rate is to be considered negligible is arbitrary.

Glasses typically are metastable substances. Like crystalline solids they exhibit macroscopic form stability, but because of their structures and some of their physical properties they must be considered as liquids with a very high viscosity. Their transition to a thermodynamically more stable structure can only be achieved by extensive atomic movements, but atom mobility is severely hindered by cross-linking.

The structures and properties of numerous substances that are thermodynamically unstable under normal conditions are only known because they are metastable and therefore can be studied under normal conditions.

4.3 Polymorphism

Molecules having the same composition but different structures are called isomers. The corresponding phenomenon for crystalline solids is called *polymorphism*. The different structures are the *modifications* or *polymorphic forms*. Modifications differ not only in the spatial arrangement of their atoms, but also in their physical and chemical properties. The structural differences may comprise anything from minor variations in the orientation of molecules up to a completely different atomic arrangement.

Different modifications of a compound are frequently designated by lower case Greek letters α, β, ..., *e.g.* α-sulfur, β-sulfur, or by roman numerals, *e.g.* tin-I, tin-II etc. Polymorphic forms of minerals have in many cases been given trivial names, like α-quartz, β-quartz, tridymite, cristobalite, coesite, keatite, and stishovite for SiO_2 forms.

More systematic (but not always unambiguous) is the designation by PEARSON symbols; their use is recommended by IUPAC (International Union of Pure and Applied Chemistry). A PEARSON symbol consists of a lower case letter for the crystal system (*cf.* the abbreviations in Table 3.1, p. 24), an upper case letter for the kind of centering of the lattice (*cf.* Fig. 2.6, p. 8) and the number of atoms in the unit cell. Example: sulfur-$oF128$ is orthorhombic, face centered and has 128 atoms per unit cell (α-sulfur).

Polymorphic forms with structures having different stacking sequences of like layers are called *polytypes*.

Which polymorphic form of a compound is formed depends on the preparation and crystallization conditions: method of synthesis, temperature, pressure, kind of solvent, cooling or heating rate, crystallization from solution, fusion or gas phase, and presence of seed crystals are some of the factors of influence.

When a compound that can form several modifications crystallizes, first a modification may form that is thermodynamically unstable under the given conditions; afterwards it converts to the more stable form (OSTWALD step rule). Selenium is an example: when elemental selenium forms by a chemical reaction in solution, it precipitates in a red modification that consists of Se_8 molecules; this then converts slowly into the stable, gray form that consists of polymeric chain molecules. Potassium nitrate is another example: at room temperature β-KNO_3 is stable, but above 128 °C α-KNO_3 is stable. From an aqueous solution at room temperature α-KNO_3 crystallizes first, then, after a short while or when triggered by the slightest mechanical stress, it transforms to β-KNO_3.

The nucleation energy governs which modification crystallizes first. This energy depends on the surface energy. As a rule, nucleation energy decreases with decreasing surface energy. The modification having the smallest nucleation energy crystallizes first. As the surface energy depends sensitively on the adsorption of extraneous particles, the sequence of crystallization of polymorphic forms can be influenced by the presence of foreign matter.

4.4 Phase Transitions

Definition: A phase transition is an event which entails a discontinuous (sudden) change of at least one property of a material.

Generally, a phase transition is triggered by an external stress which most commonly is a change in temperature or pressure. Properties that can change discontinuously include volume, density, specific heat, elasticity, compressibility, viscosity, color, electric conductivity, magnetism and solubility. As a rule, albeit not always, phase transitions involve structural changes. Therefore, a phase transition in the solid state normally involves a change from one to another modification.

If a modification is unstable at every temperature and every pressure, then its conversion into another modification is irreversible; such phase transitions are called *monotropic*. *Enantiotropic* phase transitions are reversible; they proceed under equilibrium conditions ($\Delta G = 0$). The following considerations are valid for enantiotropic phase transitions that are induced by a variation of temperature or pressure.

The first derivatives of the free enthalpy $G = U + pV - TS$ are

$$\left.\frac{\partial G}{\partial T}\right|_p = -S \quad \text{and} \quad \left.\frac{\partial G}{\partial p}\right|_T = V$$

If one of these quantities experiences a discontinuous change, *i.e.* if $\Delta S \neq 0$ or $\Delta V \neq 0$, then the phase transition is called a *first-order* transition according to EHRENFEST. It is accompanied by the exchange of conversion enthalpy $\Delta H = T\Delta S$ with the surroundings.

First-order phase transitions exhibit *hysteresis*, *i.e.* the transition takes place some time after the temperature or pressure change giving rise to it. How fast the transformation proceeds also depends on the formation or presence of sites of nucleation. The phase transition can proceed at an extremely slow rate. For this reason many thermodynamically unstable modifications are well known and can be studied in conditions under which they should already have been transformed.

In a *second-order* phase transition, volume and entropy experience a continuous variation, but at least one of the second derivatives of G exhibits a discontinuity:

$$\left.\frac{\partial^2 G}{\partial T^2}\right|_p = -\left.\frac{\partial S}{\partial T}\right|_p = -\left.\frac{1}{T}\frac{\partial H}{\partial T}\right|_p = -\frac{C_p}{T} \quad \text{and} \quad \left.\frac{\partial^2 G}{\partial p^2}\right|_T = \left.\frac{\partial V}{\partial p}\right|_T = -V\kappa$$

C_p is the specific heat at constant pressure, κ is the compressibility at constant temperature. The conversion process of a second-order phase transition can extend over a certain temperature range. If it is linked with a change of the structure (which usually is the case), this is a continuous structural change. There is no hysteresis and no metastable phases occur. A transformation that almost proceeds in a second-order manner (very small discontinuity of volume or entropy) is sometimes called 'weakly first order'.

Solid-state phase transitions can be distinguished according to BUERGER in the following manner:

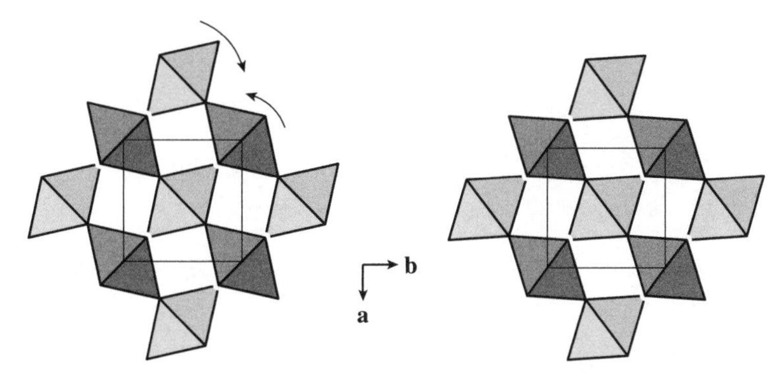

$CaCl_2$, > 217 °C (rutile type), $P4_2/mnm$ — $a = b = 637.9$ pm, $c = 419.3$ pm at 247 °C

$CaCl_2$, 17 °C, $Pnnm$ — $a = 625.9$ pm, $b = 644.4$ pm, $c = 417.0$ pm

Fig. 4.1 Rotation of the coordination octahedra during the second-order phase transition of $CaCl_2$

1. **Reconstructive phase transitions**: Chemical bonds are broken and rejoined; the reconstruction involves considerable atomic movements. Such conversions are always first-order transitions.
2. **Displacive phase transitions**: Atoms experience small shifts; if at all, only intermolecular bonds (*e.g.* hydrogen bonds) are broken and rejoined, but no primary chemical bonds. The transition may be but need not be a second-order transition.
3. **Order–disorder transitions**: Different atoms that statistically occupy the same position become ordered or vice versa. Usually this is a second-order transition.

An example of a continuous structural change is the second-order phase transition of calcium chloride (Fig. 4.1). At higher temperatures it is tetragonal (rutile type). When cooled, a mutual rotation of the coordination octahedra starts at 217 °C. As soon as the slightest rotation has taken place, the symmetry can no longer be tetragonal; at 217 °C a symmetry reduction occurs, the symmetry is 'broken'. The orthorhombic space group $Pnnm$ of the low-temperature modification has fewer symmetry operations than the space group $P4_2/mnm$ of the rutile type; $Pnnm$ is a subgroup of $P4_2/mnm$. For second-order phase transitions it is mandatory that there is a group–subgroup relation between the two space groups. For more details see Chapter 18.

With the aid of an *order parameter* one can follow the changes taking place during a second-order phase transition. A quantity that experiences changes during the phase transition and which becomes zero at the *critical temperature* T_C (or the critical pressure) can be chosen as the order parameter. T_C is the temperature at which the phase transition sets in and the symmetry is broken. For example, for $CaCl_2$ the rotation angle of the octahedra or the ratio $\eta = (b-a)/(b+a)$ of the lattice parameters can be used as the order parameter. According to LANDAU theory, below T_C the order parameter changes according to a power function of the temperature difference $T_C - T$:

$$\eta = A\left(\frac{T_C - T}{T_C}\right)^{\beta}$$

A is a constant and β is the *critical exponent* which adopts values from 0.3 to 0.5. Values around $\beta = 0.5$ are observed for long-range interactions between the particles; for short-range interactions (*e.g.* magnetic interactions) the critical exponent is closer to $\beta \approx 0.33$. As shown in the typical curve diagram in Fig. 4.2, the order parameter experiences its most relevant changes close to the critical temperature; the curve runs vertical at T_C.

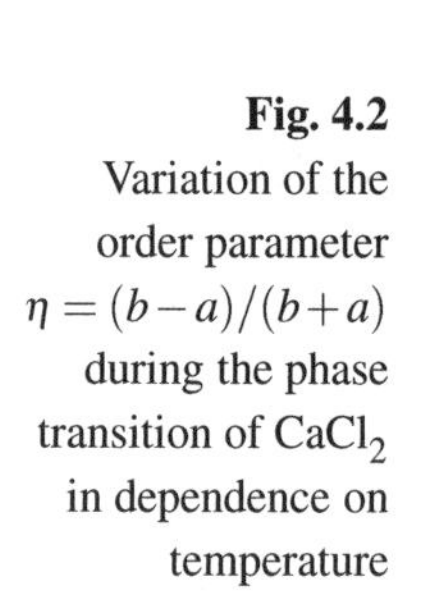
Fig. 4.2 Variation of the order parameter $\eta = (b-a)/(b+a)$ during the phase transition of $CaCl_2$ in dependence on temperature

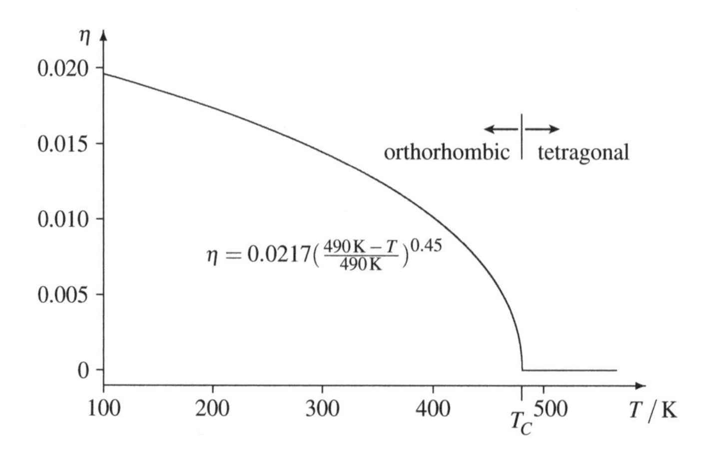

4.5 Phase Diagrams

For phases that are in equilibrium with one another, the GIBBS phase law holds:

$$F + P = C + 2$$

F is the number of degrees of freedom, *i.e.* the number of variables of state such as temperature and pressure that can be varied independently, P is the number of phases and C is the number of components. Components are to be understood as independent, pure substances (elements or compounds), from which the other compounds that eventually occur in the system can be formed. For example:

1. For pure water (one component, $C = 1$) $F + P = 3$ holds. When three phases are simultaneously in equilibrium with each other, *e.g.* vapor, liquid and ice, or vapor and two different modifications of ice, then $F = 0$; there is no degree of freedom, the three phases can coexist only at one fixed pressure and one fixed temperature ('triple point').
2. In the system iron/oxygen ($C = 2$), when two phases are present, *e.g.* Fe_3O_4 and oxygen, pressure and temperature can be varied ($F = 2$). When three phases are in equilibrium, *e.g.* Fe, Fe_2O_3 and Fe_3O_4, only one degree of freedom exists, and only the pressure *or* the temperature can be chosen freely.

A *phase diagram* in which pressure is plotted *vs.* temperature shows the existence ranges for the different phases of a system comprising only one component. Fig. 4.3 displays the phase diagram for water, in which the ranges of existence of liquid water and ten different modifications of ice are discernible, the latter being designated by Roman numerals. Within each of the marked fields only the corresponding phase is stable, but pressure and temperature can be varied independently (2 degrees of freedom). Along the delimiting lines two phases can coexist, and either the pressure or the temperature can be varied, whereas the other one has to adopt the value specified by the diagram (one degree of freedom). At triple points there is no degree of freedom, pressure and temperature have fixed values, but three phases are simultaneously in equilibrium with each other.

In phase diagrams for two-component systems the composition is plotted *vs.* one of the variables of state (pressure or temperature), the other one having a constant value. Most common are plots of the composition *vs.* temperature at ambient pressure. Such phase diagrams differ depending on whether the components form solid solutions with each other or not or whether they combine to form compounds.

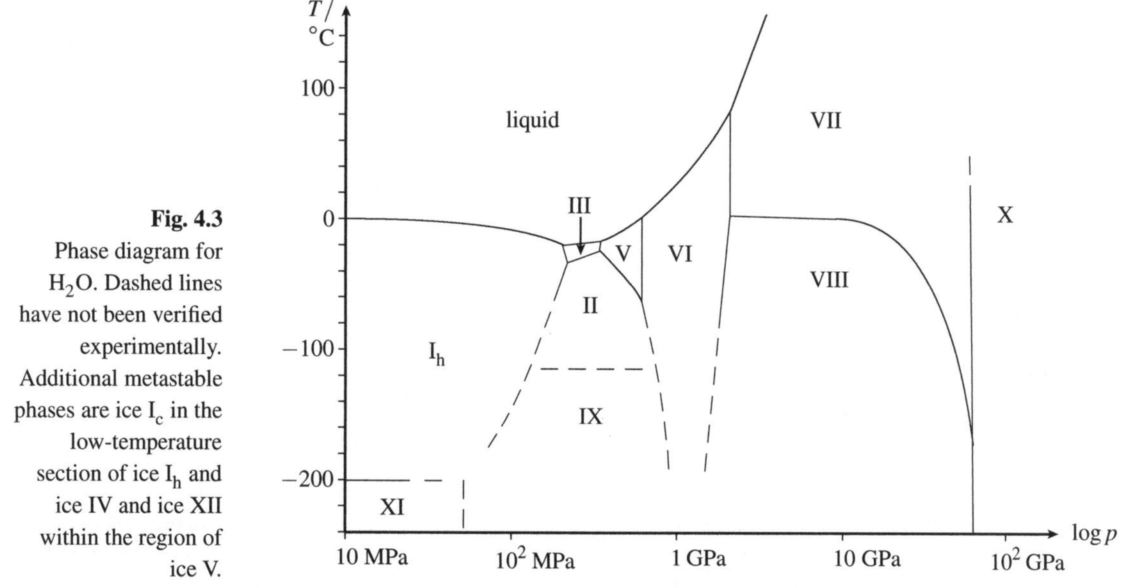

Fig. 4.3 Phase diagram for H_2O. Dashed lines have not been verified experimentally. Additional metastable phases are ice I_c in the low-temperature section of ice I_h and ice IV and ice XII within the region of ice V.
Not shown: region of gaseous H_2O at pressures below 22 MPa. The line delimiting the liquid and ice VII continues up to at a triple point at 43 GPa and 1370 °C; at this triple point it probably meets the line delimiting ice VII and ice X. The melting point continues to move upward until approximately 2150 °C at 90 GPa

The phase diagram for the antimony/bismuth system, in which mixed crystals (solid solutions) are formed, is shown in Fig. 4.4. Crystalline antimony and bismuth are isotypic, and Sb and Bi atoms can occupy the atomic positions in any proportion. The upper part of the diagram corresponds to the range of existence of the liquid phase, *i.e.* a liquid solution of antimony and bismuth. The lower part corresponds to the range of existence of the mixed crystals. In between is a range in which liquid and solid coexist. On the upper side it is delimited by the *liquidus curve*, and on the lower side by the *solidus curve*. At a given temperature, the liquid and the solid that are in equilibrium with one another have different compositions. The compositions can be read from the cross-points of the horizontal straight line marking the temperature in question with the *solidus* and the *liquidus* curves. Upon cooling an Sb/Bi melt with a composition corresponding to the point marked A in Fig. 4.4, crystallization begins when the temperature marked by the horizontal arrow is reached. The composition of the mixed crystals that form is that of point B — the mixed crystals have a higher Sb content than the melt.

The potassium/caesium phase diagram is an example of a system involving the formation of mixed crystals with a temperature minimum (Fig. 4.4). The right and left halves of the diagram are of the same type as the diagram for antimony/bismuth. The minimum corresponds to a special point for which the compositions of the solid and the liquid are the same. Other systems can have the special point at a temperature maximum.

Limited formation of mixed crystals occurs when the two components have different structures, as for example in the case of indium and cadmium. Mixed crystals containing much indium and little cadmium have the structure of indium, while those containing little indium and much cadmium have the cadmium structure. At intermediate compositions a gap is observed, *i.e.* there are no homogeneous mixed crystals, but instead a mixture of crystals rich in indium and crystals rich in cadmium is formed. This situation can even

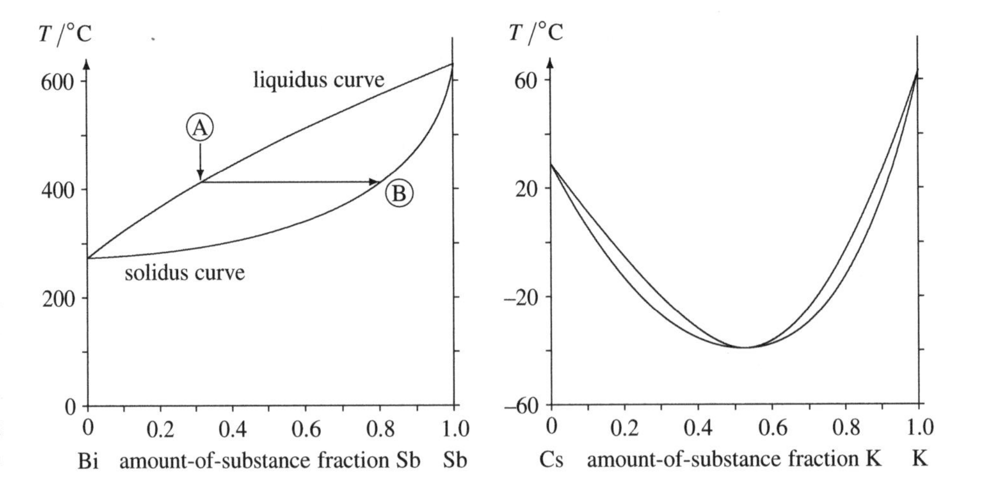

Fig. 4.4 Phase diagrams for antimony/bismuth and potassium/caesium at ambient pressure

occur when both of the pure components have the same structure, but mixed crystals may not have any arbitrary composition. Copper and silver offer an example; the corresponding phase diagram is shown in Fig. 4.5.

The phase diagram for aluminum/silicon (Fig. 4.5) is a typical example of a system of two components that form neither solid solutions (except for very low concentrations) nor a compound with one another, but are miscible in the liquid state. As a special feature an acute minimum is observed in the diagram, the *eutectic point*. It marks the melting point of the *eutectic mixture*, which is the mixture which has a lower melting point than either of the pure components or any other mixture. The *eutectic line* is the horizontal line that passes through the eutectic point. The area underneath is a region in which both components coexist as solids, *i.e.* in two phases.

A liquid solution of aluminum and silicon containing an amount-of-substance fraction of 40 % aluminum and having a temperature of 1100 °C corresponds to point A in Fig. 4.5. Upon cooling the liquid we move downwards in the diagram (as marked by the arrow). At the moment we reach the liquidus line, pure silicon begins to crystallize. As a consequence, the composition of the liquid changes as it now contains an increasing fraction of aluminum; this corresponds to a leftward movement in the diagram. There the crystal-

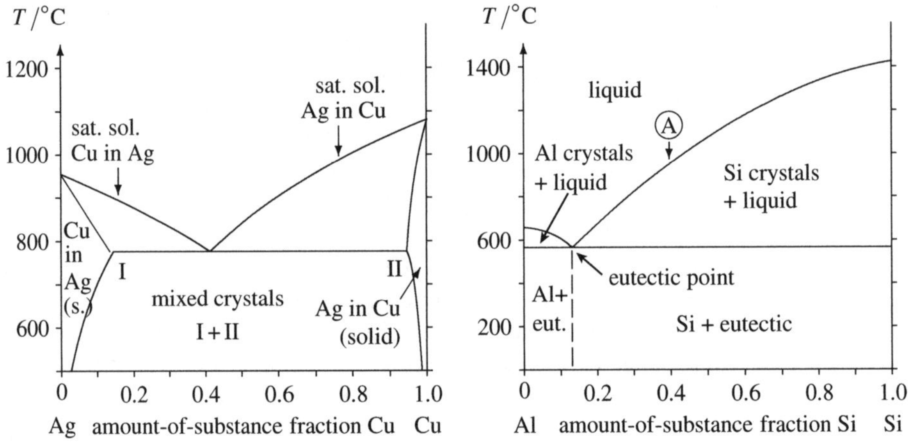

Fig. 4.5 Phase diagrams for the systems silver/copper (limited solubility in the solid) and aluminum/silicon (formation of an eutectic mixture)

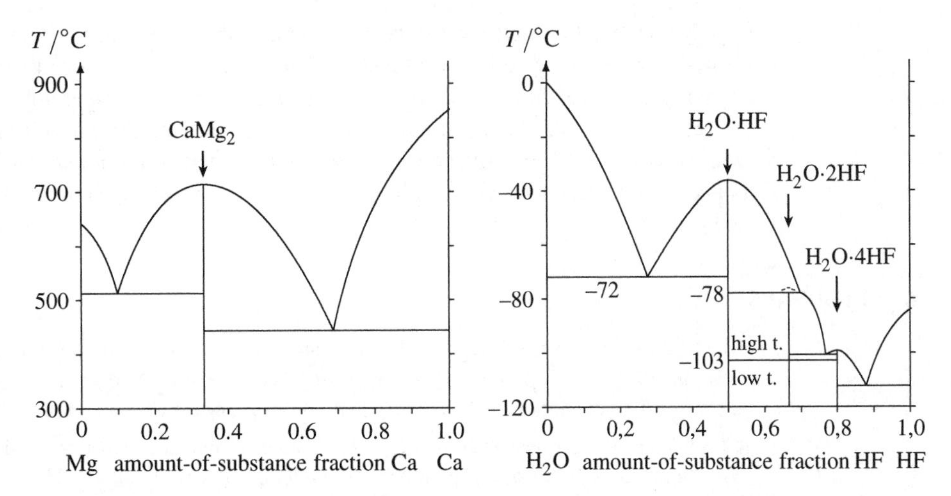

Fig. 4.6 Phase diagrams for the systems calcium/magnesium and H_2O/HF: formation of one and three compounds, respectively

lization temperature for silicon is lower. According to the amount of crystallizing silicon, the temperature for the crystallization of further silicon decreases more and more until finally the eutectic point is reached, where both aluminum and silicon solidify. The term *incongruent solidification* serves to express the continuous change of the solidification temperature.

When the two components form compounds with each other, more complicated conditions arise. Fig. 4.6 shows the phase diagram for a system in which the two components, magnesium and calcium, form a compound, $CaMg_2$. At the composition $CaMg_2$ we observe a maximum which marks the melting point of this compound. On the left there is a eutectic point, formed by the components Mg and $CaMg_2$. On the right there is another eutectic point with the components Ca and $CaMg_2$. Both the left and right parts of the phase diagram in Fig. 4.6 correspond to the phase diagram of a simple eutectic system as in Fig. 4.5.

The system H_2O/HF exhibits an even more complicated phase diagram, as three compounds occur: $H_2O{\cdot}HF$, $H_2O{\cdot}2\,HF$ and $H_2O{\cdot}4\,HF$. In these compounds H_3O^+, HF and F^- particles are joined with each other in different ways via hydrogen bridges. For two of the compounds we find maxima in the phase diagram (Fig. 4.6), in a similar way as for the compound $CaMg_2$. However, there is no maximum at the composition $H_2O{\cdot}2\,HF$ and no eutectic point between $H_2O{\cdot}2\,HF$ and $H_2O{\cdot}HF$; we can only discern a kink in the liquidus line, the *peritectic point*. The expected maximum is 'covered' (dashed line). The horizontal line running through the kink is called the *peritectic line*. The solids of composition $H_2O{\cdot}HF$ and $H_2O{\cdot}4\,HF$ show congruent melting, *i.e.* they have definite melting temperatures according to the maxima in the diagram. The solid with the composition $H_2O{\cdot}2\,HF$, however, shows incongruent melting: at $-78\,°C$ it decomposes to solid $H_2O{\cdot}HF$ and a liquid with a higher HF content. In addition, the compound $H_2O{\cdot}2\,HF$ experiences a phase transition at $-103\,°C$ from a 'high-temperature' to a low-temperature modification; in the diagram this is expressed by the horizontal line at this temperature.

Phase diagrams give valuable information about the compounds that can form in a system of components. These compounds can then be prepared and studied. For the experimental determination of phase diagrams the following methods are used. In *differential thermal analysis* (DTA) a sample of a given composition is heated or cooled slowly together with a thermally indifferent reference substance, and the temperatures of both substances are monitored continuously. When a phase transition occurs in the sample, the enthalpy of conversion is freed or absorbed and therefore a temperature difference shows up between the sample and the reference, thus indicating the phase transition. In *X-ray phase studies* an X-ray diffraction pattern is recorded continuously while the sample is being cooled or heated; when a phase transition occurs, changes in the diffraction pattern show up.

4.6 Problems

4.1 The densities of some SiO_2 modifications are: α-quartz 2.65 g cm^{-3}, β-quartz 2.53 g cm^{-3}, β-tridymite 2.27 g cm^{-3}, β-cristobalite 2.33 g cm^{-3}, vitreous SiO_2 2.20 g cm^{-3}. Should it be possible to convert β-cristobalite to some of the other modifications by applying pressure?

4.2 Silica glass is formed when molten SiO_2 is cooled rapidly. It experiences slow crystallization. Will the rate of crystallization be higher at room temperature or at 1000 °C?

4.3 BeF_2, like quartz, has a polymeric structure with F atoms linking tetrahedrally coordinated Be atoms; BF_3 is monomeric. When cooling the liquid down to solidification, which of the two is more likely to form a glass?

4.4 Is the conversion α-KNO_3 $\rightarrow$ β-KNO_3 (*cf.* p. 31) a first- or second-order phase transition?

4.5 CaO experiences a phase transition form the NaCl type to the CsCl type at a pressure of 65 GPa (images in Fig. 7.1, p. 53). What kind of a transformation is this?

4.6 Will ice at a temperature of $-10\,°C$ melt if pressure is applied to it? If so, will it refreeze if the pressure is increased even more? Which modification would have to form?

4.7 Can water at 40 °C be made to freeze? If so, what modification(s) of ice will form?

4.8 What will happen when a solution of HF and water containing an amount-of-substance fraction of 40 % HF is cooled from 0 °C to $-100\,°C$?

4.9 As shown in Fig. 12.11 (p. 126), upon heating at ambient pressure, β-quartz will experience phase transitions to β-tridymite and then to β-cristobalite at 870 °C and 1470 °C, respectively. Is it feasible to achieve a direct interconversion β-quartz $\rightleftharpoons$ β-cristobalite by temperature variation at high pressure?

5 Chemical Bonding and Lattice Energy

5.1 Chemical Bonding and Structure

Which spatial arrangement of atoms results in a stable or metastable structure depends decisively on the distribution of their electrons.

For noble gases the electronic configuration of a single atom is thermodynamically stable at normal conditions. Merely by packing the atoms closer together to form a liquid or a solid a small amount of VAN DER WAALS energy can still be released. The condensation and crystallization enthalpies being rather small, the magnitude of ΔG is governed already at relatively low temperatures by the term $T\Delta S$, and correspondingly the melting and boiling points are low.

For all other elements the electronic configuration of a single atom does not correspond to a thermodynamically stable state at normal conditions. Only at very high temperatures do single atoms occur in the vapor phase. At more ordinary temperatures atoms have to be linked to produce stable structures.

The electrons in an aggregate of atoms can only exist in certain definite energy states, just as in a single atom. These states are expressed mathematically by the eigenvalues of wave functions ψ. The wave functions result theoretically as solutions of the SCHRÖDINGER equation for the *complete set of all constituent atoms*. Although the exact mathematical solution of this equation poses insurmountable difficulties, we do have a well-founded knowledge about wave functions and thus about electrons in atomic systems. The knowledge is based on experimental data and on mathematical approximations that by now have become quite reliable; we will discuss this more broadly in Chapter 10. To begin with, we will restrict ourselves to the simplified scheme of two extreme kinds of chemical bonding, namely ionic and covalent bonds. However, we will also allow for intermediate states between these two extreme cases and we will consider the coexistence of both bonding types. As far as is relevant, we will also take into account the weaker ion–dipole, dipole–dipole and dispersion interactions.

The (localized) covalent bond is distinguished by its short range of action, which usually extends only from one atom to the next. However, within this range it is a strong bond. A near order arises around an atom; it depends on the one hand on the tight interatomic bonds and on the other hand on the mutual repulsion of the valence electrons and on the space requirements of the bonded atoms.* When atoms are linked to form larger structures, the near order can result in a long-range order in a similar way as the near order around a brick propagates to a long-range order in a brick wall.

In a nonpolymer molecule or molecular ion a limited number of atoms are linked by covalent bonds. The covalent forces within the molecule are considerably stronger than

*Frequently, *directionality* is a property attributed to the covalent bond which supposedly is taken to be the *cause* of the resulting structures. However, as the success of the valence electron pair repulsion theory shows, there exists no need to assume any orbitals directed *a priori*. The concept of directed orbitals is based on calculations in which hybridization is used as a *mathematical* aid. The popular use of hybridization models occasionally has created the false impression that hybridization is some kind of process occurring prior to bond formation and committing stereochemistry.

Inorganic Structural Chemistry, Second Edition Ulrich Müller

all forces acting outwards. For this reason, when molecular structures are being considered, one commits only a small error when one acts as if the molecule would occur by itself and had no surroundings. Common experience as well as more detailed studies by KITAIGORODSKY and calculations with force fields show that bond lengths and angles in molecules usually only undergo marginal alterations when the molecules assemble to a crystal. Only conformation angles are influenced more significantly in certain cases. Many properties of a molecular compound can therefore be explained from the molecular structure.

This is not equally valid for macromolecular compounds, in which a molecule consists of a nearly unlimited number of atoms. The interactions with surrounding molecules cannot be neglected in this case. For instance, for a substance consisting of thread-like macromolecules, it makes a difference to the physical properties whether the molecules are ordered in a crystalline manner or whether they are tangled.

Crystalline macromolecular substances can be classified according to the kind of connectivity of the covalently linked atoms as chain structures, layer structures and framework structures. The chains or layers may be electrically uncharged molecules that interact with each other only by VAN DER WAALS forces, or they can be polyanions or polycations held together by intervening counter-ions. Framework structures can also be charged, the counter-ions occupying cavities in the network. The structure of the chain, layer or framework depends to a large extent on the covalent bonds and the resulting near order around each atom.

On the other hand, the crystal structures of ionic compounds with small molecular ions depend mainly on how space can be filled most efficiently by the ions, following the principle of cations around anions and anions around cations. Geometric factors such as the relative size of the ions and the shape of molecular ions are of prime importance. More details are given in Chapter 7.

5.2 Lattice Energy

Definition: Lattice energy is the energy released when one mole of a crystalline compound is assembled at a temperature of 0 K from its infinitely separated components.

In this sense, components are taken to be:

- for molecular compounds: the molecules
- for ionic compounds: the ions
- for metals: the atoms
- for pure elements, excluding molecular species such as H_2, N_2, S_8 etc.: the atoms

For compounds that cannot be assigned uniquely to one of these substance classes, the specification of a lattice energy makes sense only if the kind of components is defined. Should SiO_2 be composed from Si and O atoms or from Si^{4+} and O^{2-} ions? For polar compounds like SiO_2, lattice energy values given in the literature usually refer to a composition formed from ions. Values calculated under this assumption should be considered with caution as the neglected covalent bonds are of considerable importance. Even in the case of an assembly from ions conditions are not always clear: should Na_2SO_4 be assembled from Na^+ and SO_4^{2-} ions or from Na^+, S^{6+} and O^{2-} ions?

Lattice Energy of Molecular Compounds

The lattice energy E of a molecular compound corresponds to the energy of sublimation at 0 K. This energy cannot be measured directly, but it is equal to the enthalpy of sublimation at a temperature T plus the thermal energy needed to warm the sample from 0 K to this temperature, minus RT. RT is the amount of energy required to expand one mole of a gas at a temperature T to an infinitely small pressure. These amounts of energy, in principle, can be measured and therefore the lattice energy can be determined experimentally in this case. However, the measurement is not simple and is subject to various uncertainties.

The following four forces acting between molecules contribute to the lattice energy of a crystal consisting of molecules:

1. The dispersion forces (LONDON forces) which are always attractive.
2. The repulsion due to the interpenetration of the electron shells of atoms that come together too closely.
3. For molecules with polar bonds, *i.e.* for molecules having the character of dipoles or multipoles, the electrostatic interaction between the dipoles or multipoles.
4. The zero point energy which always is present even at absolute zero temperature.

The zero point energy follows from quantum theory, according to which atoms do not cease to vibrate at the absolute zero point. For a DEBYE solid (that is, a homogeneous body of N equal particles) the zero point energy is

$$E_0 = N\frac{9}{8}h\nu_{\max}$$

$\nu_{\max}$ is the frequency of the highest occupied vibrational state in the crystal. For molecules with a very small mass and for molecules that are being held together via hydrogen bridges, the zero point energy makes a considerable contribution. For H_2 and He it even amounts to the predominant part of the lattice energy. For H_2O it contributes about 30 %, and for N_2, O_2 and CO about 10 %. For larger molecules the contribution of the zero point energy is marginal.

The dispersion force between two atoms results in the dispersion energy E_D, which is approximately proportional to r^{-6}, r being the distance between the atoms:

$$E_D = -\frac{C}{r^6}$$

The dispersion energy between two molecules results approximately from the sum of the contributions of atoms of one molecule to atoms of the other molecule.

For the repulsion energy E_A between two atoms that come together too closely an exponential function is usually taken:

$$E_A = B\mathrm{e}^{-\alpha r}$$

Another, equally appropriate approximation is:

$$E_A = B' r^{-n}$$

with values n ranging between 5 and 12 (BORN repulsion term).

For molecules with low polarity like hydrocarbons, electrostatic forces have only a minor influence. Molecules with highly polar bonds behave as dipoles or multipoles and exhibit corresponding interactions. For instance, hexahalide molecules like WF_6 or WCl_6 are multipoles, the halogen atoms bearing a negative partial charge $-q$, while the metal atom has a positive charge $+6q$; the partial charge q has some value between zero and one, but its exact amount is not usually known. Although the forces exerted by a multipole only have appreciable influence on close-lying molecules, they can contribute significantly to the lattice energy. The electrostatic or Coulomb energy E_C for two interacting atoms having the charges q_i and q_j is:

$$E_C = \frac{1}{4\pi\varepsilon_0} \frac{q_i q_j e^2}{r} \tag{5.1}$$

q_i, q_j being given in units of the electrical unit charge $e = 1.6022 \times 10^{-19}$ C. ε_0 = electric field constant = 8.859×10^{-12} C^{2}J^{-1}m^{-1}.

Overall, the lattice energy E can thus be calculated according to the following approximation:

$$\begin{aligned} E &= N_A \sum (E_D + E_A + E_C + E_0) \\ &= N_A \sum_{i,j} \left[-C_{ij} r_{ij}^{-6} + B_{ij} \exp(-\alpha_{ij} r_{ij}) + \frac{q_i q_j e^2}{4\pi\varepsilon_0 r_{ij}} + \frac{9}{8} h\nu_{max} \right] \end{aligned} \tag{5.2}$$

The choice of the signs gives a negative value for the lattice energy, corresponding to its definition as energy that is released upon formation of the crystal. The atoms of *one* molecule are counted with the index i, while *all* atoms of all other molecules in the crystal are counted with the index j. In this way the interaction energy of one molecule with all other molecules is calculated. The lattice energy per mole results from multiplication by AVOGADRO's number N_A. r_{ij} is the distance between the atoms i and j, q_i and q_j are their partial charges in units of the electric unit charge. B_{ij}, α_{ij} and C_{ij} are parameters that have to be determined experimentally; they are optimized to reproduce the measured sublimation enthalpies at 300 K correctly. As the contributions of the terms in equation (5.2) decrease with growing distances r_{ij}, sufficient accuracy can be obtained by considering only atoms up to some upper limit for r_{ij}. The summation can then be performed rather quickly with a computer.

Values for the partial charges of atoms can be derived from quantum mechanical calculations, from the molecular dipole moments and from rotation–vibration spectra. However, often they are not well known. If the contribution of the Coulomb energy cannot be calculated precisely, no reliable lattice energy calculations are possible.

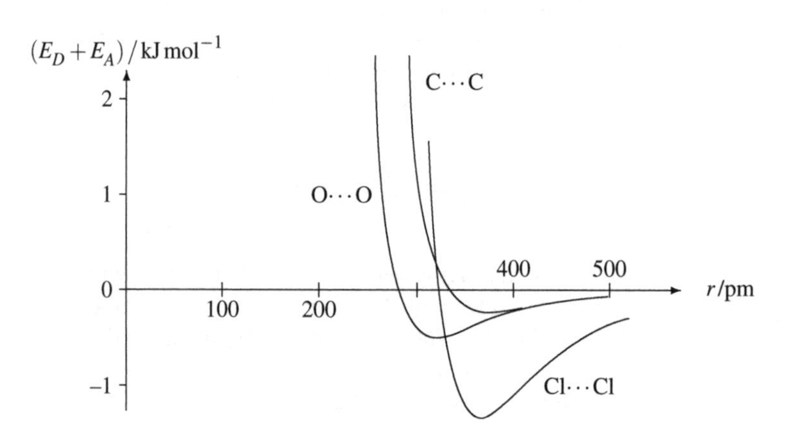

Fig. 5.1 Potential functions for the interaction energies due to repulsive and to dispersion forces between two atoms as a function of the interatomic distance

Parameters B_{ij}, α_{ij} and C_{ij} for light atoms have been listed by GAVEZZOTI [63]. Examples of the resulting potential functions are shown in Fig. 5.1. The minimum point in each graph corresponds to the interatomic equilibrium distance between *two single* atoms. In a crystal shorter distances result because a molecule contains several atoms and thus several attractive atom–atom forces are active between two molecules, and because attractive forces with further surrounding molecules cause an additional compression. All attractive forces taken together are called VAN DER WAALS forces.

Generally, increasing molecular size, heavier atoms and more polar bonds contribute to an increased lattice energy of a molecular crystal. Typical values are: argon 7.7 kJ mol^{-1}; krypton 11.1 kJ mol^{-1}; organic compounds 50 to 150 kJ mol^{-1}.

Lattice Energy of Ionic Compounds

In a molecule the partial charges of all atoms add up to zero and therefore the repulsive and the attractive electrostatic forces between two molecules are more or less balanced. Only the uneven distribution of the charges causes a certain electrostatic contribution to the lattice energy. For a polyatomic ion, however, the partial charges do not add up to zero, but to the value of the ionic charge. As a consequence, strong electrostatic interactions are present between ions, their contributions amounting to the main part of the lattice energy. Numerical values for sodium chloride illustrate this:

$$\begin{array}{rcrl}
N_A \sum q_i q_j e^2/(4\pi\varepsilon_0 r_{ij}) & = & -867 & \text{Coulomb energy} \\
N_A \sum B_{ij} \exp(-\alpha_{ij} r_{ij}) & = & 92 & \text{repulsion energy} \\
-N_A \sum C_{ij} r_{ij}^{-6} & = & -18 & \text{dispersion energy} \\
2N_A \frac{9}{8} h\nu_{max} & = & 6 & \text{zero point energy} \\
\hline
E & = & -787 & \text{kJ mol}^{-1}
\end{array}$$

The calculation of the lattice energy can be performed with the aid of equation (5.2). In the case of monoatomic ions the charges q take the values of the ionic charges. Crystals consisting of monoatomic ions like Na^+ or Cl^- have simple and symmetrical structures, which are useful for the summation according to equation (5.2). Let us take the structure of NaCl as an example. If we designate the shortest distance Na^+–Cl^- in the crystal by R, then all other interionic distances can be given as multiples of R. Which multiples occur follows from simple geometric considerations on the basis of the structure model of NaCl (cf. Fig. 7.1, p. 53). For a Na^+ ion within the crystal the Coulomb energy turns out to be:

$$\begin{aligned}
E_C &= \frac{N_A}{4\pi\varepsilon_0} \sum_j \frac{q_1 q_j e^2}{r_{1j}} \\
&= \frac{N_A e^2}{4\pi\varepsilon_0 R} \left(-\frac{6}{1} + \frac{12}{\sqrt{2}} - \frac{8}{\sqrt{3}} + \frac{6}{\sqrt{4}} - \frac{24}{\sqrt{5}} + \cdots \right) \qquad (5.3)
\end{aligned}$$

The terms in parentheses result as follows:

1. -6, because the Na^+ ion is surrounded by 6 Cl^- ions with charge -1 at a distance $r = R$ in a first sphere.
2. $+12/\sqrt{2}$, because the Na^+ ion is surrounded by 12 Na^+ ions with charge $+1$ at a distance $R\sqrt{2}$ in a second sphere.
3. $-8/\sqrt{3}$, because there are 8 Cl^- ions at a distance $R\sqrt{3}$.

Extending the series given in parentheses to infinite length, it sums to the value $-A = -1.74756$. For short, we can write:

$$E_C = -\frac{N_A e^2}{4\pi\varepsilon_0 R} A \tag{5.4}$$

The quantity A is called the MADELUNG constant. By comparison of equation (5.1) with equation (5.4) we can see that more energy is liberated by combining the ions into a crystal than by forming one mole of separate ion pairs (assuming equal interionic distances $r = R$; in fact, for a single ion pair $r < R$, so that the energy gain actually does not quite attain the factor A).

The same constant A can also be used when ions with higher charges are involved, as long as the structure type is the same:

$$E_C = -|q_1||q_2| \frac{N_A e^2}{4\pi\varepsilon_0 R} A$$

$|q_1|$ and $|q_2|$ are the absolute values of the ionic charges as multiples of the electric unit charge. The MADELUNG constant is independent of the ionic charges and of the lattice dimensions, but it is valid only for one specific structure type. Table 5.1 lists the values for some simple structure types.

Table 5.1: MADELUNG constants for some structure types

structure type	A	structure type	A
CsCl	1.76267	CaF_2	5.03879
NaCl	1.74756	TiO_2 (rutile)	4.816
ZnS (wurtzite)	1.64132	$CaCl_2$	4.730
ZnS (zinc blende)	1.63805	$CdCl_2$	4.489
		CdI_2	4.383

MADELUNG constants only cover the coulombic part of the lattice energy provided that the values of the charges q_1 and q_2 are known. A complete separation of charges between anions and cations yielding integer values for the ionic charges is met quite well only for the alkali metal halides. When some covalent bonding is present, partial charges must be assumed. The magnitudes of these partial charges are not usually known. In this case absolute values for the coulombic part of the lattice energy cannot be calculated. For ZnS, TiO_2, $CdCl_2$ and CdI_2 differing polarities have to be assumed, so that the values listed in Table 5.1 do not follow the real trend of the lattice energies. Nevertheless, MADELUNG constants are useful quantities; they can serve to estimate which structure type should be favored energetically by a compound when the Coulomb energy is the determining factor.

5.3 Problems

5.1 Derive the first four terms of the series to calculate the MADELUNG constant for CsCl (Fig. 7.1).

5.2 Calculate the contribution of the Coulomb energy to the lattice energy of:

(a) CsCl, $R = 356$ pm;

(b) CaF_2, $R = 236$ pm;

(c) BaO (NaCl type), $R = 276$ pm.

6 The Effective Size of Atoms

According to wave mechanics, the electron density in an atom decreases asymptotically towards zero with increasing distance from the atomic center. An atom therefore has no definite size. When two atoms approach each other, interaction forces between them become more and more effective.

Attractive are:

- The ever present dispersion force (LONDON attraction).
- Electronic interactions with the formation of bonding molecular orbitals (orbital energy) and the electrostatic attraction between the nuclei of atoms and electrons. These two contributions cause the bonding forces of covalent bonds.
- Electrostatic forces between the charges of ions or the partial charges of atoms having opposite signs.

Repulsive are:

- The electrostatic forces between ions or partially charged atoms having charges of the same sign.
- The electrostatic repulsion between the atomic nuclei.
- The mutual electrostatic repulsion of the electrons and the PAULI repulsion between electrons having the same spin. The PAULI repulsion contributes the principal part of the repulsion. It is based on the fact that two electrons having the same spin cannot share the same space. PAULI repulsion can only be explained by quantum mechanics, and it eludes simple model conceptions.

The effectiveness of these forces differs and, furthermore, they change to a different degree as a function of the interatomic distance. The last-mentioned repulsion force is by far the most effective at short distances, but its range is rather restricted; at somewhat bigger distances the other forces dominate. At some definite interatomic distance attractive and repulsive forces are balanced. This equilibrium distance corresponds to the minimum in a graph in which the potential energy is plotted as a function of the atomic distance ('potential curve', *cf.* Fig. 5.1, p. 42).

The equilibrium distance that always occurs between atoms conveys the impression of atoms being spheres of a definite size. In fact, in many cases atoms can be treated as if they were more or less hard spheres.

Since the attractive forces between the atoms differ depending on the type of bonding forces, for every kind of atom several different sphere radii have to be assigned according to the bonding types. From experience we know that for one specific kind of bonding the atomic radius of an element has a fairly constant value. We distinguish the following radius types: VAN DER WAALS radii, metallic radii, several ionic radii depending on the ionic charges, and covalent radii for single, double and triple bonds. Furthermore, the values vary depending on coordination numbers: the larger the coordination number, the bigger is the radius.

Inorganic Structural Chemistry, Second Edition Ulrich Müller

6.1 Van der Waals Radii

In a crystalline compound consisting of molecules, the molecules usually are packed as close as possible, but with atoms of neighboring molecules not coming closer than the sums of their VAN DER WAALS radii. The shortest commonly observed distance between atoms of the same element in adjacent molecules is taken to calculate the VAN DER WAALS radius for this element. Some values are listed in Table 6.1. A more detailed study reveals that covalently bonded atoms are not exactly spherical. For instance, a halogen atom bonded to a carbon atom is flattened to some degree, *i.e.* its VAN DER WAALS radius is shorter in the direction of the extension of the C–halogen bond than in transverse directions (*cf.* Table 6.1). If the covalent bond is more polar, as in metal halides, then the deviation from the spherical form is less pronounced. The kind of bonding also can have some influence; for example, carbon atoms in acetylenes have a slightly bigger radius than in other compounds.

Distances that are shorter than the sums of the corresponding listed values of the VAN DER WAALS radii occur when there exist special attractive forces. For example, in a solvated ion the distances between the ion and atoms of the solvent molecules cannot be calculated with the aid of VAN DER WAALS radii. The same applies in the presence of hydrogen bonding.

Table 6.1: Van der Waals radii / pm

H	120

spherical approximation [65, 67]

								He	140
C	170	N	155	O	152	F	147	Ne	154
Si	210	P	180	S	180	Cl	175	Ar	188
Ge		As	185	Se	190	Br	185	Kr	202
Sn		Sb	200	Te	206	I	198	Xe	216

flattened atoms bonded to C [66]

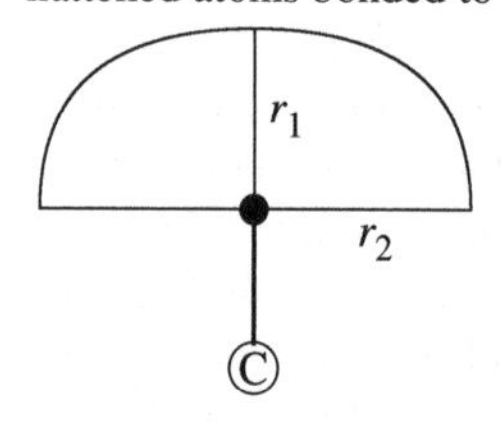

	r_1	r_2		r_1	r_2		r_1	r_2
N	160	160	O	154	154	F	130	138
			S	160	203	Cl	158	178
			Se	170	215	Br	154	184
H	101	126				I	176	213

6.2 Atomic Radii in Metals

The degree of cohesion of the atoms in metals is governed by the extent to which occupation of bonding electron states outweighs antibonding states in the electronic energy bands (*cf.* Section 10.8). Metals belonging to groups in the left part of the periodic table have few valence electrons; the numbers of occupied bonding energy states are low. Metals in the right part of the periodic table have many valence electrons; a fraction of them has to be accommodated in antibonding states. In both cases we have relatively weak metallic bonding. When many bonding but few antibonding states are occupied, the resulting bond forces between the metal atoms are large. This is valid for the metals belonging to the central part of the block of transition elements. Atomic radii in metals therefore decrease from

Table 6.2: Atomic radii in metals / pm. All values refer to coordination number 12, except for the alkali metals (c.n. 8), Ga (c.n. 1+6), Sn (c.n. 4+2), Pa (c.n. 10), U, Np and Pu

Li	Be												
152	112												
Na	Mg											Al	
186	160											143	
K	Ca	Sc	Ti	V	Cr	Mn	Fe	Co	Ni	Cu	Zn	Ga	
230	197	162	146	134	128	137	126	125	125	128	134	135	
Rb	Sr	Y	Zr	Nb	Mo	Tc	Ru	Rh	Pd	Ag	Cd	In	Sn
247	215	180	160	146	139	135	134	134	137	144	151	167	154
Cs	Ba	La	Hf	Ta	W	Re	Os	Ir	Pt	Au	Hg	Tl	Pb
267	222	187	158	146	139	137	135	136	139	144	151	171	175

Ce	Pr	Nd	Pm	Sm	Eu	Gd	Tb	Dy	Ho	Er	Tm	Yb	Lu
182	182	182	181	180	204	179	178	177	176	175	174	193	174
Th	Pa	U	Np	Pu	Am	Cm	Bk	Cf	Es	Fm	Md	No	Lr
180	161	156	155	159	173	174	170	169					

the alkali metals up to the metals of the groups six to eight, and then they increase. Superimposed on this sequence is the general tendency of decreasing atomic sizes observed in all periods from the alkali metals to the noble gases, which is due to the increasing nuclear charge (Table 6.2). For intermetallic compounds the ratio of the total number of available valence electrons to the number of atoms (the 'valence electron concentration') is a decisive factor affecting the effective atomic size.

6.3 Covalent Radii

Covalent radii are derived from the observed distances between covalently bonded atoms of the same element. For example, the C–C bond length in diamond and in alkanes is 154 pm; half of this value, 77 pm, is the covalent radius for a single bond at a carbon atom having coordination number 4 (sp^3 C atom). In the same way we calculate the covalent radii for chlorine (100 pm) from the Cl–Cl distance in a Cl_2 molecule, for oxygen (73 pm) from the O–O distance in H_2O_2 and for silicon (118 pm) from the bond length in elemental silicon. If we add the covalent radii for C and Cl, we obtain $77 + 100 = 177$ pm; this value corresponds rather well to the distances observed in C–Cl compounds. However, if we add the covalent radii for Si and O, $118 + 73 = 191$ pm, the value obtained does not agree satisfactorily with the distances observed in SiO_2 (158 to 162 pm). Generally we must state: the more polar a bond is, the more its length deviates to lower values compared with the sum of the covalent radii. To take this into account, SHOMAKER and STEVENSON derived the following correction formula:

$$d(\text{AX}) = r(\text{A}) + r(\text{X}) - c|x(\text{A}) - x(\text{X})|$$

$d(\text{AX})$ = bond length, $r(\text{A})$ und $r(\text{X})$ = covalent radii of the atoms A and X, $x(\text{A})$ and $x(\text{X})$ = electronegativities of A and X.

The correction parameter c depends on the atoms concerned and has values between 2 and 9 pm. For C–X bonds no correction is necessary when X is an element of the 5th,

6th or 7th main groups, except for N, O and F. The influence of bond polarity also shows up in the fact that the bond lengths depend on the oxidation states; for instance, the P–O bonds in P_4O_6 (164 pm) are longer than in P_4O_{10} (160 pm; sum of the covalent radii 183 pm). Deviations of this kind are larger for 'soft' atoms, *i.e.* for atoms that can be polarized easily.

Calculated bond lengths also are uncertain when it is not known to exactly what degree multiple bonding is present, what influence lone electron pairs exercise on adjacent bonds, and to what extent the ionic charge and the coordination number have an effect. The range of Cl–O bond lengths illustrates this: HOCl 170 pm, ClO_2^- 156 pm, ClO_3^- 149 pm, ClO_4^- 143 pm, $HClO_3$ one at 164 and three at 141 pm, ClO_2 147 pm, ClO_2^+ 131 pm. Problems related to bond lengths are also dealt with on pages 60 and 67 – 71.

6.4 Ionic Radii

The shortest cation–anion distance in an ionic compound corresponds to the sum of the ionic radii. This distance can be determined experimentally. However, there is no straightforward way to obtain values for the radii themselves. Data taken from carefully performed X-ray diffraction experiments allow the calculation of the electron density in the crystal; the point having the minimum electron density along the connection line between a cation and an adjacent anion can be taken as the contact point of the ions. As shown in the example of sodium fluoride in Fig. 6.1, the ions in the crystal show certain deviations from spherical shape, *i.e.* the electron shell is polarized. This indicates the presence of some degree of covalent bonding, which can be interpreted as a partial backflow of electron density from the anion to the cation. The electron density minimum therefore does not necessarily represent the ideal place for the limit between cation and anion.

The commonly used values for ionic radii are based on an arbitrarily assigned standard radius for a certain ion. In this way, a consistent set of radii for other ions can be

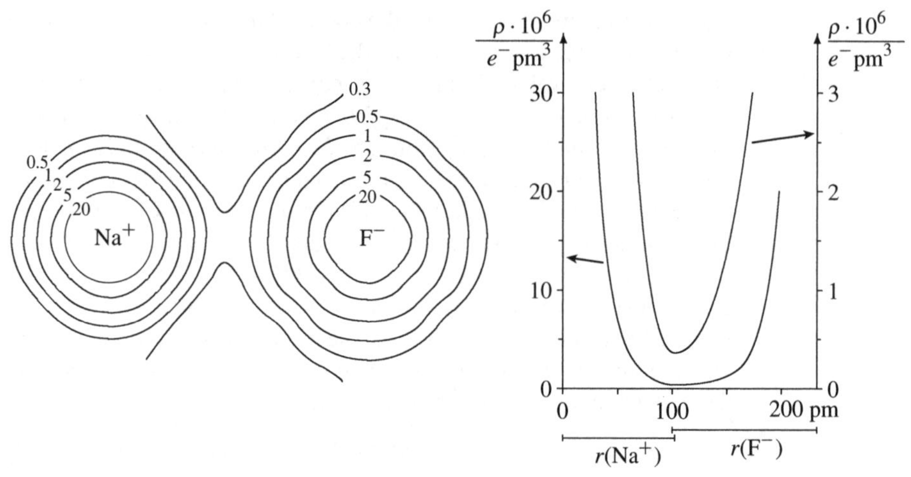

Fig. 6.1
Experimentally determined electron density ρ (multiples of $10^{-6} e^-/pm^3$) in crystalline sodium fluoride [68]). Left: plane of intersection through adjacent ions; right: electron density along the connecting line Na^+—F^- and marks for the ionic radii values according to Table 6.3

derived. Several tables have been published, each of which was derived using a different standard value for one ionic radius (ionic radii by GOLDSCHMIDT, PAULING, AHRENS, SHANNON). The values by SHANNON are based on a critical evaluation of experimentally determined interatomic distances and on the standard radius of 140 pm for the O^{2-} ion with sixfold coordination. They are listed in Tables 6.3 and 6.4.

Ionic radii can also be used when considerable covalent bonding is involved. The higher the charge of a cation, the greater is its polarizing effect on a neighboring anion, *i.e.* the covalent character of the bond increases. Nevertheless, arithmetically one can still assume a constant radius for the anion and assign a radius for the cation that will yield the correct interatomic distance. A value like $r(Nb^{5+}) = 64$ pm therefore does not imply the existence of an Nb^{5+} ion with this radius, but means that in niobium(V) compounds the bond length between an Nb atom and a more electronegative atom X can be calculated as the sum of $r(Nb^{5+})$ plus the anionic radius of X. However, the values are not completely independent of the nature of X; for instance, the values given in Table 6.4 cannot be used readily for sulfur compounds; for this purpose another set of slightly different ionic radii has been derived [70]. Conversely, one can deduce the oxidation states of the atoms from observed bond lengths.

Ionic radii of soft (easily polarized) ions depend on the counter-ion. The H^- ion is an example; its radius is 130 pm in MgH_2, 137 pm in LiH, 146 pm in NaH and 152 pm in KH.

The ionic radii listed in Tables 6.3 and 6.4 in most cases apply to ions which have coordination number 6. For other coordination numbers slightly different values have to be taken. For every unit by which the coordination number increases or decreases, the ionic radius increases or decreases by 1.5 to 2 %. For coordination number 4 the values are approximately 4 % smaller, and for coordination number 8 about 3 % greater than for coordination number 6. The reason for this is the mutual repulsion of the coordinated ions,

Table 6.3: Ionic radii for main group elements according to SHANNON [69], based on $r(O^{2-}) = 140$ pm. Numbers with signs: oxidation states. All values refer to coordination number 6 (except c.n. 4 for N^{3-})

H −1 ~150	Li +1 76	Be +2 45	B +3 27	C +4 16	N −3 146 +3 16	O −2 140	F −1 133
	Na +1 102	Mg +2 72	Al +3 54	Si +4 40	P +3 44 +5 38	S −2 184 +6 29	Cl −1 181
	K +1 138	Ca +2 100	Ga +3 62	Ge +2 73 +4 53	As +3 58 +5 46	Se −2 198 +4 50	Br −1 196
	Rb +1 152	Sr +2 118	In +3 80	Sn +2 118 +4 69	Sb +3 76 +5 60	Te −2 221 +4 97 +6 56	I −1 220 +5 95 +7 53
	Cs +1 167	Ba +2 135	Tl +1 150 +3 89	Pb +2 119 +4 78	Bi +3 103 +5 76	Po +4 94 +6 67	

Table 6.4: Ionic radii for transition elements according to SHANNON [69], based on $r(O^{2-}) =$ 140 pm. Numbers with signs: oxidation states; ls = low spin, hs = high spin; roman numerals: coordination numbers if other than 6

	Sc	Ti	V	Cr	Mn	Fe	Co	Ni	Cu	Zn	
+2				ls 73	ls 67	ls 61	ls 65		+1 77		
+2		86	79	hs 80	hs 83	hs 78	hs 75	69	73	74	+2
+3	75	67	64	62	ls 58	ls 55	ls 55	ls 56	ls 54		+3
+3					hs 65	hs 65	hs 61	hs 60			+3
+4		61	58	55	53	59	hs 53	ls 48			+4
+5			54	49	IV 26						+5
+6				44	IV 25	IV 25					+6
	Y	Zr	Nb	Mo	Tc	Ru	Rh	Pd	Ag	Cd	
+1									115		+1
+2								86	94	95	+2
+3	90		72	69		68	67	76	75		+3
+4		72	68	65	65	62	60	62			+4
+5			64	61	60	57	55				+5
+6				59							+6
	La	Hf	Ta	W	Re	Os	Ir	Pt	Au	Hg	
+1									137	119	+1
+2								80		102	+2
+3	103		72				68		85		+3
+4		71	68	66	63	63	63	63			+4
+5			64	62	58	58	57	57	57		+5
+6				60	55	55					+6
	Ac										
+3	112										

	Ce	Pr	Nd	Pm	Sm	Eu	Gd	Tb	Dy	Ho	Er	Tm	Yb	Lu
+2						117			107			103	102	
+3	101	99	98	97	96	95	94	92	91	90	89	88	87	86
+4	87	85						76						
	Th	Pa	U	Np	Pu	Am	Cm	Bk	Cf	Es	Fm	Md	No	Lr
+3		104	103	101	100	98	97	96	95					
+4	94	90	89	87	86	85	85	83	82					
+5		78	76	75	74									
+6			73	72	71									

the effect of which increases when more of them are present. The size of the coordinated ions also has some influence: a cation that is surrounded by six small anions appears to be slightly smaller than the same cation surrounded by six large anions because in the latter case the anions repel each other more. To account for this, a correction function was derived by PAULING [73]. When covalent bonding is involved, the ionic radii depend to a larger extent on the coordination number. For instance, increasing the coordination number from 6 to 8 entails an increase of the ionic radii of lanthanoid ions of about 13 %, and for Ti^{4+} and Pb^{4+} of about 21 %. An ionic radius decrease of 20 to 35 % is observed when the coordination number of a transition element decreases from 6 to 4.

6.5 Problems

6.1 In the following tetrahedral molecules the bond lengths are:
SiF_4 155 pm; $SiCl_4$ 202 pm; SiI_4 243 pm.
Calculate the halogen–halogen distances and compare them with the VAN DER WAALS distances. What do you conclude?

6.2 Use ionic radii to calculate expected bond lengths for:
Molecules WF_6, WCl_6, PCl_6^-, PBr_6^-, SbF_6^-, MnO_4^{2-};
Solids (metal atom has c.n. 6) TiO_2, ReO_3, EuO, $CdCl_2$.

7 Ionic Compounds

7.1 Radius Ratios

In an energetically favorable packing of cations and anions only anions are directly adjacent to a cation and vice versa. In this way, the attractive forces between ions of opposite charges outweigh the repulsive forces between ions of like charges. Packing many ions together into a crystal frees an amount of energy which is larger by the factor A than in the formation of separate ion pairs (assuming equal interionic distances R). A is the MADELUNG constant discussed in Section 5.2 (p. 43), which has a definite value for a given crystal structure type. One might now think that the structure type having the largest MADELUNG constant for a given chemical composition should always be favored. However, this is not the case.

The stability of a certain structure type depends essentially on the relative sizes of cations and anions. Even with a larger MADELUNG constant a structure type can be less stable than another structure type in which cations and anions can approach each other more closely; this is so because the lattice energy also depends on the interionic distances [*cf.* equation (5.4), p. 44]. The relative size of the ions is quantified by the *radius ratio* r_M/r_X, r_M being the cation radius and r_X the anion radius. In the following the ions are taken to be hard spheres having specific radii.

For compounds of the composition MX (M = cation, X = anion) the CsCl type has the largest MADELUNG constant. In this structure type a Cs^+ ion is in contact with eight Cl^- ions in a cubic arrangement (Fig. 7.1). The Cl^- ions have no contact with one another. With cations smaller than Cs^+ the Cl^- ions come closer together and when the radius ratio has the value of $r_M/r_X = 0.732$, the Cl^- ions are in contact with each other. When $r_M/r_X < 0.732$, the Cl^- ions remain in contact, but there is no more contact between anions and cations. Now another structure type is favored: its MADELUNG constant is indeed smaller, but it again allows contact of cations with anions. This is achieved by the smaller coordination number 6 of the ions that is fulfilled in the NaCl type (Fig. 7.1). When the radius ratio becomes even smaller, the zinc blende (sphalerite) or the wurtzite type should occur, in which the ions only have the coordination number 4 (Fig. 7.1; zinc blende and wurtzite are two modifications of ZnS).

The geometric considerations leading to the following values are outlined in Fig. 7.2:

r_M/r_X	coordination number and polyhedron	structure type
> 0.732	8 cube	CsCl
0.414 to 0.732	6 octahedron	NaCl
< 0.414	4 tetrahedron	zinc blende

The purely geometric approach we have considered so far is still too simple. The really determining factor is the lattice energy, the calculation of which is somewhat more

Inorganic Structural Chemistry, Second Edition Ulrich Müller

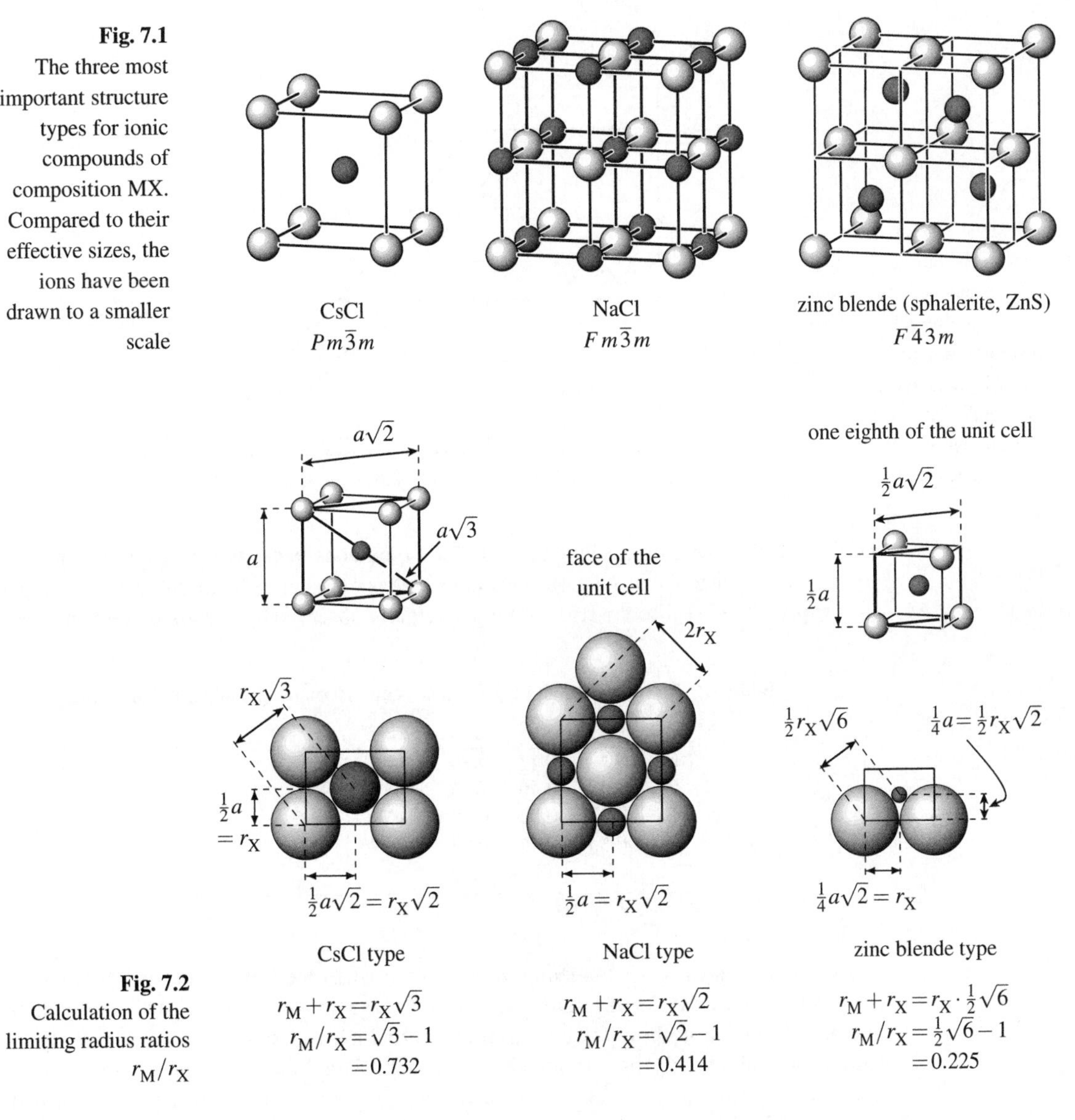

Fig. 7.1 The three most important structure types for ionic compounds of composition MX. Compared to their effective sizes, the ions have been drawn to a smaller scale

Fig. 7.2 Calculation of the limiting radius ratios r_M/r_X

complicated. If we only take into account the electrostatic part of the lattice energy, then the relevant magnitude in equation (5.4) is the ratio A/R (A = MADELUNG constant, R = shortest cation–anion distance). Fig. 7.3 shows how the electrostatic part of the lattice energy depends on the radius ratio for chlorides. The transition from the NaCl type to the zinc blende type is to be expected at the crossing point of the curves at $r_M/r_X \approx 0.3$ instead of $r_M/r_X = 0.414$. The transition from the NaCl type to the CsCl type is to be expected at $r_M/r_X \approx 0.71$. The curves were calculated assuming hard Cl^- ions with r_{Cl^-} = 181 pm. If, in addition, we take into account the increase of the ionic radius for an increased coordination number, then we obtain the dotted line in Fig. 7.3 for the CsCl type. As a consequence, the CsCl type should not occur at all, as the dotted line always runs below

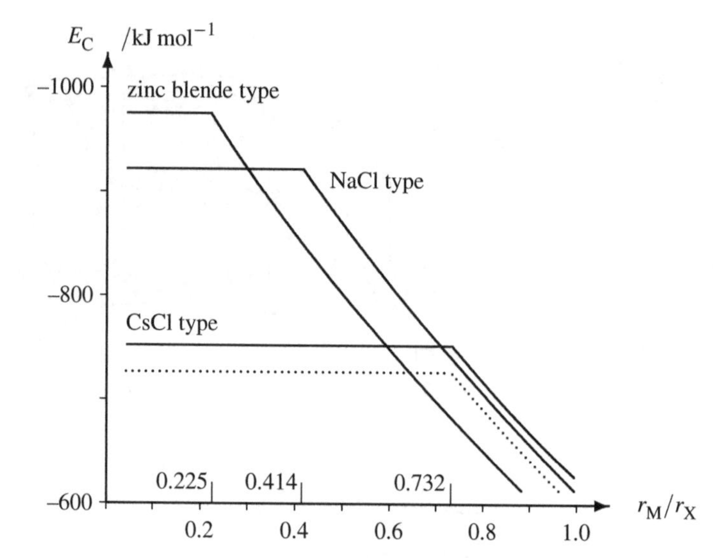

Fig. 7.3 The electrostatic part of the lattice energy for chlorides crystallizing in the CsCl, NaCl and zinc blende type as a function of the radius ratio

the line for the NaCl type. However, the CsCl type does occur when heavy ions are involved; this is due to the dispersion energy, which has a larger contribution in the case of the CsCl type. Table 7.1 lists the structure types actually observed for the alkali metal halides.

Table 7.1: Radius ratios and observed structure types for the alkali metal halides

	Li	Na	K	Rb	Cs	
F	0.57	0.77	0.96*	0.88*	0.80*	
Cl	0.42	0.56	0.76	0.84	0.92	
Br	0.39	0.52	0.70	0.78	0.85	CsCl
I	0.35	0.46	0.63	0.69	0.76	type
		NaCl type				

* r_X/r_M

Twelve anions can be arranged around a cation when the radius ratio is 0.95 to 1.00. However, unlike the three structure types considered so far, geometrically the coordination number 12 does not allow for any arrangement which has cations surrounded only by anions and anions only by cations simultaneously. This kind of coordination therefore does not occur among ionic compounds. When r_M/r_X becomes larger than 1, as for RbF and CsF, the relations are reversed: in this case the cations are larger than the anions and the contacts among the cations determine the limiting radius ratios; the same numerical values and structure types apply, but the inverse radius ratios have to be taken, *i.e.* r_X/r_M.

The zinc blende type is unknown for truly ionic compounds because there exists no pair of ions having the appropriate radius ratio. However, it is well known for compounds with considerable covalent bonding even when the zinc blende type is not to be expected according to the relative sizes of the atoms in the sense of the above-mentioned considerations. Examples are CuCl, AgI, ZnS, SiC, and GaAs. We focus in more detail on this structure type in Chapter 12.

In the structure types for compounds MX so far considered, both anions and cations have the same coordination numbers. In compounds MX_2 the coordination number of the cations must be twice that of the anions. The geometric considerations concerning

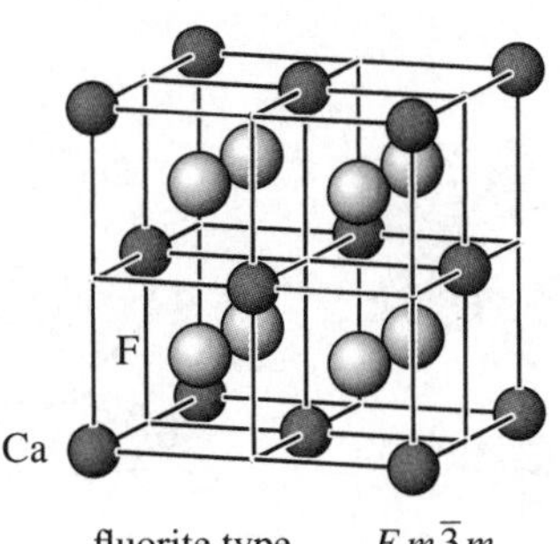

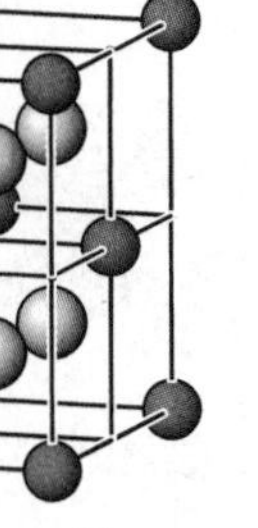

Fig. 7.4 Fluorite and rutile type

fluorite type $Fm\bar{3}m$ rutile type $P4_2/mnm$ $c/a \approx 0.65$

the relations of radius ratios and coordination polyhedra are the same. First of all, two structure types fulfill the conditions and are of special importance (Fig. 7.4):

r_M/r_X	coordination number and polyhedron cation		anion		structure type	examples
> 0.732	8	cube	4	tetrahedron	fluorite (CaF_2)	SrF_2, BaF_2, EuF_2, $SrCl_2$, $BaCl_2$, ThO_2
0.414 to 0.732	6	octahedron	3	triangle	rutile (TiO_2)	MgF_2, FeF_2, ZnF_2, SiO_2*, SnO_2, RuO_2

* stishovite

When the positions of cations and anions are interchanged, the same structure types result for the CsCl, NaCl and zinc blende type. In the case of the fluorite type the interchange also involves an interchange of the coordination numbers, *i.e.* the anions obtain coordination number 8 and the cations 4. This structure type sometimes is called 'anti-fluorite' type; it is known for the alkali metal oxides (Li_2O, ... , Rb_2O).

The structure types discussed so far have a favorable arrangement of cations and anions and are well suited for ionic compounds consisting of spherical ions. However, their occurrence is by no means restricted to ionic compounds. The majority of their representatives are found among compounds with considerable covalent bonding and among intermetallic compounds.

Several additional, more complicated structure types are known for ionic compounds. For example, according to the radius ratio, one could expect the rutile type for strontium iodide ($r_{Sr^{2+}}/r_{I^-} = 0.54$). In fact, the structure consists of Sr^{2+} ions with a coordination number of 7 and anions having two different coordination numbers, 3 and 4.

7.2 Ternary Ionic Compounds

When three different kinds of spherical ions are present, their relative sizes are also an important factor that controls the stability of a structure. The PbFCl type is an example having anions packed with different densities according to their sizes. As shown in Fig. 7.5, the Cl^- ions form a layer with a square pattern. On top of that there is a layer of F^- ions, also with a square pattern, but rotated through 45°. The F^- ions are situated above the edges of the squares of the Cl^- layer (dotted line in Fig. 7.5). With this arrangement the F^-–F^- distances are smaller by a factor of 0.707 ($= \frac{1}{2}\sqrt{2}$) than the Cl^-–Cl^- distances; this matches the ionic radius ratio of $r_{F^-}/r_{Cl^-} = 0.73$. An F^- layer contains twice as many ions as a Cl^- layer. Every Pb^{2+} ion is located in an antiprism having as vertices four F^- and four

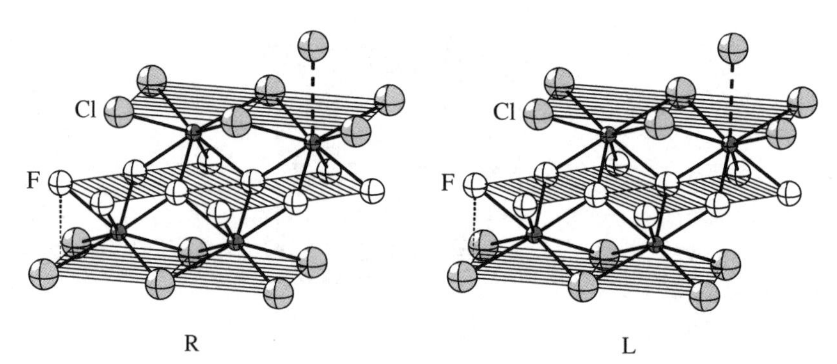

Fig. 7.5
The PbFCl type (stereo view).

How to view the stereo picture: the left picture is to be viewed with the right eye, and the right one with the left eye. It requires some practice to orient the eyes crosswise for this purpose. As an aid, one can hold a finger tip halfway between the figure and the eyes. Aim at the finger tip and move the finger towards the paper or away from it until the images in the center merge to one image. Then focus the image without turning the eyes

Cl^- ions that form two square faces of different sizes. Pb^{2+} ions are located under one half of the squares of the F^- ions; an equal number of Pb^{2+} ions are situated above the other half of the squares, which in turn form the base faces of further antiprisms that are completed by another layer of Cl^- ions. In this way, the total number of Pb^{2+} ions is the same as the number of F^- ions; the number of Cl^- ions also is the same because there are two Cl^- layers for every F^- layer. Together, these layers form a slab that is limited by Cl^- ions on either side. In the crystal these slabs are stacked with staggered adjacent Cl^- layers. As a consequence, the coordination sphere of each Pb^{2+} ion is completed by a fifth Cl^- ion (dashed in Fig. 7.5).

Numerous compounds adopt the PbFCl structure. These include, apart from fluoride chlorides, oxide halides MOX (M = Bi, lanthanoids, actinoids; X = Cl, Br, I), hydride halides like CaHCl and many compounds with metallic properties like ZrSiS or NbSiAs.

Further ternary compounds for which the relative sizes of the ions are an important factor for their stability are the perovskites and the spinels, which are discussed in Sections 17.4 and 17.6.

7.3 Compounds with Complex Ions

The structures of ionic compounds comprising complex ions can in many cases be derived from the structures of simple ionic compounds. A spherical ion is substituted by the complex ion and the crystal lattice is distorted in a manner adequate to account for the shape of this ion.

Rod-like ions like CN^-, C_2^{2-} or N_3^- can substitute the Cl^- ions in the NaCl type when all of them are oriented parallel and the lattice is stretched in the corresponding direction. In CaC_2 the acetylide ions are oriented parallel to one of the edges of the unit cell; as a consequence, the symmetry is no longer cubic, but tetragonal (Fig. 7.6). The hyperoxides KO_2, RbO_2 and CsO_2 as well as peroxides like BaO_2 crystallize in the CaC_2 type. In CsCN and NaN_3 the cyanide and azide ions, respectively, are oriented along one of the space diagonals of the unit cell, and the symmetry is rhombohedral (Fig. 7.6).

The structure of calcite ($CaCO_3$) can be derived from the NaCl structure by substituting the Cl^- ions for CO_3^{2-} ions. These are oriented perpendicular to one of the space diago-

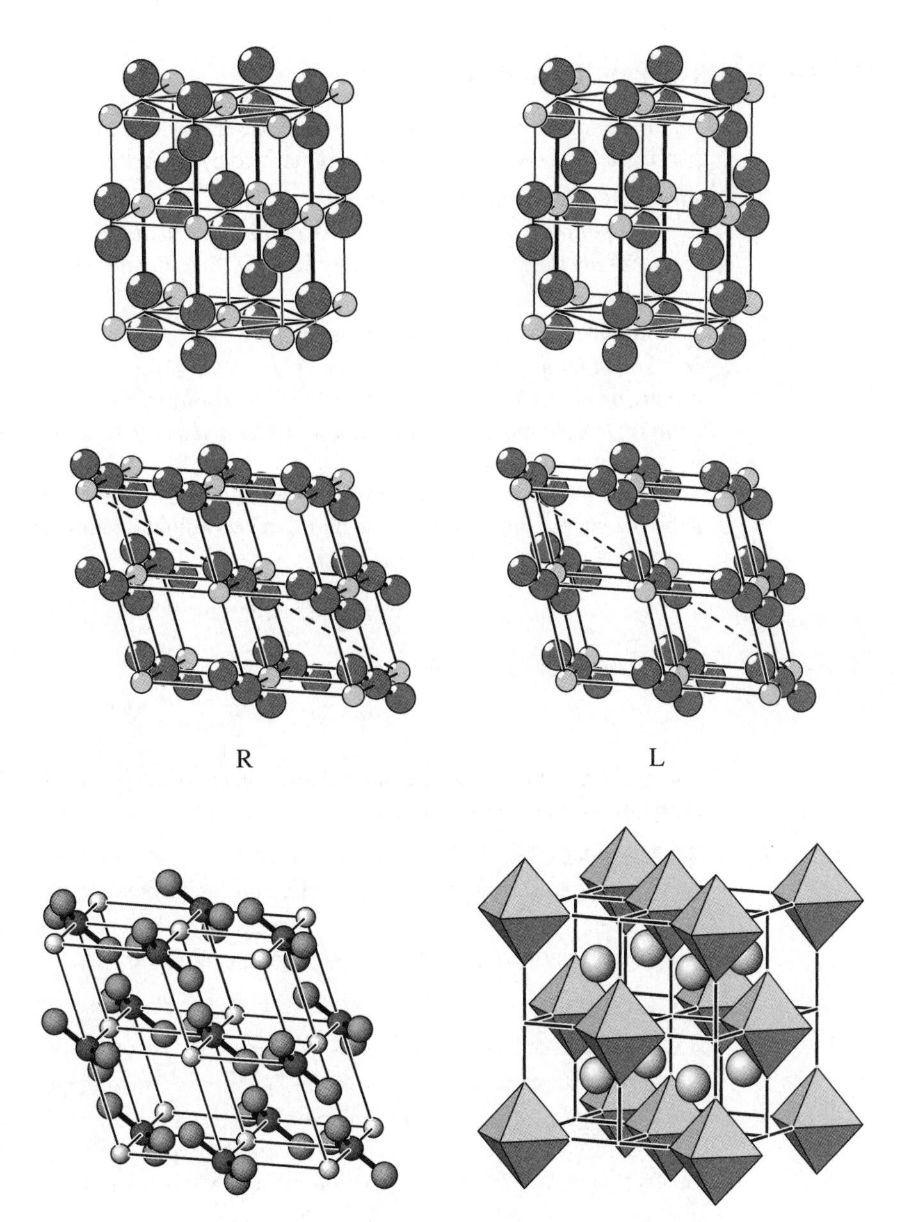

Fig. 7.6 The structures of CaC_2 and NaN_3 (stereo views). Heavy outlines: body-centered tetragonal unit cell of CaC_2. Dashed line at NaN_3: direction of the elongation of the NaCl cell

Fig. 7.7 The structures of $CaCO_3$ (calcite) and K_2PtCl_6. The section of the calcite structure shown does not correspond to the unit cell (as can be seen from the orientations of the CO_3^{2-} groups on opposite edges)

nals of the unit cell and require an expansion of the lattice perpendicular to this diagonal (Fig. 7.7). The calcite type is also encountered among borates (*e.g.* $AlBO_3$) and nitrates ($NaNO_3$). Another way of regarding this structure is discussed on p. 171.

By substituting the Ca^{2+} ions in the CaF_2 type for $PtCl_6^{2-}$ ions and the F^- ion for K^+ ions, one obtains the K_2PtCl_6 type (Fig. 7.7). It occurs among numerous hexahalo salts. In this structure type each of a group of four $PtCl_6^{2-}$ ions has one octahedron face in contact with one K^+ ion, which therefore has coordination number 12. How this structure can be regarded as derived from the perovskite type with a close packing of Cl and K particles is discussed on p. 204.

7.4 The Rules of Pauling and Baur

Important structural principles for ionic crystals, which had already been recognized in part by V. GOLDSCHMIDT, were summarized by L. PAULING in the following rules.

First rule: Coordination polyhedra

A coordination polyhedron of anions is formed around every cation. The cation–anion distances are determined by the sum of the ionic radii, and the coordination number of the cation by the radius ratio.

Second rule: The electrostatic valence rule

In a stable ionic structure the valence (ionic charge) of each anion with changed sign is exactly or nearly equal to the sum of the electrostatic bond strengths to it from adjacent cations. The electrostatic bond strength is defined as the ratio of the charge on a cation to its coordination number.

Let a be the coordination number of an anion. Of the set of its a adjacent cations, let n_i be the charge on the i-th cation and k_i its coordination number. The electrostatic bond strength of this cation is:

$$s_i = \frac{n_i}{k_i} \tag{7.1}$$

The charge z_j of the j-th anion is:

$$z_j \approx -p_j = -\sum_{i=1}^{a} s_i = -\sum_{1}^{a} \frac{n_i}{k_i} \tag{7.2}$$

The rule states that the electrostatic charges in an ionic crystal are balanced locally around every ion as evenly as possible.

Example 7.1

Let the cation M^{2+} in a compound MX_2 have coordination number 6. Its electrostatic bond strength is $s = 2/6 = \frac{1}{3}$. The correct charge for the anion, $z = -1$, can only be obtained when the anion has the coordination number $a = 3$.

Example 7.2

Let the cation M^{4+} in a compound MX_4 also have coordination number 6; its electrostatic bond strength is $s = 4/6 = \frac{2}{3}$. For an anion X^- having coordination number $a = 2$ we obtain $\sum s_i = \frac{2}{3} + \frac{2}{3} = \frac{4}{3}$; for an anion with $a = 1$ the sum is $\sum s_i = \frac{2}{3}$. For other values of a the resulting p_j deviate even more from the expected value $z = -1$. The most favorable structure will have anions with $a = 2$ and with $a = 1$, and these in a ratio of 1:1, so that the correct value for z results in the mean.

The electrostatic valence rule usually is met rather well by polar compounds, even when considerable covalent bonding is present. For instance, in calcite ($CaCO_3$) the Ca^{2+} ion has coordination number 6 and thus an electrostatic bond strength of $s(Ca^{2+}) = \frac{1}{3}$. For the C atom, taken as C^{4+} ion, it is $s(C^{4+}) = \frac{4}{3}$. We obtain the correct value of z for the oxygen atoms, considering them as O^{2-} ions, if every one of them is surrounded by one C and two Ca particles, $z = -[2s(Ca^{2+}) + s(C^{4+})] = -[2 \cdot \frac{1}{3} + \frac{4}{3}] = -2$. This corresponds to the actual structure. $NaNO_3$ and YBO_3 have the same structure; in these cases the rule also is fulfilled when the ions are taken to be Na^+, N^{5+}, Y^{3+}, B^{3+} and O^{2-}. For the numerous silicates no or only marginal deviations result when the calculation is performed with metal ions, Si^{4+} and O^{2-} ions.

The electrostatic valence rule has turned out to be a valuable tool for the distinction of the particles O^{2-}, OH^- and OH_2. Because H atoms often cannot be localized reliably by X-ray diffraction, which is the most common method for structure determination, O^{2-}, OH^- and OH_2 cannot be distinguished unequivocally at first. However, their charges must harmonize with the sums p_j of the electrostatic bond strengths of the adjacent cations.

Example 7.3
Kaolinite, $Al_2Si_2O_5(OH)_4$ or "$Al_2O_3 \cdot 2SiO_2 \cdot 2H_2O$", is a sheet silicate with Al atoms in octahedral and Si atoms in tetrahedral coordination; the corresponding electrostatic bond strengths are:

$$s(Al^{3+}) = \tfrac{3}{6} = 0.5 \qquad s(Si^{4+}) = \tfrac{4}{4} = 1.0$$

The atoms in a sheet are situated in planes with the sequence O(1)–Al–O(2)–Si–O(3) (*cf.* Fig. 16.21e, p. 183). The particles O(2), which are shared by octahedra and tetrahedra, have c.n. 3 (2×Al, 1×Si), the other O particles have c.n. 2. We calculate the following sums of electrostatic bond strengths:

$$\text{O(1)}: \quad p_1 = 2 \cdot s(Al^{3+}) = 2 \cdot 0.5 = 1$$
$$\text{O(2)}: \quad p_2 = 2 \cdot s(Al^{3+}) + 1 \cdot s(Si^{4+}) = 2 \cdot 0.5 + 1 = 2$$
$$\text{O(3)}: \quad p_3 = 2 \cdot s(Si^{4+}) = 2 \cdot 1 = 2$$

Therefore, the OH^- ions must take the O(1) positions and the O^{2-} ions the remaining positions.

Third rule: Linking of polyhedra

An ionic crystal can be described as a set of linked polyhedra. The electrostatic valence rule allows the deduction of the number of polyhedra that share a common vertex, but not how many vertices are common to two adjacent polyhedra. Two shared vertices are equivalent to one shared edge, three or more common vertices are equivalent to a shared face. In the four modifications of TiO_2, rutile, high-pressure TiO_2 (α-PbO_2 type), brookite, and anatase, the Ti atoms have octahedral coordination. As required by the electrostatic valence rule, every O atom is shared by three octahedra. In rutile and high-pressure TiO_2 every octahedron has two common edges with other octahedra, in brookite there are three and in anatase four shared edges per octahedron. The third rule states in what way the kind of polyhedron linkage affects the stability of the structure:

The presence of shared edges and especially of shared faces in a structure decreases its stability; this effect is large for cations with high charge and low coordination number.

The stability decrease is due to the electrostatic repulsion between the cations. The centers of two polyhedra are closest to each other in the case of a shared face and they are relatively distant when only one vertex is shared (*cf.* Fig. 2.3, p. 6, and Table 16.1, p. 167).

According to this rule, rutile and, at high pressures, the modification with the α-PbO_2 structure are the most stable forms of TiO_2. Numerous compounds crystallize in the rutile type and some in the α-PbO_2 type, whereas scarcely any examples are known for the brookite and the anatase structures.

Exceptions to the rule are observed for compounds with low polarity, *i.e.* when covalent bonds predominate. Fluorides and oxides (including silicates) usually fulfill the rule, whereas it is inapplicable to chlorides, bromides, iodides, and sulfides. For instance, in metal trifluorides like FeF_3 octahedra sharing vertices are present, while in most other trihalides octahedra usually share edges or even faces.

In some cases a tendency exactly opposite to the rule is observed, *i.e.* decreasing stability in the sequence face-sharing > edge-sharing > vertex-sharing. This applies when it is favorable to allow the atoms in the centers of the polyhedra to come close to one another. This is observed when the metal atoms in transition metal compounds have d electrons and tend to form metal–metal bonds. For example, $TiBr_3$, TiI_3, $ZrBr_3$ and ZrI_3 form columns consisting of octahedra sharing opposite faces, with metal atoms forming M–M bonds in pairs between adjacent octahedra (*cf.* Fig. 16.10, p. 175).

Fourth rule: Linking of polyhedra having different cations

In a crystal containing different cations, those with large charge and small coordination number tend not to share polyhedron elements with each other, i.e. they tend to keep as far apart as possible.

Silicates having an O : Si ratio larger than or equal to 4 are orthosilicates, *i.e.* the SiO_4 tetrahedra do not share atoms with each other, but with the polyhedra about the other cations. Examples: olivines, M_2SiO_4 ($M = Mg^{2+}, Fe^{2+}$) and garnets, $M_3M'_2[SiO_4]_3$ ($M = Mg^{2+}, Ca^{2+}, Fe^{2+}$; $M' = Al^{3+}, Y^{3+}, Cr^{3+}, Fe^{3+}$).

The extended electrostatic valence rules

Two additional rules, put forward by W. H. BAUR, deal with the bond lengths d(MX) in ionic compounds:

The distances d(MX) *within the coordination polyhedron of a cation* M *vary in the same manner as the values* p_j *corresponding to the anions* X [*cf.* equation (7.2)],

and

For a given pair of ions the average value of the distances d(MX) *within a coordination polyhedron,* $\overline{d(\mathrm{MX})}$*, is approximately constant and independent of the sum of the* p_j *values received by all the anions in the polyhedron. The deviation of an individual bond length from the average value is proportional to* $\Delta p_j = p_j - \overline{p}$ ($\overline{p}$ = mean value of the p_j for the polyhedron). Therefore, the bond lengths can be predicted from the equation:

$$d(\mathrm{MX}(j)) = \overline{d(\mathrm{MX})} + b\Delta p_j \qquad (7.3)$$

$\overline{d(\mathrm{MX})}$ and b are empirically derived values for given pairs of M and X in a given coordination.

Example 7.4

In baddeleyite, a modification of ZrO_2, Zr^{4+} has coordination number 7 in the sense of the formula $ZrO_{3/3}O_{4/4}$; *i.e.* there are two kinds of O^{2-} ions, O(1) with c.n. 3 and O(2) with c.n. 4. The electrostatic valence strength of a Zr^{4+} ion is:

$$s = \tfrac{4}{7}$$

For O(1) and O(2) we calculate:

$$\mathrm{O}(1): \quad p_1 = 3 \cdot \tfrac{4}{7} = 1.714 \qquad \mathrm{O}(2): \quad p_2 = 4 \cdot \tfrac{4}{7} = 2.286$$

We expect shorter distances for O(1); the observed mean distances are:

$$d(\text{Zr–O(1)}) = 209 \text{ pm} \quad \text{and} \quad d(\text{Zr–O(2)}) = 221 \text{ pm}$$

The average values are:

$$\overline{d(\mathrm{ZrO})} = \tfrac{1}{7}(3 \cdot 209 + 4 \cdot 221) = 216 \text{ pm} \quad \text{and} \quad \overline{p} = \tfrac{1}{7}(3 \cdot 1.714 + 4 \cdot 2.286) = 2.041$$

With $b = 20.4$ pm the actual distances can be calculated according to equation (7.3).

Table 7.2 lists values for $\overline{d(\mathrm{MX})}$ and b that have been derived from extensive data. They can be used to calculate the bond lengths in oxides, usually with deviations of less than ±2 pm from the actual values.

Table 7.2: Some average values $\overline{d(\mathrm{MO})}$ and parameters b for the calculation of bond lengths in oxides according to equation (7.3) [74]

bond	ox. state	c.n.	$\overline{d(\mathrm{MO})}$ /pm	b /pm	bond	ox. state	c.n.	$\overline{d(\mathrm{MO})}$ /pm	b /pm
Li–O	+1	4	198	33	Si–O	+4	4	162	9
Na–O	+1	6	244	24	P–O	+5	4	154	13
Na–O	+1	8	251	31	S–O	+6	4	147	13
K–O	+1	8	285	11					
Mg–O	+2	6	209	12	Ti–O	+4	6	197	20
Ca–O	+2	8	250	33	V–O	+5	4	172	16
B–O	+3	3	137	11	Cr–O	+3	6	200	16
B–O	+3	4	148	13	Fe–O	+2	6	214	30
Al–O	+3	4	175	9	Fe–O	+3	6	201	22
Al–O	+3	6	191	24	Zn–O	+2	4	196	18

7.5 Problems

7.1 Use ionic radius ratios (Tables 6.3 and 6.4) to decide whether the CaF_2 or the rutile type is more likely to be adopted by: NiF_2, CdF_2, GeO_2, K_2S.

7.2 In garnet, $Mg_3Al_2Si_3O_{12}$, an O^{2-} ion is surrounded by 2 Mg^{2+}, 1 Al^{3+} and 1 Si^{4+} particle. There are cation sites having coordination numbers of 4, 6 and 8. Use PAULING's second rule to decide which cations go in which sites.

7.3 YIG (yttrium iron garnet), $Y_3Fe_5O_{12}$, has the same structure as garnet. Which are the appropriate sites for the Y^{3+} and Fe^{3+} ions? If the electrostatic valence rule is insufficient for you to come to a decision, take ionic radii as an additional criterion.

7.4 In crednerite, $\mathrm{Cu}^{[2l]}\mathrm{Mn}^{[6o]}\mathrm{O}_2^t$, every oxygen atom is surrounded by 1 Cu and 3 Mn. Can the electrostatic valence rule help to decide whether the oxidation states are Cu^+ and Mn^{3+}, or Cu^{2+} and Mn^{2+}?

7.5 Silver cyanate, AgNCO, consists of infinite chains of alternating Ag^+ and NCO^- ions. Ag^+ has c.n. 2 and only one of the terminal atoms of the cyanate group is part of the chain skeleton, being coordinated to 2 Ag^+. Decide with the aid of PAULING's second rule which of the cyanate atoms (N or O) is the coordinated one. (Decompose the NCO^- to N^{3-}, C^{4+} and O^{2-}).

7.6 In $Rb_2V_3O_8$ the Rb^+ ions have coordination number 10; there are two kinds of vanadium ions, V^{4+} with c.n. 5 and V^{5+} with c.n. 4, and four kinds of O^{2-} ions. The mutual coordination of these particles is given in the table, the first value referring to the number of O^{2-} ions per cation, the second to the number of cations per O^{2-} ion (the sums of the first numbers per row and of the second numbers per column correspond to the c.n.s).

	O(1)	O(2)	O(3)	O(4)	c.n.
Rb^+	2;4	4;2	1;2	3;3	10
V^{4+}	1;1	4;1	–	–	5
V^{5+}	–	2;1	1;2	1;1	4
c.n.	5	4	4	4	

Calculate the electrostatic bond strengths of the cations and determine how well the electrostatic valence rule is fulfilled. Calculate the expected individual V–O bond lengths using data from Table 7.2 and the values $\overline{d(\mathrm{V^{4+}O})} = 189$ pm and $b(\mathrm{V^{4+}O}) = 36$ pm.

8 Molecular Structures I: Compounds of Main Group Elements

Molecules and molecular ions consist of atoms that are connected by covalent bonds. With few exceptions, molecules and molecular ions only exist when hydrogen or elements of the fourth to seventh main groups of the periodic table are involved (the exceptions are molecules such as Li_2 in the gas phase). These elements tend to attain the electron configuration of the noble gas that follows them in the periodic table. For every covalent bond in which one of their atoms participates, it gains one electron. The **8 − *N* rule** holds: *an electron configuration corresponding to a noble gas is attained when the atom takes part in* $8-N$ *covalent bonds;* N = main group number = 4 to 7 (except for hydrogen).

Usually a molecule consists of atoms with different electronegativities, and the more electronegative atoms have smaller coordination numbers (we only count covalently bonded atoms as belonging to the coordination sphere of an atom). The more electronegative atoms normally fulfill the $8-N$ rule; in many cases they are 'terminal atoms', *i.e.* they have coordination number 1. Elements of the second period of the periodic table almost never surpass the coordination number 4 *in molecules.* However, for elements of higher periods this is quite common, the $8-N$ rule being violated in this case.

The structure of a molecule depends essentially on the covalent bond forces acting between its atoms. In the first place, they determine the *constitution* of the molecule, that is, the sequence of the linkage of the atoms. The constitution can be expressed in a simple way by means of the valence bond formula. For a given constitution the atoms arrange themselves in space according to certain principles. These include: atoms not bonded directly with one another may not come too close (repulsion of interpenetrating electron shells); and the valence electron pairs of an atom keep as far apart as possible from each other.

8.1 Valence Shell Electron-Pair Repulsion

The structures of numerous molecules can be understood and predicted with the *valence shell electron-pair repulsion theory* (VSEPR theory) of GILLESPIE and NYHOLM. In the first place, it is applicable to compounds of main group elements. The special aspects concerning transition group elements are dealt with in Chapter 9. However, transition group elements having the electron configurations d^0, high-spin d^5, and d^{10} generally can be treated in the same way as main group elements; the d electrons need not be taken into account in these cases.

In order to apply the theory, one first draws a valence bond formula with the correct constitution, including all lone electron pairs. This formula shows how many valence electron pairs are to be considered at an atom. Every electron pair is taken as one unit (orbital). The electron pairs are being attracted by the corresponding atomic nucleus, but they exercise a mutual repulsion. A function proportional to $1/r^n$ can be used to approximate the repulsion energy between two electron pairs; r is the distance between the centers of gravity of their charges, and n has some value between 5 and 12. $n = 1$ would correspond to a purely electrostatic repulsion. In fact, the contribution of the PAULI repulsion between electrons having the same spin is more important (*cf.* p. 45), and with $n = 6$ one obtains good agreement with experimental data.

The next step is to consider how the electron pairs have to be arranged to achieve a minimum energy for their mutual repulsion. If the centers of the charge of all orbitals are equidistant from the atomic nucleus, every orbital can be represented by a point on the surface of a sphere. The problem thus amounts to deducing what distribution of the points on the sphere corresponds to a minimum for the sum $\sum(1/r_i^n)$, covering all distances r_i between the points. As a result, we obtain a definite polyhedron for every number of points (Fig. 8.1). The resulting polyhedron is independent of the value of the exponent n only for 2, 3, 4, 6, 8, 9, and 12 points. For five points the trigonal bipyramid is only slightly more favorable than the square pyramid. A model to visualize the mutual arrangement of the orbitals about a common center consists of tightly joined balloons; the pressure in the balloons simulates the value of n.

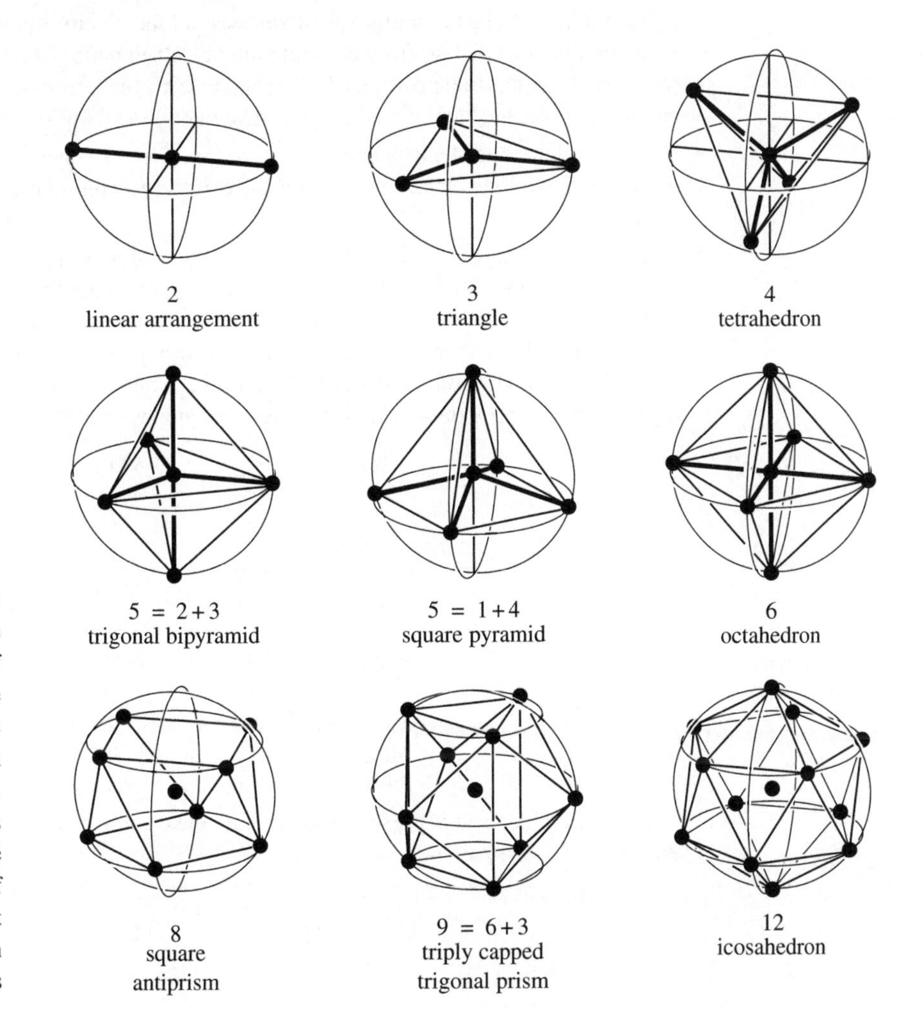

Fig. 8.1 Possible arrangements of points on the surface of a sphere with minimum repulsion energy. When not all points are equivalent, the numbers of equivalent positions are given as sums

Molecules having no lone electron pairs at the central atom and having only equal atoms bonded to this atom usually have structures that correspond to the polyhedra shown in Fig. 8.1.

Some polyhedra have vertices that are not equivalent in the first place. In any case, non-equivalent points always result when the corresponding orbitals belong to bonds with atoms of different elements or when some of the points represent lone electron pairs. In these cases the charge centers of the orbitals have different distances from the atomic center. A charge center that is closer to the atomic nucleus also has shorter distances to the remaining orbitals and thus exerts stronger repulsions on them. In the balloon model an electron pair close to the atomic nucleus corresponds to a larger balloon. This has important consequences for the molecular structure. The following aspects have to be taken into account:

1. A lone electron pair is under the direct influence of only one atomic nucleus, and its charge center therefore is located significantly closer to the nucleus than the centers of bonding electron pairs. Lone electron pairs are especially effective sterically in the following manner:

- If the polyhedron has non-equivalent vertices, a lone electron pair will take the position that is most distant from the remaining electron pairs. At a trigonal bipyramid these are the equatorial positions. In SF_4 and in ClF_3 the lone electron pairs thus take equatorial positions, and fluorine atoms take the two axial positions (*cf.* Table 8.1).
- Two equivalent lone electron pairs take those positions which are farthest apart. In an octahedron two lone electron pairs thus have the *trans* configuration, as for example in XeF_4.
- Due to their stronger repulsion lone electron pairs press other electron pairs closer together. The more lone electron pairs are present, the more this effect is noticeable, and the more the real molecular shape deviates from the ideal polyhedron. For larger central atoms the charge centers of bonding electron pairs are more distant from each other, their mutual repulsion is reduced and the lone electron pairs can press them together to a larger extent. The following bond angles exemplify this:

$$\begin{array}{ccccc}
\mathrm{CH_4} & & \mathrm{NH_3} & & \mathrm{OH_2} \\
109.5^\circ & > & 107.3^\circ & > & 104.5^\circ \\
 & & \vee & & \vee \\
\mathrm{SiH_4} & & \mathrm{PH_3} & & \mathrm{SH_2} \\
109.5^\circ & > & 93.5^\circ & > & 92.3^\circ \\
 & & \vee & & \vee \\
\mathrm{GeH_4} & & \mathrm{AsH_3} & & \mathrm{SeH_2} \\
109.5^\circ & > & 92.0^\circ & > & 91.0^\circ \\
 & & \vee & & \vee \\
\mathrm{SnH_4} & & \mathrm{SbH_3} & & \mathrm{TeH_2} \\
109.5^\circ & > & 91.5^\circ & > & 89.5^\circ
\end{array}$$

If only an unpaired electron is present instead of an electron pair, its influence is reduced, for example:

$[O{=}N{=}O]^+$ 180° $\quad$ NO_2 134° $\quad$ NO_2^- 115°

Table 8.1: Molecular structures of compounds AX_nE_m. A = main group element, E = lone electron pair

composition	structure	angle XAX	examples
AX_2E	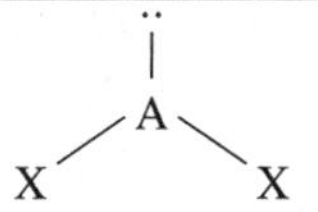	$< 120°$	$SnCl_2$(g), $GeBr_2$(g)
AX_2E_2	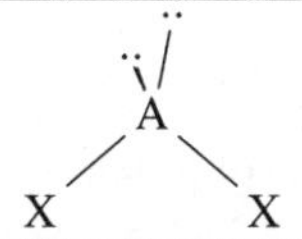	$< 109,5°$	H_2O, F_2O, Cl_2O, H_2S, H_2N^-
AX_2E_3	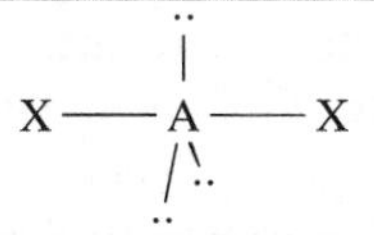	$180°$	XeF_2, I_3^-
AX_3E	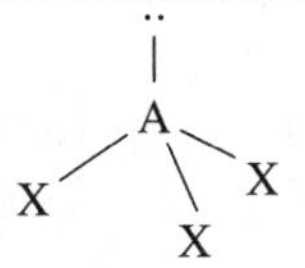	$< 109,5°$	NH_3, NF_3, PH_3, PCl_3, OH_3^+, SCl_3^+, $SnCl_3^-$
AX_3E_2	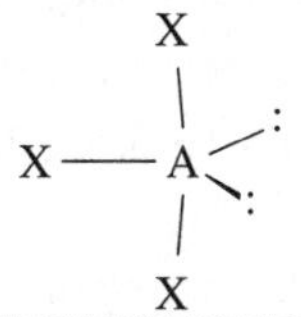	$< 90°$	ClF_3
AX_4E	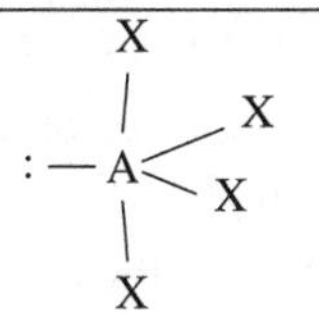	$< 90°$ and $< 120°$	SF_4
AX_4E_2	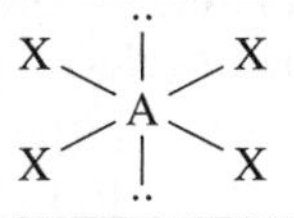	$90°$	XeF_4, BrF_4^-, ICl_4^-
AX_5E	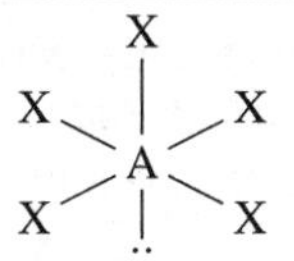	$< 90°$	$SbCl_5^{2-}$, SF_5^-, BrF_5
AX_5E_2	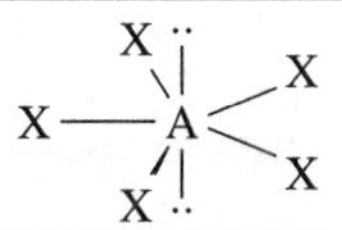	$72°$	XeF_5^-

By definition, the coordination number includes the adjacent atoms, and lone electron pairs are not counted. On the other hand, now we are considering lone pairs as occupying polyhedron vertices. To take account of this, we regard the coordination sphere as including the lone pairs, but we designate them with a ψ, for example: ψ_2-octahedral = octahedron with two lone electron pairs and four ligands.

2. Decreasing **electronegativity of the ligand atoms** causes the charge centers of bonding electron pairs to shift toward the central atom; their repulsive activity increases. Ligands having low electronegativity therefore have a similar influence, albeit less effective, as lone electron pairs. Correspondingly the bond angles in the following pairs increase:

F_2O		H_2O		NF_3		NH_3
103.2°	<	104.5°		102.1°	<	107.3°

However, bond angles cannot be understood satisfactorily by considering only the electron-pair repulsion because they also depend on another factor:

3. The effective size of the ligands. In most cases (unless the central atom is very large) the ligands come closer to each other than the corresponding VAN DER WAALS distance; the electron shells of the ligand atoms interpenetrate one another and additional repulsive forces become active. This is especially valid for large ligand atoms. Within a group of the periodic table decreasing electronegativities and increasing atom sizes go hand in hand, so that they act in the same sense. The increase of the following bond angles is due to both effects:

	HCF_3		$HCCl_3$		$HCBr_3$		HCI_3
Hal–C–Hal	108.8°	<	110.4°	<	110.8°	<	113.0°
	PF_3		PCl_3		PBr_3		PI_3
Hal–P–Hal	97.8°	<	100.1°	<	101.0°	<	102°

When the electronegativity and the size of the ligand atoms have opposing influence, no safe predictions can be made:

F_2O		H_2O	influence of the
103.2°	<	104.5°	electronegativity predominates
Cl_2O		H_2O	influence of the
110.8°	>	104.5°	ligand size predominates

Sometimes the opposing influence of the two effects is just balanced. For example, the steric influence of chlorine atoms and methyl groups is often the same (the carbon atom of a methyl group is smaller, but it is less electronegative than a chlorine atom):

Cl_2O		Me_2O		PCl_3		PMe_3
110.8°	≈	111°		100.1°	≈	99.1°

4. A **pre-existing distortion** is found when certain bond angles deviate from the ideal values of the corresponding polyhedron for geometric reasons. In this case the remaining angles adapt themselves. Forced angle deviations result mainly in small rings. For example:

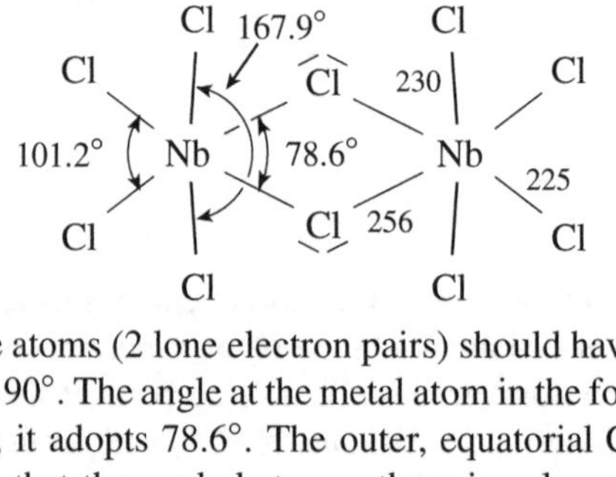

Cl 167.9° Cl
Cl Cl 230 Cl
101.2° Nb 78.6° Nb
225
Cl Cl 256 Cl
Cl Cl

The bridging chlorine atoms (2 lone electron pairs) should have bond angles smaller than 109.5° but larger than 90°. The angle at the metal atom in the four-membered ring is forced to a value under 90°; it adopts 78.6°. The outer, equatorial Cl atoms now experience a reduced repulsion, so that the angle between them is enlarged from 90° to 101.2°. Due to this distortion the axial Cl atoms should be inclined slightly outwards; however, as the Nb–Cl bonds in the ring are longer and their charge centers therefore are situated farther away from the centers of the niobium atoms, they are less repulsive, and the axial Cl atoms are inclined inwards. The increased Nb–Cl distances in the ring are a consequence of the higher coordination number (2 instead of 1) at the bridging Cl atoms (*cf.* point 6).

5. Multiple bonds can be treated as ring structures with bent bonds. The distortions dealt with in the preceding paragraph must be taken into account. For example, in ethylene every C atom is surrounded tetrahedrally by four electron pairs; two pairs mediate the double bond between the C atoms via two bent bonds. The tension in the bent bonds reduces the angle between them and decreases their repulsion toward the C–H bonds, and the HCH bond angle is therefore bigger than 109.5°.

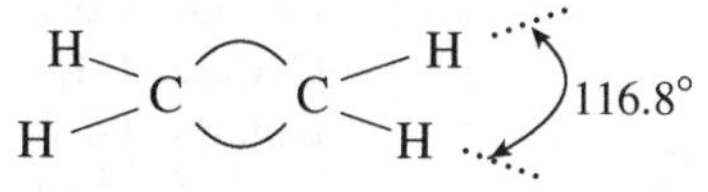

116.8°

Usually, it is more straightforward to treat double and triple bonds as if they form a single orbital that is occupied by four and six electrons, respectively. The increased repulsive power of this orbital corresponds to its high charge. In this way, the structure of the ethylene molecule can be considered as having triangularly coordinated C atoms, but with angles deviating from 120°; the two angles between the double bond and the C–H bonds will be larger than 120°, and the H–C–H will be smaller than 120°.

Molecules like $OPCl_3$ and O_2SCl_2 used to be (and often continue to be) formulated with double bonds, in violation of the octet rule at the P or S atom. This would require the participation of *d* orbitals of these atoms; however, this is not the case according to more recent quantum mechanical calculations. The formulae with formal charges seem to be closer to the true conditions, and if the more electronegative atom obtains a negative formal charge, this qualitatively corresponds to the actual charge distribution.

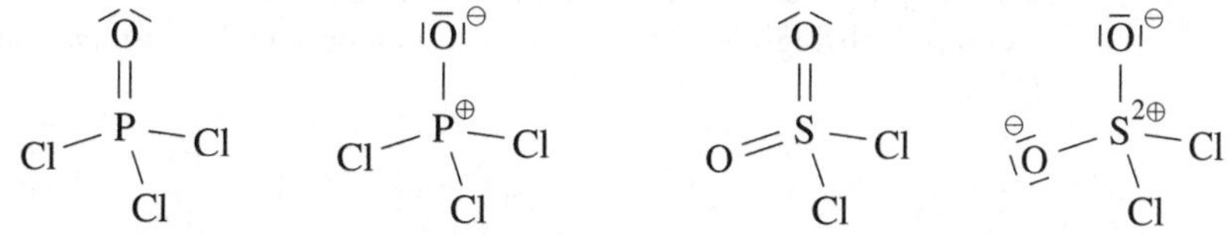

To estimate the bond angles it does not matter which kind of formula one uses. A double bond as well as an excess negative charge act repulsively on the other bonds. In any case, one expects a tetrahedral $OPCl_3$ molecule, but with widened OPCl angles. In O_2SCl_2 the OSO angle will be the largest. In Table 8.2 some examples are listed; they also show the influence of electronegativity and atomic size.

6. Bond lengths are affected like bond angles. The more electron pairs are present, the more they repel each other and the longer the bonds become. The increase in bond length with increasing coordination number is also mentioned in the discussion of ionic radii (p. 49). For example:

distance Sn–Cl: $SnCl_4$ 228 pm $Cl_4Sn(OPCl_3)_2$ 233 pm

The polarity of bonds indeed has a much more marked influence on their lengths. With increasing negative charge of a particle the repulsive forces gain importance. Examples:

distance Sn–Cl: $Cl_4Sn(OPCl_3)_2$ 233 pm $SnCl_6^{2-}$ 244 pm

Table 8.2: Bond angles in degrees for some molecules having multiple bonds. X = singly, Z = double bonded ligand atom

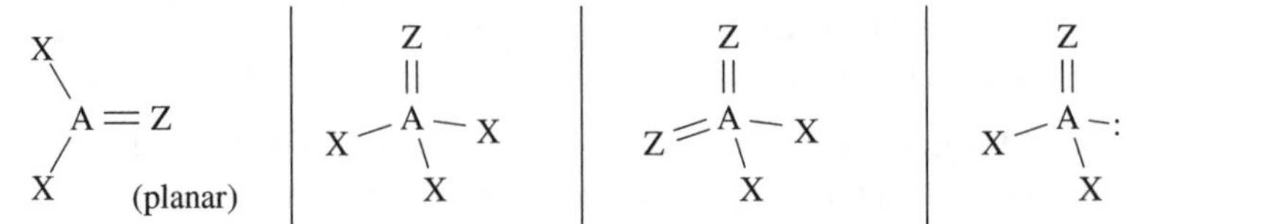

X, A=Z, X (planar)		X–A(=Z)–X, X		Z=A(=Z)–X, X			X–A(=Z)–:, X		
	XAX		XAX		XAX	ZAZ		XAX	XAZ
$F_2C{=}O$	107.7	$F_3P{=}O$	101.3	F_2SO_2	98.6	124.6	$F_2S{=}O$	92.2	106.2
$Cl_2C{=}O$	111.3	$Cl_3P{=}O$	103.3	Cl_2SO_2	101.8	122.4	$Cl_2S{=}O$	96.3	107.4
$Br_2C{=}O$	112.3	$Br_3P{=}O$	105.4				$Br_2S{=}O$	96.7	106.3
$Me_2C{=}O$	112.4	$Me_3P{=}O$	104.1	Me_2SO_2	102.6	119.7	$Me_2S{=}O$	96.6	106.6

	P–O / pm	P–F / pm	O–P–O / °	F–P–F / °
POF_3	144	152	–	101.3
$PO_2F_2^-$	147	157	122	97
PO_3F^{2-}	151	159	114	–
PO_4^{3-}	155	–	109.5	–

The bond between two atoms of different electronegativities is polar. The opposite partial charges of the atoms cause them to attract each other. A change in polarity affects the bond length. This can be noted especially when the more electronegative atom participates in more bonds than it should have according to the $8-N$ rule: contrary to its electronegativity it must supply electrons for the bonds, its negative partial charge is lowered or even becomes positive, and the attraction to the partner atom is decreased. This effect is conspicuous for bridging halogen atoms, as can be seen by comparison with the non-bridging atoms:

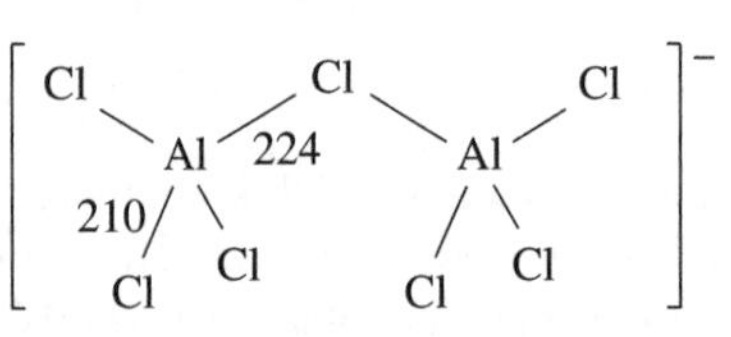

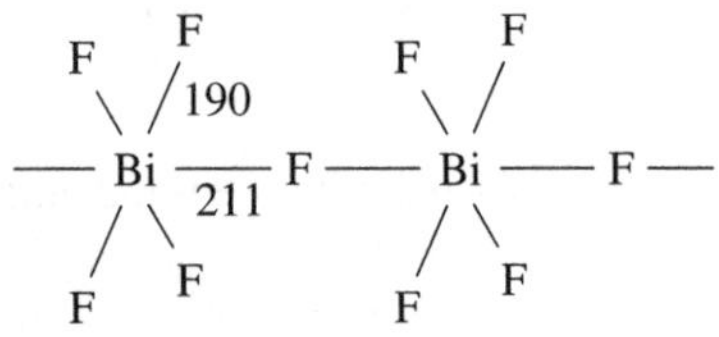

$[Cl_3Al–Cl–AlCl_3]^-$ (Al–Cl terminal 210, Al–Cl bridging 224)

BiF_5 chain: Bi–F terminal 190, Bi–F bridging 211

Niobium pentachloride mentioned on page 66 is another example. The BAUR rules also express these facts (*cf.* p. 60).

7. Influence of a partial valence shell. Atoms of elements of the third period, such as Si, P, and S, and of higher periods can accommodate more than four valence electron pairs in their valence shell (hypervalent atoms). In fact, most main group elements only tolerate a maximum of six electron pairs (for example the S atom in SF_6). Compounds having more than six valence electron pairs per atom are known only for the heavy elements, the iodine atom in IF_7 being an example. Obviously an increased repulsion between the electron pairs comes into effect when the bond angles become smaller than 90°; this would have to be the case for coordination numbers higher than six. However, crowding electron pairs down to angles of 90° is possible without large resistance (*cf.* hydrogen compounds listed on page 64; note the marked jump for the angles between the second and third periods).

If the central atom can still take over electrons and if a ligand has lone electron pairs, then these tend to pass over to the central atom to some degree. In other words, the electron pairs of the ligand reduce their mutual repulsion by shifting partially towards the central atom. This applies especially for small ligand atoms like O and N, particularly when high formal charges have to be allocated to them. For this reason terminal O and N atoms tend to form multiple bonds with the central atom, for example:

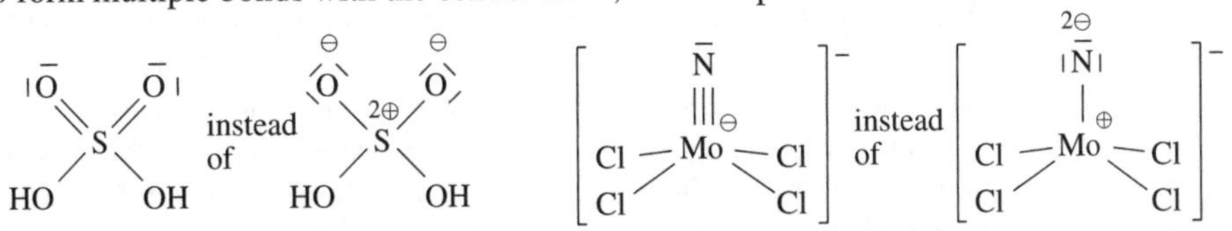

$(HO)_2S(=O)_2$ instead of $(HO)_2S^{2\oplus}(O^{\ominus})_2$ $\qquad [Cl_4Mo{\equiv}N]^-$ (Mo: $\ominus$) instead of $[Cl_4Mo^{\oplus}{-}N^{2\ominus}]^-$

In the case of the molecule of sulfuric acid the left resonance formula would indeed require the participation of *d* orbitals at the sulfur atom; according to recent theoretical calculations this is not justified (for H_2SO_4 the formal charges even seem to reflect the real charge distribution quite well). However, the resonance formula with the formal charges can also explain the bond lengths and angles: the negative charges at the O atoms cause a mutual repulsion of these atoms and an angle widening; simultaneously the S–O bonds are shortened due to the $S^{2\oplus}$–$O^{\ominus}$ attraction.

A similar explanation can be given for the larger Si–O–Si bond angles as compared to C–O–C. Electron density is given over from the oxygen atom into the valence shells of the silicon atoms, but not of the carbon atoms, in the sense of the resonance formulas:

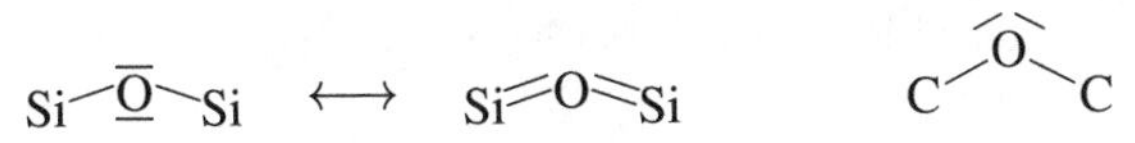

$$Si{-}\overline{\underline{O}}{-}Si \longleftrightarrow Si{=}\overline{O}{=}Si \qquad C{-}O{-}C$$

Examples:

angle SiOSi		angle COC	
$O(SiH_3)_2$	144°	$O(CH_3)_2$	111°
α-quartz	142°	$O(C_6H_5)_2$	124°
α-cristobalite	147°		

That the COC angle is larger in diphenyl ether than in diethyl ether can be explained by the electron acceptor capacity of the phenyl groups. When two strongly electron-accepting atoms are bonded to an oxygen atom, the transition of the lone electron pairs can go so far that a completely linear group of atoms M=O=M results, as for example in the $[Cl_3FeOFeCl_3]^{2-}$ ion.

The electron transition and the resulting multiple bonds should reveal themselves by shortened bond lengths. In fact, bridging oxygen atoms between metal atoms in a linear arrangement exhibit rather short metal–oxygen bonds. Because of the high electronegativity of oxygen the charge centers of the bonding electron pairs will be located more towards the side of the oxygen atom, *i.e.* the bonds will be polar. This can be expressed by the following resonance formulas:

$$[\ Cl_3\overset{2\ominus}{Fe}{=}\overset{2\oplus}{O}{=}\overset{2\ominus}{Fe}Cl_3 \longleftrightarrow Cl_3Fe\ |\overset{2\ominus}{\overline{\underline{O}}}|\ FeCl_3\ ]^{2-}$$

The lowest formal charges result when both formulas have equal weight. In the case of bridging fluorine atoms the ionic formula should be more important to achieve lower formal charges, *e.g.*:

$$[\ F_5\overset{2\ominus}{Sb}{=}\overset{3\oplus}{F}{=}\overset{2\ominus}{Sb}F_5 \longleftrightarrow F_5Sb\ |\overset{\ominus}{\overline{\underline{F}}}|\ SbF_5\ ]^{-}$$

In fact, bridging fluorine atoms usually exhibit bond angles between 140 and 180° and rather long bonds. The significance of the right formula, representing weak interactions between the central F^- ion and two SbF_5 molecules, also shows up in the chemical reactivity: fluorine bridges are cleaved easily.

Restrictions

The consideration of the mutual valence electron-pair repulsion as a rule results in correct qualitative models for molecular structures. In spite of the simple concept the theory is well founded and compatible with the more complicated and less illustrative MO theory (Chapter 10). The results often are by no means inferior to those of sophisticated calculations. Nevertheless, in some cases the model fails. Examples are the ions $SbBr_6^{3-}$, $SeBr_6^{2-}$, and $TeCl_6^{2-}$, which have undistorted octahedral structures, although the central atom still has a lone electron pair. This electron pair is said to be 'stereochemically inactive', although this is not quite true, because its influence still shows up in increased bond lengths. This phenomenon is observed only for higher coordination numbers (≥ 6), when the central atom is a heavy atom and when the ligands belong to a higher period of the periodic table, *i.e.* when the ligands are easily polarized. The decreasing influence of the lone electron pairs can also be seen by comparing the solid-state structures of AsI_3, SbI_3 and BiI_3. AsI_3 forms pyramidal molecules (bond angles 100.2°), but in the solid they are associated in that three iodine atoms of adjacent molecules are coordinated to an arsenic atom. In all, the coordination is distorted octahedral, with three intramolecular As–I lengths of 259 pm and three intermolecular lengths of 347 pm. In BiI_3 the coordination is octahedral with six equal Bi–I distances (307 pm). SbI_3 takes an intermediate position (3× 287 pm, 3× 332 pm).

The theory also cannot explain the '*trans*-influence' that is observed between ligands that are located on opposite sides of the central atom on a straight line, as for two ligands in *trans* arrangement in octahedral coordination. The more tightly one ligand is bonded to the central atom, as evidenced by a short bond, the longer is the bond to the ligand in the *trans* position. Particularly multiple bonds are strongly effective in this way. This can also be noted in the reactivity: the weakly bonded ligand is easily displaced. On the other hand, a lone electron pair usually does not cause a lengthened bond of the ligand *trans* to it; on the contrary, this bond tends to be shorter (distances in pm):

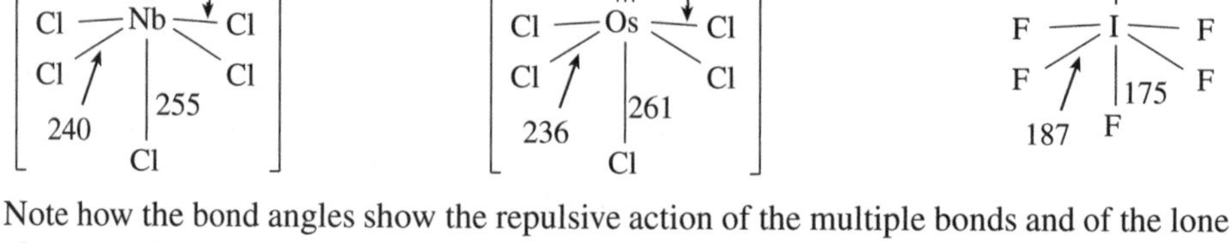

$[NbOCl_5]^{2-}$: Nb=O 197, O–Nb–Cl 95°, Nb–Cl 240, Nb–Cl (*trans*) 255

$[OsNCl_5]^{2-}$: Os≡N 161, N–Os–Cl 96°, Os–Cl 236, Os–Cl (*trans*) 261

IF_5: I–F 187, I–F (*trans* to lone pair) 175

Note how the bond angles show the repulsive action of the multiple bonds and of the lone electron pair.

Transition metal compounds with ligands of low electronegativity also show deviations, in spite of a d^0 electron configuration. For example, $W(CH_3)_6$ does not have the expected octahedral structure, but is trigonal-prismatic.

In one respect the valence shell electron-pair repulsion theory is no better (and no worse) than other theories of molecular structure. Predictions can only be made when the constitution is known, *i.e.* when it is already known which and how many atoms are joined

Table 8.3: Axial and equatorial bond lengths /pm for trigonal-bipyramidal distribution of the valence electrons

AX_5	AX_{ax}	AX_{eq}	AX_4E	AX_{ax}	AX_{eq}	AX_3E_2	AX_{ax}	AX_{eq}
PF_5	158	153	SF_4	165	155	ClF_3	170	160
AsF_5	171	166	SeF_4	177	168	BrF_3	181	172
PCl_5	212	202						

to each other. For example, it cannot be explained why the following pentahalides consist of so different kinds of molecules or ions in the solid state: $SbCl_5$ monomeric, $(NbCl_5)_2$ dimeric, $(PaCl_5)_\infty$ polymeric, $PCl_4^+PCl_6^-$ ionic, $PBr_4^+Br^-$ ionic; PCl_2F_3 monomeric, $AsCl_4^+AsF_6^-$ (= $AsCl_2F_3$) ionic, $SbCl_4^+[F_4ClSb\text{–}F\text{–}SbClF_4]^-$ (= $SbCl_2F_3$) ionic.

8.2 Structures with Five Valence Electron Pairs

The features described in the preceding section can be studied well with molecules for which five valence electron pairs are to be considered. As they also show some peculiarities, we deal with them separately. The favored arrangement for five points on a sphere is the trigonal bipyramid. Its two axial and three equatorial positions are not equivalent, a greater repulsive force being exercised on the axial positions. Therefore, lone electron pairs as well as ligands with lower electronegativities prefer the equatorial sites. If the five ligands are of the same kind, the bonds to the axial ligands are longer (in other words, the covalent radius is larger in the axial direction). *Cf.* Table 8.3.

The molecular parameters for CH_3PF_4 and $(CH_3)_2PF_3$ illustrate the influence of the lower electronegativity of the methyl groups and the corresponding increased repulsive effect of the electron pairs of the P–C bonds:

Energetically, the tetragonal pyramid is almost as favorable as the trigonal bipyramid. With a bond angle of 104° between the apical and a basal position the repulsion energy is only 0.14 % more when a purely coulombic repulsion is assumed; for the PAULI repulsion the difference is even less. Furthermore, the transformation of a trigonal bipyramid to a tetragonal pyramid requires only a low activation energy; as a consequence, a fast exchange of positions of the ligands from one trigonal bipyramid to a tetragonal pyramid and on to a differently oriented trigonal bipyramid occurs ('BERRY rotation', Fig. 8.2). This explains why only a doublet peak is observed in the ^{19}F-NMR spectrum of PF_5 even

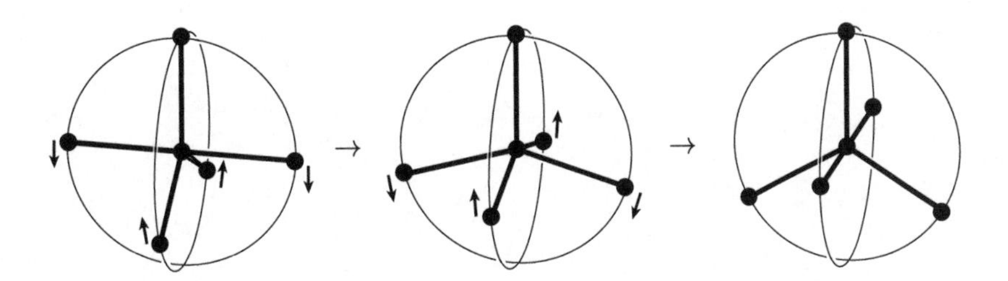

Fig. 8.2 Change of ligand positions between trigonal bipyramids and a tetragonal pyramid

at low temperatures; the doublet is due to the P–F spin-spin coupling. If the fast positional exchange did not occur, two doublets with an intensity ratio of 2 : 3 would be expected.

The tetragonal pyramid is often favored when one double bond is present, especially with compounds of transition metals having a d^0 configuration. Molecules or ions such as $O{=}CrF_4$, $O{=}WCl_4$ (as monomers in the gas phase), $O{=}TiCl_4^{2-}$ and $S{=}NbCl_4^-$ have this structure. However, $O{=}SF_4$ has a trigonal-bipyramidal structure with the oxygen atom in an equatorial position.

Very low energy differences also result for different polyhedra with higher coordination numbers, including coordination number 7. In these cases the electron pair repulsion theory no longer allows reliable predictions.

8.3 Problems

8.1 What structures will the following molecules have according to VSEPR theory?
$BeCl_2$(g), BF_3, PF_3, BrF_3, $TeCl_3^+$, XeF_3^+, $GeBr_4$, $AsCl_4^+$, SbF_4^-, ICl_4^-, BrF_4^+, $TiBr_4$, $SbCl_5$, $SnCl_5^-$, TeF_5^-, $ClSF_5$, O_3^-, Cl_3^-, S_3^{2-}, O_2ClF_3, $O_2ClF_2^-$, $OClF_4^-$, O_3BrF, O_3XeF_2.

8.2 The following dimeric species are associated via two bridging halogen atoms. What are their structures?
Be_2Cl_4, Al_2Br_6, I_2Cl_6, $As_2Cl_8^{2-}$, Ta_2I_{10}.

8.3 What structure is to be expected for $H_2C{=}SF_4$?

8.4 Arrange the following molecules in the order of increasing bond angles.
(a) OF_2, SF_2, SCl_2, S_3^-, S_3^{2-};
(b) Angle H–N–H in H_3CNH_2, $[(H_3C)_2NH_2]^+$;
(c) Angle F_{ax}–P–F_{ax} in PCl_2F_3, PCl_3F_2.

8.5 There are two bridging chlorine atoms in Al_2Cl_6. Give the sequence of increasing bond lengths and bond angles and estimate approximate values for the angles.

8.6 Which of the following species should have the longer bond lengths?
$SnCl_3^-$ or $SnCl_5^-$; PF_5 or PF_6^-; $SnCl_6^{2-}$ or $SbCl_6^-$.

8.7 Which of the following species are the most likely to violate the VSEPR rules?
SbF_5^{2-}, $BiBr_5^{2-}$, TeI_6^{2-}, ClF_5, IF_7, IF_8^-.

9 Molecular Structures II: Compounds of Transition Metals

9.1 Ligand Field Theory

The mutual interaction between *bonding* electron pairs is the same for transition metal compounds as for compounds of main group elements. All statements concerning molecular structure apply equally. However, *nonbonding* valence electrons behave differently. For transition metal atoms these generally are *d* electrons that can be accommodated in five *d* orbitals. In what manner the electrons are distributed among these orbitals and in what way they become active stereochemically can be judged with the aid of *ligand field theory*. The concept of ligand field theory is equivalent to that of the valence shell electron-pair repulsion theory: it considers how the *d* electrons have to be distributed so that they attain a minimum repulsion with each other *and* with the bonding electron pairs. In its original version by H. BETHE it was formulated as crystal field theory; it considered the electrostatic repulsion between the *d* electrons and the ligands, which were treated as point-like ions.* After the success of the valence shell electron-pair repulsion theory it appears more appropriate to consider the interactions between nonbonding *d* electrons and bonding electron pairs; thus the same notions apply for both theories. This way, one obtains qualitatively correct structural statements with relatively simple models. The more exact molecular orbital theory draws the same conclusions.

The relative orientations of the regions with high charge density of *d* electrons and of bonding electrons about an atom can be described with the aid of a coordinate system that has its origin in the center of the atom. Two sets of *d* orbitals are to be distinguished (Fig. 9.1): the first set consists of two orbitals oriented along the coordinate axes, and the second set consists of three orbitals oriented toward the centers of the edges of a circumscribed cube.

Octahedral Coordination

If an atom has six ligands, then the mutual repulsion of the six bonding electron pairs results in an octahedral coordination. The positions of the ligands correspond to points on the axes of the coordinate system. If nonbonding electrons are present, these will prefer the orbitals d_{xy}, d_{yz} and d_{xz} because the regions of high charge density of the other two *d* orbitals are especially close to the bonding electron pairs (Fig. 9.1). The three orbitals favored energetically are termed t_{2g} orbitals (this is a symbol for the orbital symmetry; the *t* designates a triply degenerate state); the other two are e_g orbitals (*e* = doubly degenerate; from German *entartet* = degenerate). *Cf.* the diagram in the margin on the next page.

*The terms crystal field theory and ligand field theory are not used in a uniform way. As only interactions between adjacent atoms are being considered, without referring to crystal influences, the term crystal field theory does not seem adequate. Some authors consider certain electronic interactions (like π bonds) as part of ligand field theory, although they originate from MO theory.

Inorganic Structural Chemistry, Second Edition Ulrich Müller

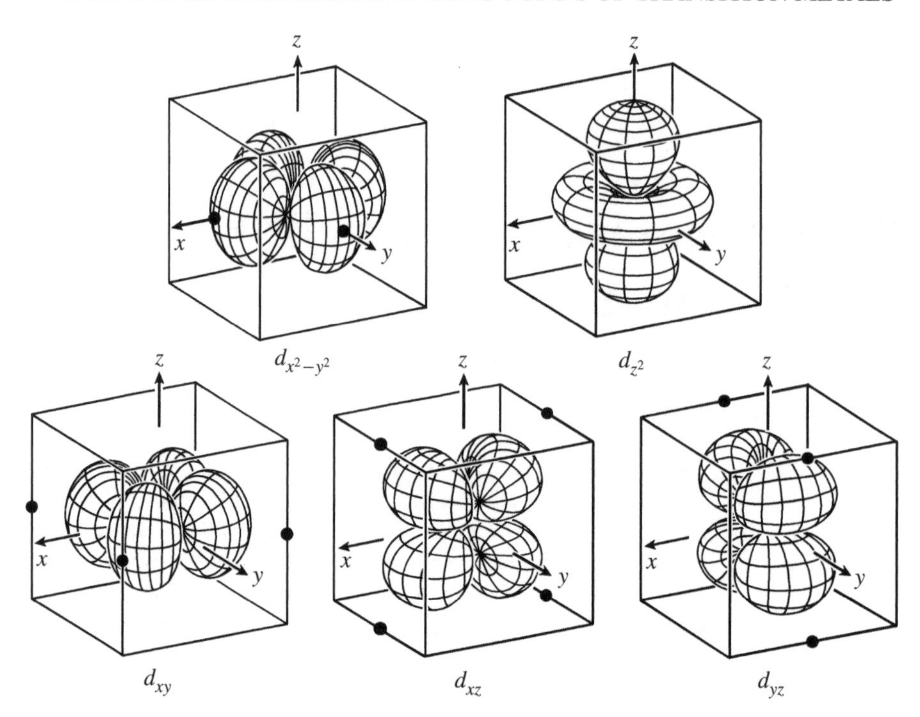

Fig. 9.1 Orientation of the regions of high electron density for 3*d* orbitals. True-to-scale drawings of areas with constant value for the wave functions. The dots ● on the circumscribed cubes mark the directions of preferential orientation of the 'partial clouds'

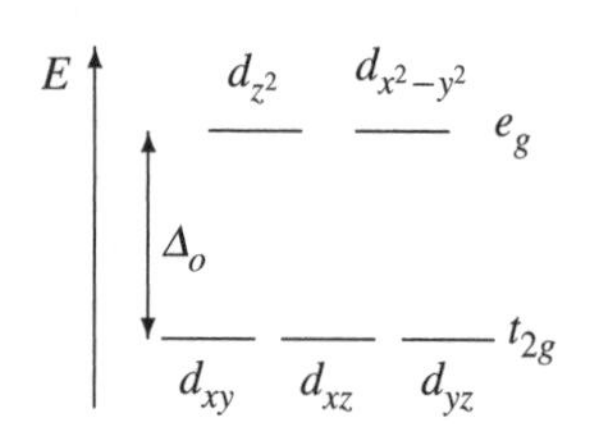

The energy difference between the occupation of a t_{2g} and an e_g orbital is termed Δ_o. The value of Δ_o depends on the repulsion exercised by the bonding electron pairs on the *d* electrons. Compared to a transition metal atom the bonded ligand atoms are usually much more electronegative. The centers of charge of the bonding electron pairs are much closer to them, especially when they are strongly electronegative. Therefore, one can expect a decreasing influence on the *d* electrons and thus a decrease of Δ_o with increasing ligand electronegativity. Decreasing Δ_o values also result with increasing sizes of the ligand atoms; in this case the electron pairs are distributed over a larger space so that the difference of their repulsive action on a t_{2g} and an e_g orbital is less marked. In the presence of multiple bonds between the metal atom and the ligands, as for example in metal carbonyls, the electron density of the bonds is especially high and their action is correspondingly large. Δ_o is a value that can be measured directly with spectroscopic methods: by photoexcitation of an electron from the t_{2g} to the e_g level we have $\Delta_o = h\nu$. The *spectrochemical series* is obtained by ordering different ligands according to decreasing Δ_o:

$$\begin{aligned} &CO > CN^- > PR_3 > NO_2^- > NH_3 > \mathbf{N}CS^- > H_2O > RCO_2^- \approx OH^- \\ &> F^- > NO_3^- > Cl^- \approx \mathbf{S}CN^- > S^{2-} > Br^- > I^- \end{aligned}$$

When two or three nonbonding electrons are present, they will occupy two or three of the t_{2g} orbitals (HUND's rule). This is more favorable than pairing electrons in one orbital because the pairing requires that the electrostatic repulsion between the two electrons be overcome. The energy necessary to include a second electron in an already occupied orbital is called the electron pairing energy *P*. When four nonbonding electrons are present, there are two alternatives for the placement of the fourth electron. If $P > \Delta_o$, then it will be an e_g orbital and all four electrons will have parallel spin: we call this a *high-spin complex*. If $P < \Delta_o$, then it is more favorable to form a *low-spin complex* leaving the e_g orbitals unoccupied and having two paired electrons:

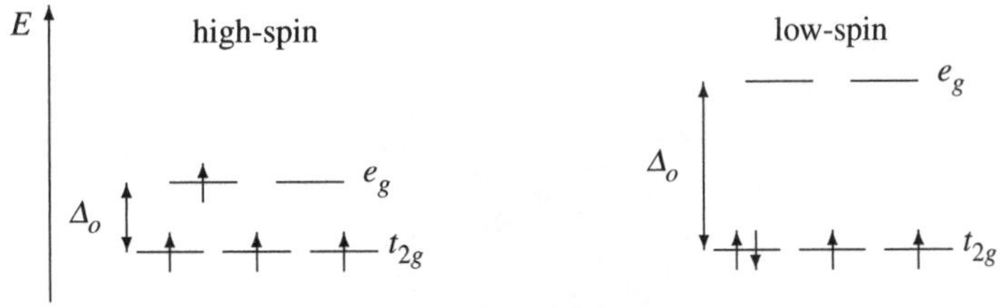

In a high-spin d^4 complex only one of the two e_g orbitals is occupied. If it is the d_{z^2} orbital then it exerts a strong repulsion on the bonding electrons of the two ligands on the z axis. These ligands are forced outwards; the coordination octahedron suffers an elongation along the z axis. This effect is known as the *Jahn–Teller effect*. Instead of the d_{z^2} orbital the $d_{x^2-y^2}$ orbital could have been occupied, which would have produced elongations along the x and y axes. However, a higher force is needed to stretch four bonds; stretching only two bonds is energetically more favorable, and consequently only examples with octahedra elongated in one direction are known.

The JAHN–TELLER effect is always to be expected when degenerate orbitals are unevenly occupied with electrons. In fact, it is observed for the following electronic configurations:

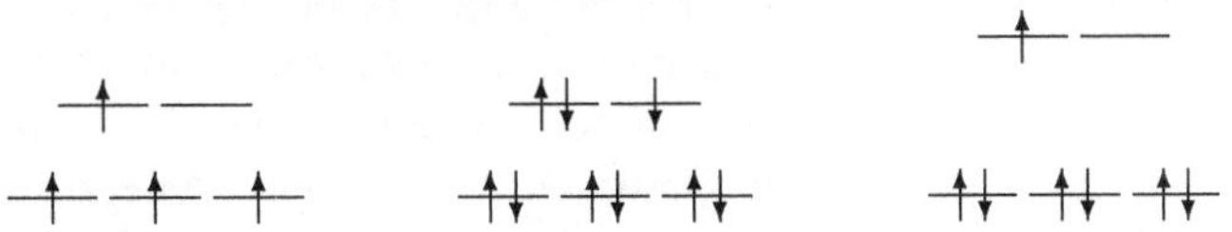

	d^4 high-spin	d^9	d^7 low-spin
Examples	Cr(II), Mn(III)	Cu(II)	Ni(III)

A JAHN–TELLER distortion should also occur for configuration d^1. However, in this case the occupied orbital is a t_{2g} orbital, for example d_{xy}; this exerts a repulsion on the ligands on the axes x and y which is only slightly larger than the force exerted along the z axis. The distorting force is usually not sufficient to produce a perceptible effect. Ions like TiF_6^{3-} or $MoCl_6^-$ show no detectable deviation from octahedral symmetry.

Not even the slightest JAHN–TELLER distortion and therefore no deviation from the ideal octahedral symmetry are to be expected when the t_{2g} and e_g orbitals are occupied evenly. This applies to the following electronic configurations:

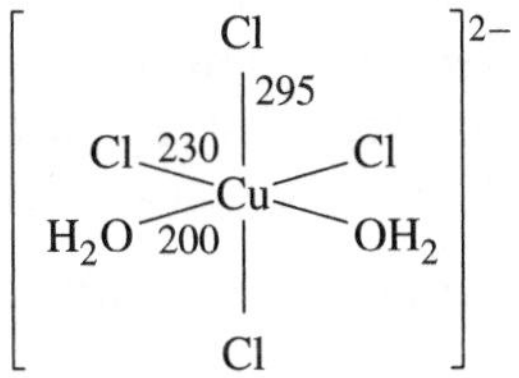

d^0, d^3, d^5 high-spin, d^6 low-spin, d^8 and d^{10}. However, for configuration d^8, octahedral coordination occurs only rarely (see below, square coordination).

If there are different kinds of ligands, those which have the smaller influence according to the spectrochemical series prefer the positions with the stretched bonds. For example, in the $[CuCl_4(OH_2)_2]^{2-}$ ion two of the Cl atoms take the positions in the vertices of the elongated axis of the coordination polyhedron.

Tetrahedral Coordination

We can imagine the four ligands of a tetrahedrally coordinated atom to be placed in four of the eight vertices of a cube. The orbitals d_{xy}, d_{yz} and d_{xz} (t_2 orbitals), which are oriented toward the cube edges, are closer to the bonding electron pairs than the orbitals $d_{x^2-y^2}$ and

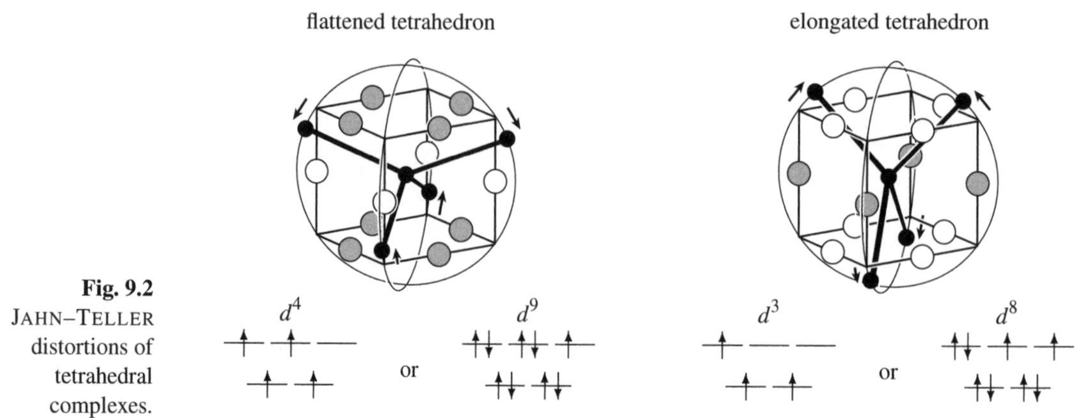

Fig. 9.2 JAHN–TELLER distortions of tetrahedral complexes.
The arrows indicate the directions of displacement of the ligands due to repulsion by the nonbonding d electrons. The spheres on the cube edges mark the centers of gravity of the charges of the t_2 orbitals; a gray sphere means occupation by one electron more than a white sphere

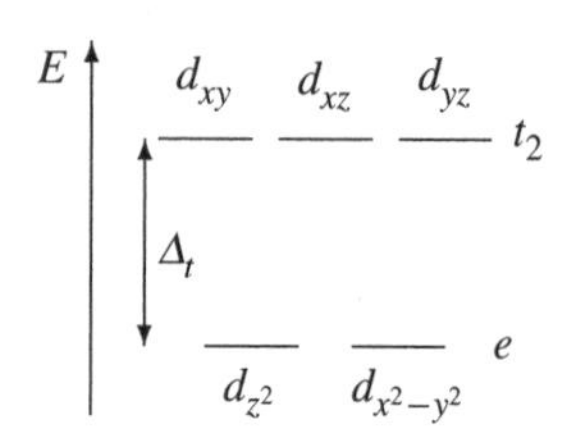

d_{z^2} (e orbitals). Consequently, the t_2 orbitals experience a larger repulsion and become energetically higher than the e orbitals; the sequence is opposite to that of octahedral coordination. The energy difference is termed Δ_t. Since none of the d orbitals is oriented toward a cube vertex, $\Delta_t < \Delta_o$ is expected or, more specifically, $\Delta_t \approx \frac{4}{9}\Delta_o$ (for equal ligands, equal central atom and equal bond lengths). Δ_t is always smaller than the spin pairing energy; tetrahedral complexes are always high-spin complexes.

If the t_2 orbitals are occupied unevenly, JAHN–TELLER distortions occur. For configuration d^4, one of the t_2 orbitals is unoccupied; for d^9, one has single occupation and the rest double. As a consequence, the ligands experience differing repulsions, and a flattened tetrahedron results (Fig. 9.2). Typical bond angles are, for example in the $CuCl_4^{2-}$ ion, $2 \times 116°$ and $4 \times 106°$.

For the configurations d^3 and d^8 one t_2 orbital has one electron more than the others; in this case an elongated tetrahedron is to be expected; however, the deformation turns out to be smaller than for d^4 and d^9, because the deforming repulsion force is being exerted by only one electron (instead of two; Fig. 9.2). Since the deformation force is small and the requirements of the packing in the crystal sometimes cause opposite deformations, observations do not always conform to expectations. For example, $NiCl_4^{2-}$ (d^8) has been observed to have undistorted, slightly elongated or slightly flattened tetrahedra depending on the cation. For uneven occupation of the e orbitals distortions could also be expected, but the effect is even smaller and usually it is not detectable; VCl_4 (d^1) for example has undistorted tetrahedra.

Square Coordination

When the two ligands on the z axis of an octahedral complex are removed, the remaining ligands form a square. The repulsion between bonding electrons on the z axis ceases for the d_{z^2}, the d_{xz}, and the d_{yz} electrons. Only one orbital, namely $d_{x^2-y^2}$, still experiences a strong repulsion from the remaining bond electrons and is energetically unfavorable (Fig. 9.3). Square coordination is the preferential coordination for d^8 configuration, as for Ni(II) and especially for Pd(II), Pt(II), and Au(III), in particular with ligands that cause a strong

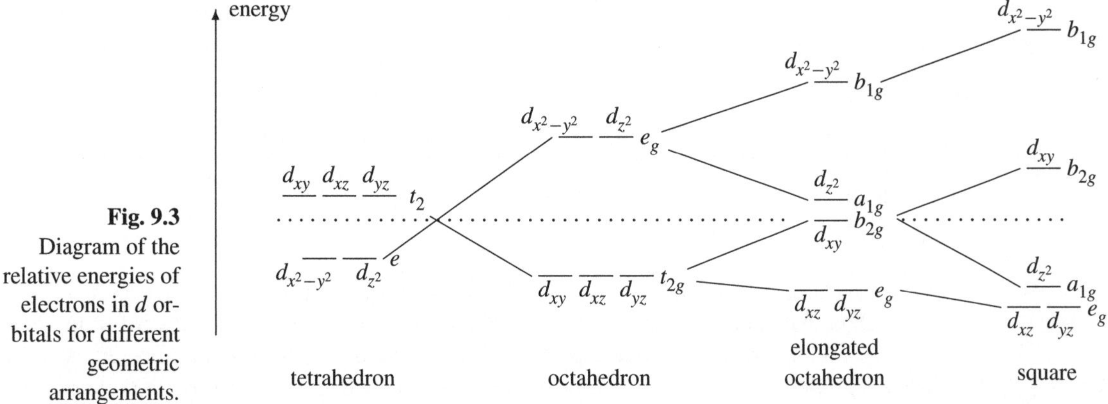

Fig. 9.3 Diagram of the relative energies of electrons in d orbitals for different geometric arrangements. The 'centers of gravity' (mean values of the energy levels) for all term sequences were positioned on the dotted line

splitting of the energy levels. Both an octahedral complex (two electrons in e_g orbitals) and a tetrahedral complex (four electrons in t_2 orbitals) are less favorable in this case.

9.2 Ligand Field Stabilization Energy

When ligands approach a central atom or ion, the following energetic contributions become effective:

- Energy gain (freed energy) by the formation of covalent bonds.
- Energy expenditure due to the mutual repulsion of the bonding electron pairs and due to the repulsion between ligands that approach each other too closely.
- Energy expenditure due to the repulsion exerted by bonding electron pairs on nonbonding electrons of the central atom.

Ligand field theory mainly considers the last contribution. For this contribution the geometric distribution of the ligands is irrelevant as long as the electrons of the central atom have a spherical distribution; the repulsion energy is always the same in this case. All half and fully occupied electron shells of an atom are spherical, namely d^5 high-spin and d^{10} (and naturally d^0). This is not so for other d electron configurations.

In order to compare the structural options for transition metal compounds and to estimate which of them are most favorable energetically, the *ligand field stabilization energy* (LFSE) is a useful parameter. This is defined as the difference between the repulsion energy of the bonding electrons toward the d electrons as compared to a notional repulsion energy that would exist if the d electron distribution were spherical.

In an octahedral complex a d_{z^2} electron is oriented toward the ligands (the same applies for $d_{x^2-y^2}$); it exercises more repulsion than if it were distributed spherically. Compared to this imaginary distribution it has a higher energy state. On the other hand, a d_{xy} electron is lowered energetically: it is being repelled less than an electron with spherical distribution. The *principle of the weighted mean* holds: the sum of the energies of the raised and the lowered states must be equal to the energy of the spherical state. Since there are two raised and three lowered states for an octahedron, the following scheme results:

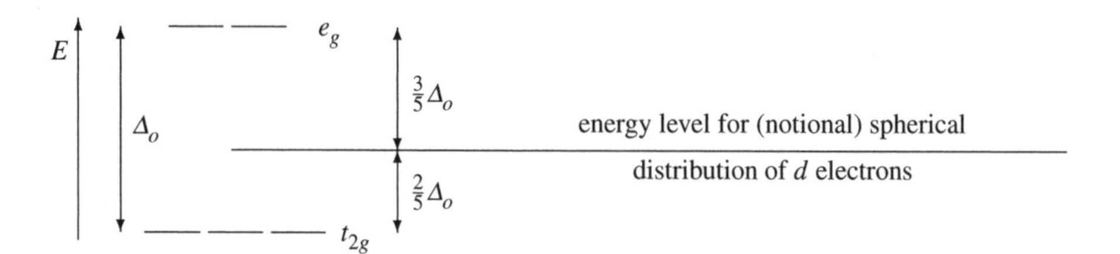

The energy level diagrams in Fig. 9.3 have been drawn according to the principle of the weighted mean energy. They show how the energy levels are placed relative to the level of the notional state of a spherical d electron distribution. They do not represent absolute energy values, as the absolute level of the notional state also depends on the other energy contributions mentioned above. Even when the central atoms and the ligands are the same, the level of the notional state differs on an absolute scale for different ligand arrangements, *i.e.* the different term schemes are shifted mutually.

Table 9.1 lists the values for the ligand field stabilization energies for octahedral and tetrahedral complexes. The values are given as multiples of Δ_o and Δ_t. In Fig. 9.4 the values have been plotted; the curves also show the influence of the other energy contributions for 3d elements. In the series from Ca^{2+} to Zn^{2+} the ionic radii decrease and the bond energies increase; correspondingly the curves run downwards from left to right. The dashed lines apply for the notional ions with spherical electron distributions. The actual energy values for the truly spherical electron distributions d^0, d^5 high-spin and d^{10} are situated on these lines. Due to the decreasing ionic radii octahedral complexes become less stable than tetrahedral complexes toward the end of the series (because of increasing repulsive forces between the bonding electron pairs and due to the more crowded ligand atoms); for this reason the dashed line for octahedra bends upwards at the end. The ligand field stabilization energy is the reason for the occurrence of two minima in the curves for high-spin complexes. The minima correspond to the configurations d^3 and d^8 for octahedral and

Table 9.1: Ligand field stabilization energies (LFSE) for octahedral and tetrahedral ligand distributions

	number of d electrons										
	0	1	2	3	4	5	6	7	8	9	10
octahedra, high-spin	electron distribution × energy $/\Delta_o$										
$\frac{3}{5}\Delta_o$ e_g	0	0	0	0	$1\times\frac{3}{5}$	$2\times\frac{3}{5}$	$2\times\frac{3}{5}$	$2\times\frac{3}{5}$	$2\times\frac{3}{5}$	$3\times\frac{3}{5}$	$4\times\frac{3}{5}$
$-\frac{2}{5}\Delta_o$ t_{2g}	0	$-1\times\frac{2}{5}$	$-2\times\frac{2}{5}$	$-3\times\frac{2}{5}$	$-3\times\frac{2}{5}$	$-3\times\frac{2}{5}$	$-4\times\frac{2}{5}$	$-5\times\frac{2}{5}$	$-6\times\frac{2}{5}$	$-6\times\frac{2}{5}$	$-6\times\frac{2}{5}$
sum = LFSE $/\Delta_o$	0	$-\frac{2}{5}$	$-\frac{4}{5}$	$-\frac{6}{5}$	$-\frac{3}{5}$	0	$-\frac{2}{5}$	$-\frac{4}{5}$	$-\frac{6}{5}$	$-\frac{3}{5}$	0
octahedra, low-spin	electron distribution × energy $/\Delta_o$										
$\frac{3}{5}\Delta_o$ e_g	0	0	0	0	0	0	0	$1\times\frac{3}{5}$	$2\times\frac{3}{5}$	$3\times\frac{3}{5}$	$4\times\frac{3}{5}$
$-\frac{2}{5}\Delta_o$ t_{2g}	0	$-1\times\frac{2}{5}$	$-2\times\frac{2}{5}$	$-3\times\frac{2}{5}$	$-4\times\frac{2}{5}$	$-5\times\frac{2}{5}$	$-6\times\frac{2}{5}$	$-6\times\frac{2}{5}$	$-6\times\frac{2}{5}$	$-6\times\frac{2}{5}$	$-6\times\frac{2}{5}$
sum = LFSE $/\Delta_o$	0	$-\frac{2}{5}$	$-\frac{4}{5}$	$-\frac{6}{5}$	$-\frac{8}{5}$	$-\frac{10}{5}$	$-\frac{12}{5}$	$-\frac{9}{5}$	$-\frac{6}{5}$	$-\frac{3}{5}$	0
tetrahedra, high-spin	electron distribution × energy $/\Delta_t$										
$\frac{2}{5}\Delta_t$ t_2	0	0	0	$1\times\frac{2}{5}$	$2\times\frac{2}{5}$	$3\times\frac{2}{5}$	$3\times\frac{2}{5}$	$3\times\frac{2}{5}$	$4\times\frac{2}{5}$	$5\times\frac{2}{5}$	$6\times\frac{2}{5}$
$-\frac{3}{5}\Delta_t$ e	0	$-1\times\frac{3}{5}$	$-2\times\frac{3}{5}$	$-2\times\frac{3}{5}$	$-2\times\frac{3}{5}$	$-2\times\frac{3}{5}$	$-3\times\frac{3}{5}$	$-4\times\frac{3}{5}$	$-4\times\frac{3}{5}$	$-4\times\frac{3}{5}$	$-4\times\frac{3}{5}$
sum = LFSE $/\Delta_t$	0	$-\frac{3}{5}$	$-\frac{6}{5}$	$-\frac{4}{5}$	$-\frac{2}{5}$	0	$-\frac{3}{5}$	$-\frac{6}{5}$	$-\frac{4}{5}$	$-\frac{2}{5}$	0

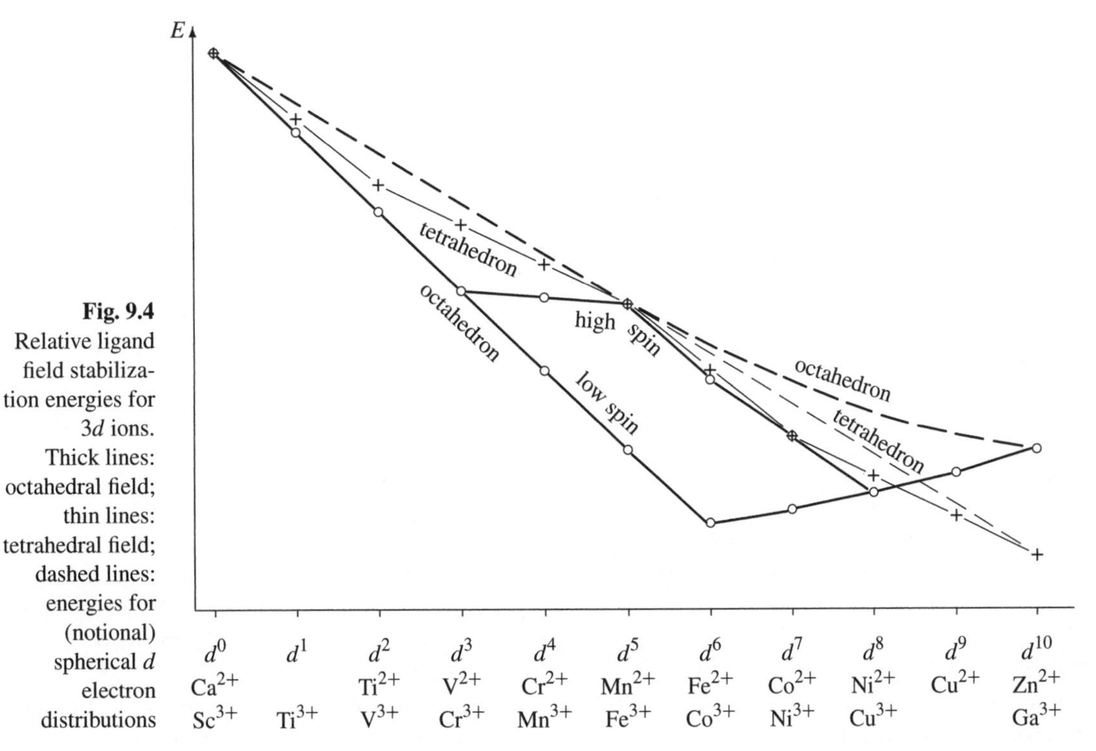

Fig. 9.4 Relative ligand field stabilization energies for 3*d* ions. Thick lines: octahedral field; thin lines: tetrahedral field; dashed lines: energies for (notional) spherical *d* electron distributions

to d^2 and d^7 for tetrahedral complexes. The stabilization energies are less for tetrahedral ligand fields, since generally $\Delta_o > \Delta_t$ (in Fig. 9.4 $\Delta_t = \frac{4}{9}\Delta_o$ was assumed). For octahedral low-spin complexes there is only one minimum at d^6.

For high-spin compounds only rather small stabilization differences result between octahedral and tetrahedral coordination for the configurations d^7 and d^8 (Fig. 9.4). Co^{2+} shows a tendency to tetrahedral coordination, whereas this tendency is overcompensated for Ni^{2+} by the larger ligand field stabilization for octahedra, so that Ni^{2+} prefers octahedral coordination. Here the different locations of the maxima of the ligand field stabilization energies takes effect (Table 9.1): it is largest for tetrahedra at configuration d^7 (Co^{2+}) and for octahedra at d^8 (Ni^{2+}). With increasing ligand sizes the tendency toward tetrahedral coordination becomes more marked; in other words, the octahedral arrangement becomes relatively less stable; in Fig. 9.4 this would be expressed by an earlier upwards bending of the thick dashed line. Fe^{2+} and Mn^{2+} also form tetrahedral complexes with larger ligands like Cl^- or Br^-.

In Fig. 9.4 the additional stabilization by the JAHN–TELLER effect has not been taken into account. Its inclusion brings the point for the (distorted) octahedral coordination for Cu^{2+} further down, thus rendering this arrangement more favorable.

The ligand field stabilization is expressed in the lattice energies of the halides MX_2. The values obtained by the BORN–HABER cycle from experimental data are plotted *vs.* the *d* electron configuration in Fig. 9.5. The ligand field stabilization energy contribution is no more than 200 kJ mol^{-1}, which is less than 8% of the total lattice energy. The ionic radii also show a similar dependence (Fig. 9.6; Table 6.4, p. 50).

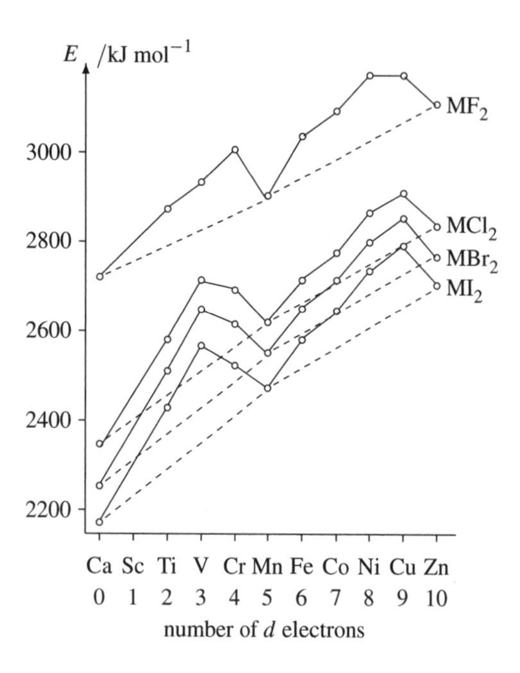

Fig. 9.5
Lattice energies of the dihalides of elements of the first transition metal period

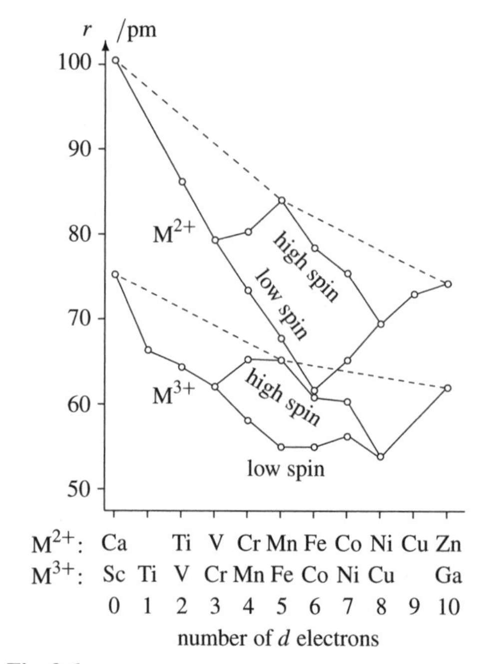

Fig. 9.6
Ionic radii of the elements of the first transition metal period in octahedral coordination

9.3 Coordination Polyhedra for Transition Metals

According to the preceding statements certain coordination polyhedra occur preferentially for compounds of transition metals, depending on the central atom, the oxidation state, and the kind of ligand. The general tendencies can be summarized as follows:

The series of 3*d* elements from scandium to iron as well as nickel preferably form octahedral complexes in the oxidation states I, II, III, and IV. Octahedra and tetrahedra are known for cobalt, and tetrahedra for zinc and copper(I). Copper(II) (d^9) forms JAHN-TELLER distorted octahedra and tetrahedra. With higher oxidation states (= smaller ionic radii) and larger ligands the tendency to form tetrahedra increases. For vanadium(V), chromium(VI) and manganese(VII) almost only tetrahedral coordination is known (VF_5 is an exception). Nickel(II) low-spin complexes (d^8) can be either octahedral or square.

Among the heavier 4*d* and 5*d* elements, tetrahedral coordination only occurs for silver, cadmium, and mercury and when the oxidation states are very high as in MoO_4^{2-}, ReO_4^- or OsO_4. Octahedra are very common, and higher coordination numbers, especially 7, 8, and 9, are not unusual, as for example in ZrO_2 (c.n. 7), $Mo(CN)_8^{4-}$ or $LaCl_3$ (c.n. 9). A special situation arises for the electronic configuration d^8, namely for Pd(II), Pt(II), Ag(III), and Au(III), which almost always have square coordination. Pd(0), Pt(0), Ag(I), Au(I), and Hg(II) (d^{10}) frequently show linear coordination (c.n. 2). In Table 9.2 the most important coordination polyhedra are summarized with corresponding examples.

Table 9.2 Most common coordination polyhedra for coordination numbers 2 to 6 for transition metal compounds

polyhedron	c.n.	electron config.	central atom	examples
linear arrangement	2	d^{10}	Cu(I), Ag(I), Au(I), Hg(II)	Cu_2O, $Ag(CN)_2^-$, $AuCN^*$, $AuCl_2^-$, $HgCl_2$, HgO^*
triangle	3	d^{10}	Cu(I), Ag(I), Au(I), Hg(II)	$Cu(CN)_3^{2-}$, $Ag_2Cl_5^{3-}$, $Au(PPh_3)_3^+$, HgI_3^-
square	4	d^8	Ni(II), Pd(II), Pt(II), Au(III)	$Ni(CN)_4^{2-}$, $PdCl_2{}^*$, PtH_4^{2-}, $Pt(NH_3)_2Cl_2$, $AuCl_4^-$
tetrahedron	4	d^0	Ti(IV), V(V), Cr(VI), Mo(VI), Mn(VII), Re(VII) Ru(VIII), Os(VIII)	$TiCl_4$, VO_4^{3-}, $CrO_3{}^*$, CrO_4^{2-}, MoO_4^{2-}, WO_4^{2-} Mn_2O_7, ReO_4^- RuO_4, OsO_4
		d^1	V(IV), Cr(V), Mn(VI), Ru(VII)	VCl_4, CrO_4^{3-}, MnO_4^{2-}, RuO_4^-
		d^5	Mn(II), Fe(III)	$MnBr_4^{2-}$, Fe_2Cl_6
		d^6	Fe(II)	$FeCl_4^{2-}$
		d^7	Co(II)	$CoCl_4^{2-}$
		d^8	Ni(II)	$NiCl_4^{2-}$
		d^9	Cu(II)	$CuCl_4^{2-}$ †
		d^{10}	Ni(0), Cu(I), Zn(II), Hg(II)	$Ni(CO)_4$, $Cu(CN)_4^{3-}$ $Zn(CN)_4^{2-}$, HgI_4^{2-}
square pyramid	5	d^0	Ti(IV), V(V), Nb(V), Mo(VI), W(VI),	$TiOCl_4^{2-}$, VOF_4^-, $NbSCl_4^-$, $MoNCl_4^-$, $WNCl_4^-$
		d^1	V(IV), Cr(V), Mo(V), W(V), Re(VI)	$VO(NCS)_4^{2-}$, $CrOCl_4^-$, $MoOCl_4^-$, $WSCl_4^-$, $ReOCl_4$
		d^2	Os(VI)	$OsNCl_4^-$
		d^4	Mn(III), Re(III)	$MnCl_5^{2-}$, Re_2Cl_8
		d^7	Co(II)	$Co(CN)_5^{3-}$
trigonal bipyramid	5	d^2	V(IV)	$VCl_3(NMe_3)_2$
		d^8	Fe(0)	$Fe(CO)_5$
octahedron	6		nearly all; rarely Pd(II), Pt(II), Au(III), Cu(I)	

* endless chain † Jahn–Teller distorted

9.4 Isomerism

Two compounds are *isomers* when they have the same chemical composition but different molecular structures. Isomers have different physical and chemical properties.

Constitution isomers have molecules with different *constitutions*, *i.e.* the atoms linked with one another differ. For example:

F–S–S–F

S=SF₂

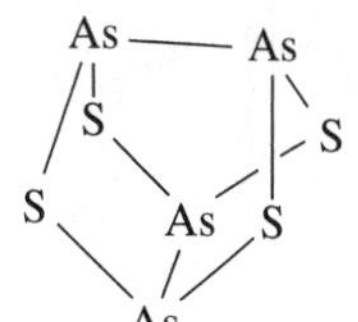

Transition metal complexes in particular show several kinds of constitution isomers, namely:

Bonding isomers, differing by the kind of ligand atom bonded to the central atom, for example:

Ph_3As $AsPh_3$ Pt $N{\equiv}C{-}S$ $S{-}C{\equiv}N$ Ph_3As $AsPh_3$ Pt N C S N C S

Further ligands that can be bonded by different atoms include OCN^- and NO_2^-. Cyanide ions always are linked with their C atoms in isolated complexes, but in polymeric structures as in Prussian blue they can be coordinated via both atoms ($Fe{-}C{\equiv}N{-}Fe$).

Coordination isomers occur when complex cations and complex anions are present and ligands are exchanged between anions and cations, for example:

$[Cu(NH_3)_4][PtCl_4]$ $[Pt(NH_3)_4][CuCl_4]$
$[Pt(NH_3)_4][PtCl_6]$ $[Pt(NH_3)_4Cl_2][PtCl_4]$

Further variations are:

Hydrate isomers, e.g. $[Cr(OH_2)_6]Cl_3$, $[Cr(OH_2)_5Cl]Cl_2{\cdot}H_2O$, $[Cr(OH_2)_4Cl_2]Cl{\cdot}2H_2O$

Ionization isomers, e.g. $[Pt(NH_3)_4Cl_2]Br_2$, $[Pt(NH_3)_4Br_2]Cl_2$

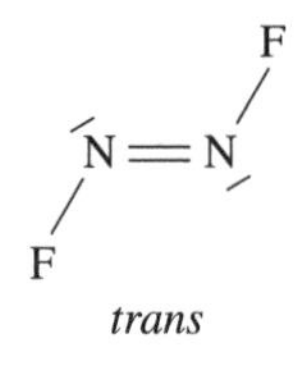

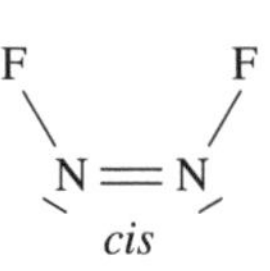

Stereo isomers have the same constitution, but a different spatial arrangement of their atoms; they differ in their *configuration*. Two cases have to be distinguished: geometric isomers (diastereomers) and enantiomers.

Geometric isomers occur as *cis–trans* isomers in compounds with double bonds like in N_2F_2 and especially when coordination polyhedra have different kinds of ligands. The most important types are square and octahedral complexes with two or more different ligands (Fig. 9.7). To designate them in more complicated cases, the polyhedron vertices are numbered alphabetically, for example *abf*-triaqua-*cde*-tribromoplatinum(IV) for *mer*-$[PtBr_3(OH_2)_3]^+$. No geometric isomers exist for tetrahedral complexes. With other

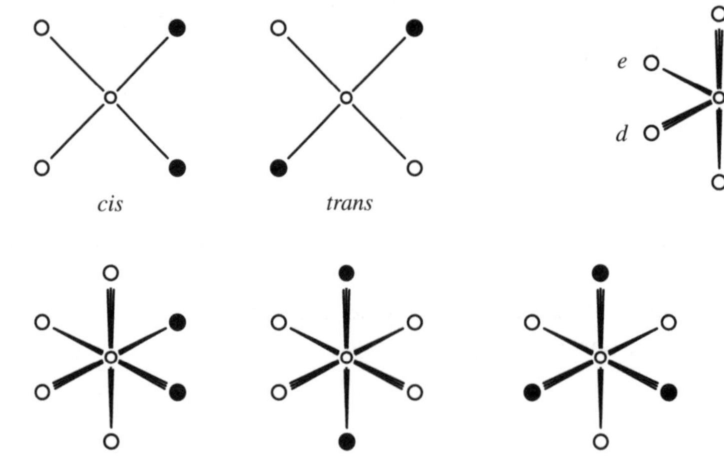

Fig. 9.7 Geometric isomers for square and octahedral coordination with two different ligands. Top right: designation of ligand positions in an octahedral complex

Table 9.3: Number of possible geometric isomers depending on the number of different ligands (designated by A, B, C, ...) for some coordination polyhedra (excluding chelate complexes). Of every pair of enantiomers only one representative was counted

polyhedron	ligands	total number	chiral number	polyhedron	ligands	total number	chiral number
tetrahedron	unrestricted	1	ABCD	octahedron	AB_5	1	0
square	AB_3	1	0		A_2B_4	2	0
	A_2B_2	2	0		A_3B_3	2	0
	ABC_2	2	0		ABC_4	2	0
	ABCD	3	0		AB_2C_3	3	0
trigonal bipyramid	AB_4	2	0		$A_2B_2C_2$	5	1
	A_2B_3	3	0		$ABCD_3$	4	1
	ABC_3	4	0		ABC_2D_2	6	2
	AB_2C_2	5	1		$ABCDE_2$	9	6
	$ABCD_2$	7	3		ABCDEF	15	15
	ABCDE	10	10				

coordination polyhedra the number of possible isomers increases with the number of different ligands (Table 9.3); however, usually only one or two of them are known.

Enantiomers have structures of exactly the same kind and yet are different. Their structures correspond to mirror images. In their physical properties they differ only with respect to phenomena that are polar, *i.e.* that have some kind of a preferred direction. This especially includes polarized light, the polarization plane of which experiences a rotation when it passes through a solution of the substance. For this reason enantiomers have also been called optical isomers. In their chemical properties enantiomers differ only when they react with a compound that is an enantiomer itself.

The requirement for the existence of enantiomers is a *chiral* structure. Chirality is solely a symmetry property: a rigid object is chiral if it is not superposable by pure rotation or translation on its image formed by inversion. Such an object contains no rotoinversion axis (or rotoreflection axis; *cf.* Section 3.1). Since the reflection plane and the inversion center are special cases of rotoinversion axes ($\bar{2}$ and $\bar{1}$), they are excluded.

A chiral object and the opposite object formed by inversion form a pair of *enantiomorphs.* If an enantiomorph is a molecular entity, it is called an enantiomer. An equimolar mixture of enantiomers is a *racemate.*

In crystals, in addition, no glide planes may occur. Rotation axes and screw axes are permitted. As a consequence, only 65 out of the 230 space-group types may occur; these are called *Sohncke space-group types* after L. SOHNCKE who was the first to describe them. Among the 65 SOHNCKE space group types there are 11 enantiomorphic pairs which have only one kind of right- or left-handed screw axis (*e.g.* $P4_1$ and $P4_3$). Only these 22 space-group types are chiral themselves. The remaining 43 SOHNCKE space-group types do permit chiral crystal structures, but their space groups are not chiral.*

The great majority of known chiral compounds are naturally occurring organic substances, their molecules having one or more *asymmetrically substituted carbon atoms* (stereogenic atoms). Chirality is present when a tetrahedrally coordinated atom has

*In literature, SOHNCKE space-group types are often termed 'chiral space groups', which is not correct. Most chiral molecular compounds do not crystallize in a chiral (enantiomorphic) space group. For details see [86].

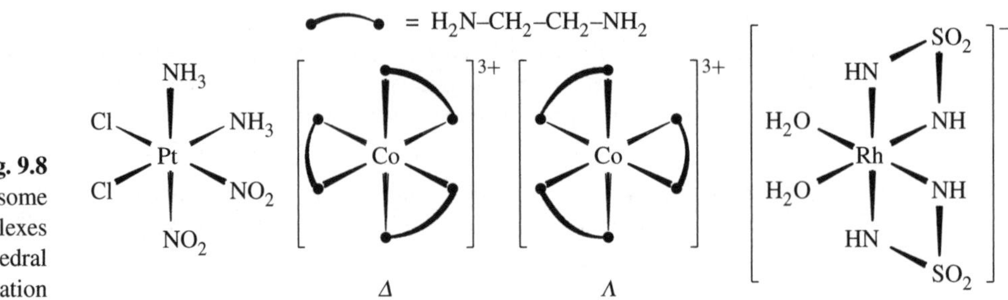

Fig. 9.8 Examples of some chiral complexes with octahedral coordination

four different ligands.* Known inorganic enantiomers are mainly complex compounds, mostly with octahedral coordination. In Table 9.3 ligand combinations are listed for which chiral molecules are possible. Well known chiral complexes are chelate complexes, some examples being shown in Fig. 9.8. The configuration of trichelate complexes like $[Co(H_2N(CH_2)_2NH_2)_3]^{3+}$ can be designated by Δ or Λ: view the structure along the threefold rotation axis, as shown in Fig. 9.8; if the chelate groups are oriented like the turns of a right-handed screw, then the symbol is Δ.

9.5 Problems

9.1 State which of the following octahedral high-spin complexes should be JAHN–TELLER distorted.
TiF_6^{2-}, MoF_6, $[Cr(OH_2)_6]^{2+}$, $[Mn(OH_2)_6]^{2+}$, $[Mn(OH_2)_6]^{3+}$, $FeCl_6^{3-}$, $[Ni(NH_3)_6]^{2+}$, $[Cu(NH_3)_6]^{2+}$.

9.2 State which of the following tetrahedral complexes should be JAHN–TELLER distorted, and what kind of a distortion it should be.
$CrCl_4^-$, $MnBr_4^{2-}$, $FeCl_4^-$, $FeCl_4^{2-}$, $NiBr_4^{2-}$, $CuBr_4^{2-}$, $Cu(CN)_4^{3-}$, $Zn(NH_3)_4^{2+}$.

9.3 Decide whether the following complexes are tetrahedral or square.
$Co(CO)_4^-$, $Ni(PF_3)_4$, $PtCl_2(NH_3)_2$, $Pt(NH_3)_4^{2+}$, $Cu(OH)_4^{2-}$, Au_2Cl_6 (dimeric via chloro bridges).

9.4 What are the point groups of the complexes shown in Fig. 9.8 and why are they chiral?

9.5 How many isomers do you expect for the following complexes?
(a) $PtCl_2(NH_3)_2$; (b) $ZnCl_2(NH_3)_2$; (c) $[OsCl_4F_2]^{2-}$; (d) $[CrCl_3(OH_2)_3]^{3-}$; (e) $Mo(CO)_5OR_2$.

*In organic stereochemistry the terms 'center of chirality' or 'center of asymmetry' are often used; usually they refer to an asymmetrically substituted C atom. These terms should be avoided since they are contradictions in themselves: a chiral object by definition has no center (the only kind of center existing in symmetry is the inversion center).

10 Molecular Orbital Theory and Chemical Bonding in Solids

10.1 Molecular Orbitals

Molecular orbital (MO) theory currently offers the most accurate description of the bonding within a molecule. The term *orbital* is a neologism reminiscent of the concept of an orbiting electron, but it also expresses the inadequacy of this concept for the precise characterization of the behavior of an electron. Mathematically an electron is treated as a standing wave by the formulation of a wave function ψ. For the hydrogen atom the wave functions for the ground state and all excited states are known exactly; they can be calculated as solutions of the SCHRÖDINGER equation. Hydrogen-like wave functions are assumed for other atoms, and their calculation is performed with sophisticated approximation methods.

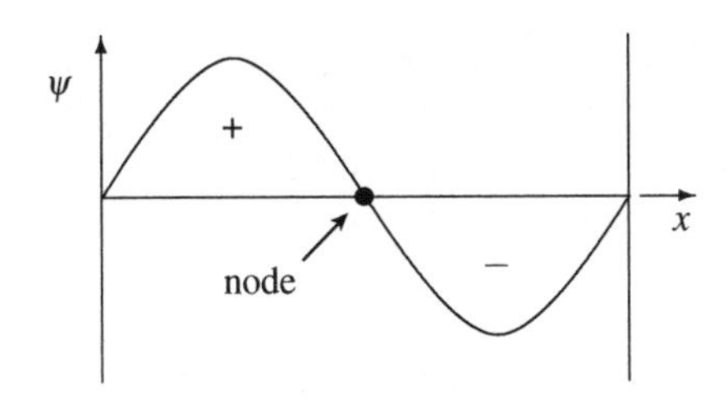

The wave function of an electron corresponds to the expression used to describe the amplitude of a vibrating chord as a function of the position x. The opposite direction of the motion of the chord on the two sides of a vibrational node is expressed by opposite signs of the wave function. Similarly, the wave function of an electron has opposite signs on the two sides of a nodal surface. The wave function is a function of the site x, y, z, referred to a coordinate system that has its origin in the center of the atomic nucleus.

Wave functions for the orbitals of molecules are calculated by linear combinations of *all* wave functions of *all* atoms involved. The total number of orbitals remains unaltered, *i.e.* the total number of contributing atomic orbitals must be equal to the number of molecular orbitals. Furthermore, certain conditions have to be obeyed in the calculation; these include linear independence of the molecular orbital functions and normalization. In the following we will designate wave functions of atoms by χ and wave functions of molecules by ψ. We obtain the wave functions of an H_2 molecule by linear combination of the $1s$ functions χ_1 and χ_2 of the two hydrogen atoms:

$$\psi_1 = \tfrac{1}{2}\sqrt{2}(\chi_1 + \chi_2) \qquad \psi_2 = \tfrac{1}{2}\sqrt{2}(\chi_1 - \chi_2)$$

+ χ_1 + χ_2 — bonding

+ χ_1 − χ_2 — antibonding

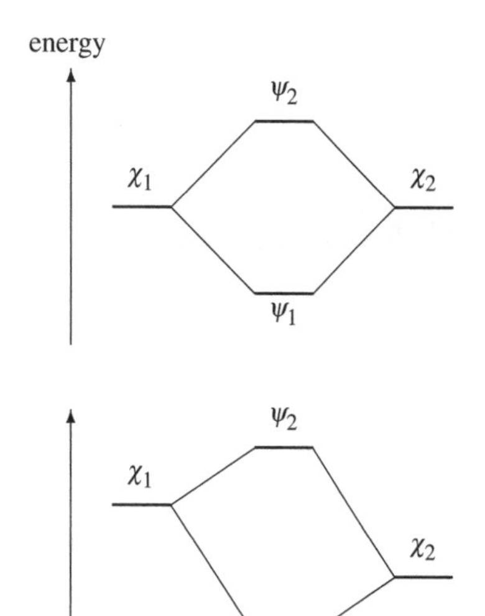

Compared to an H atom, electrons with the function ψ_1 are less energetic, and those with the function ψ_2 are more energetic. When the two available electrons 'occupy' the molecular orbital ψ_1, this is energetically favorable; ψ_1 is the wave function of a *bonding* molecular orbital. ψ_2 belongs to an *antibonding* molecular orbital; its occupation by electrons requires the input of energy.

When calculating the wave functions for the bonds between two atoms of different elements, the functions of the atoms contribute with different coefficients c_1 and c_2:

$$\psi_1 = c_1\chi_1 + c_2\chi_2 \quad (10.1)$$

$$\psi_2 = c_2\chi_1 - c_1\chi_2 \quad (10.2)$$

The probability of finding an electron at a site x, y, z is given by ψ^2. Integrated over all space, the probability must be equal to 1:

$$1 = \int \psi_1^2 \, \mathrm{d}V = \int |c_1\chi_1 + c_2\chi_2|^2 \, \mathrm{d}V = c_1^2 + c_2^2 + 2c_1c_2S_{12} \quad (10.3)$$

S_{12} is the *overlap integral* between χ_1 and χ_2. The term $2c_1c_2S_{12}$ is the *overlap population*; it expresses the electronic interaction between the atoms. The contributions c_1^2 and c_2^2 can be assigned to the atoms 1 and 2, respectively.

Equation (10.3) is fulfilled when $c_1^2 \approx 1$ and $c_2^2 \approx 0$; in this case the electron is localized essentially at atom 1 and the overlap population is approximately zero. This is the situation of a minor electronic interaction, either because the corresponding orbitals are too far apart or because they differ considerably in energy. Such an electron does not contribute to bonding.

For ψ_1 the overlap population $2c_1c_2S_{12}$ is positive, and the electron is bonding; for ψ_2 it is negative, and the electron is antibonding. The sum of the values $2c_1c_2S_{12}$ of all occupied orbitals of the molecule, the MULLIKEN overlap population, is a measure of the bond strength or bond order (b.o.):

$$\text{b.o.} = \tfrac{1}{2}[(\text{number of bonding electrons}) - (\text{number of antibonding electrons})]$$

Despite the given formula, the calculation of the bond order is not always clear in the case of the occupation of orbitals having only a minor bonding or minor antibonding effect; should they be counted or not? Nevertheless, the bond order is a simple and useful concept.† In valence-bond formulas it corresponds to the number of bonding lines.

Orbitals other than *s* orbitals can also be combined to give bonding, antibonding or nonbonding molecular orbitals. Nonbonding are those orbitals for which bonding and antibonding components cancel each other. Some possibilities are shown in Fig. 10.1. Note the signs of the wave functions. A bonding molecular orbital having no nodal surface is a σ orbital; if it has one nodal plane parallel to the connecting line between the atomic centers it is a π orbital, and with two such nodal planes it is a δ orbital. Antibonding orbitals usually are designated by an asterisk *.

† Chemists very successfully use many concepts in a more intuitive manner, although they generally tend not to define their concepts clearly (like the bond order), and to ignore definitions if they happen not to be convenient.

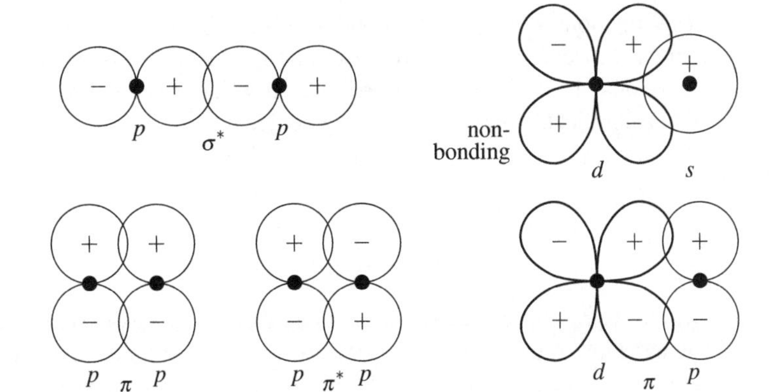

Fig. 10.1 Some combinations of atomic orbitals to give molecular orbitals. Asterisks denote antibonding orbitals

10.2 Hybridization

In order to calculate the orbitals for a methane molecule, the four $1s$ functions of the four hydrogen atoms and the functions $2s$, $2p_x$, $2p_y$ and $2p_z$ of the carbon atom are combined to give eight wave functions, four of which are bonding and four of which are antibonding. The four bonding wave functions are:

$$\psi_1 = \frac{c_1}{2}(s+p_x+p_y+p_z)+c_2\chi_{H1}+c_3(\chi_{H2}+\chi_{H3}+\chi_{H4})$$
$$\psi_2 = \frac{c_1}{2}(s+p_x-p_y-p_z)+c_2\chi_{H2}+c_3(\chi_{H3}+\chi_{H4}+\chi_{H1})$$
$$\psi_3 = \frac{c_1}{2}(s-p_x+p_y-p_z)+c_2\chi_{H3}+c_3(\chi_{H4}+\chi_{H1}+\chi_{H2})$$
$$\psi_4 = \frac{c_1}{2}(s-p_x-p_y+p_z)+c_2\chi_{H4}+c_3(\chi_{H1}+\chi_{H2}+\chi_{H3})$$

$\psi_1, \psi_2, \ldots$ are wave functions of the CH_4 molecule, s, p_x, p_y and p_z designate the wave functions of the C atom, and $\chi_{H1}, \chi_{H2}, \ldots$ correspond to the H atoms. Among the coefficients c_1, c_2 and c_3, one is negligible: $c_3 \approx 0$.

Formulated as in the preceding paragraph, the functions are not especially illustrative. They do not correspond to the idea a chemist associates with the formation of a bond between two atoms: in his or her imagination the atoms approach each other and their atomic orbitals merge into a bonding molecular orbital. To match this kind of mental picture it is expedient to start from atomic orbitals whose spatial orientations correspond to the orientations of the bonds of the molecule that is formed. Such orbitals can be obtained by *hybridization* of atomic orbitals. Instead of calculating the molecular orbitals of the methane molecule in one step according to the equations mentioned above, one proceeds in two steps. First, only the wave functions of the C atom are combined to give sp^3 hybrid orbitals:

$$\chi_1 = \tfrac{1}{2}(s+p_x+p_y+p_z)$$
$$\chi_2 = \tfrac{1}{2}(s+p_x-p_y-p_z)$$
$$\chi_3 = \tfrac{1}{2}(s-p_x+p_y-p_z)$$
$$\chi_4 = \tfrac{1}{2}(s-p_x-p_y+p_z)$$

The functions χ_1 to χ_4 correspond to orbitals having preferential alignments oriented towards the vertices of an circumscribed tetrahedron. Their combinations with the wave functions of four hydrogen atoms placed in these vertices yield the following functions, the insignificant coefficient c_3 being neglected:

$$\psi_1 = c_1\chi_1 + c_2\chi_{\mathrm{H1}}$$
$$\psi_2 = c_1\chi_2 + c_2\chi_{\mathrm{H2}}$$
$$\text{etc.}$$

ψ_1 corresponds to a bonding orbital that essentially involves the interaction of the C atom with the first H atom; its charge density ψ_1^2 is concentrated in the region between these two atoms. This matches the idea of a localized C–H bond: The electron pair of this orbital is assigned to a bond between these two atoms and symbolized by a dash in the valence bond formula.

To be more exact, every bond is a 'multi-center bond' with contributions of the wave functions of all atoms. However, due to the charge concentration in the region between two atoms and because of the inferior contributions χ_{H2}, χ_{H3}, and χ_{H4}, the bond can be taken to a good approximation to be a 'two-center-two-electron bond' (2*c*2*e* bond) between the atoms C and H1. From the mathematical point of view the hybridization is not necessary for the calculation, and in the usual molecular orbital calculations it is not performed. It is, however, a helpful mathematical trick for adapting the wave functions to a chemist's mental picture.

For molecules with different structures different hybridization functions are appropriate. An infinity of hybridization functions can be formulated by linear combinations of s and p orbitals:

$$\chi_i = \alpha_i s + \beta_i p_x + \gamma_i p_y + \delta_i p_z$$

The coefficients must be normalized, *i.e.* $\alpha_i^2 + \beta_i^2 + \gamma_i^2 + \delta_i^2 = 1$. Their values determine the preferential directions of the hybrid orbitals. For example, the functions

$$\begin{aligned}
\chi_1 &= 0.833s + 0.32(p_x + p_y + p_z) \\
\chi_2 &= 0.32s + 0.547(p_x - p_y - p_z) \\
\chi_3 &= 0.32s + 0.547(-p_x + p_y - p_z) \\
\chi_4 &= 0.32s + 0.547(-p_x - p_y + p_z)
\end{aligned}$$

define an orbital χ_1 having contributions of 69 % (= $0.833^2 \times 100$ %) s and 31 % p and three orbitals χ_2, χ_3 and χ_4, each with contributions of 10 % s and 90 % p. They are adequate to calculate the wave functions for a molecule $|AX_3$ that has a lone electron pair (χ_1) with a larger s contribution and bonds with larger p orbital contributions as compared to sp^3 hybridization. The corresponding bond angles are between 90° and 109.5°, namely 96.5°.

To derive the values of the coefficients α_i, β_i, γ_i, and δ_i so that the bond energy is maximized and the correct molecular structure results, the mutual interactions between the electrons have to be considered. This requires a great deal of computational expenditure. However, in a qualitative manner the interactions can be estimated rather well: that is exactly what the valence shell electron-pair repulsion theory accomplishes.

10.3 The Electron Localization Function

Wave functions can be calculated rather reliably with quantum-chemical approximations. The sum of the squares of all wave functions ψ_i of the occupied orbitals at a site x, y, z is the electron density $\rho(x, y, z) = \sum \psi_i^2$. It can also be determined experimentally by X-ray diffraction (with high expenditure). The electron density is not very appropriate to visualize chemical bonds. It shows an accumulation of electrons close to the atomic nuclei. The enhanced electron density in the region of chemical bonds can be displayed after the contribution of the inner atomic electrons has been subtracted. But even then it remains difficult to discern and to distinguish the electron pairs.

Redress can be obtained by the *electron localization function* (ELF). It decomposes the electron density spatially into regions that correspond to the notion of electron pairs, and its results are compatible with the valence shell electron-pair repulsion theory. An electron has a certain electron density $\rho_1(x, y, z)$ at a site x, y, z; this can be calculated with quantum mechanics. Take a small, spherical volume element ΔV around this site. The product $n_1(x, y, z) = \rho_1(x, y, z)\Delta V$ corresponds to the number of electrons in this volume element. For a given number of electrons the size of the sphere ΔV adapts itself to the electron density. For this given number of electrons one can calculate the probability $w(x, y, z)$ of finding a second electron with the same spin within this very volume element. According to the PAULI principle this electron must belong to another electron pair. The electron localization function is defined with the aid of this probability:

$$\mathrm{ELF}(x, y, z) = \frac{1}{1 + [c - w(x, y, z)]^2}$$

c is a positive constant that is arbitrarily chosen as to yield ELF = 0.5 for a homogeneous electron gas.

The properties of the thus defined function are:

- ELF is a function of the spatial coordinates x, y, z.
- ELF adopts values between 0 and 1.
- In a region where an electron pair is present, where therefore the probability of coming across a second electron pair is low, ELF adopts high values. Low values of ELF separate the regions of different electron pairs.
- The symmetry of ELF corresponds to the symmetry of the molecule or crystal.

ELF can be visualized with different kinds of images. Colored sections through a molecule are popular, using white for high values of ELF, followed by yellow–red–violet–blue–dark blue for decreasing values; simultaneously, the electron density can be depicted by the density of colored points. Contour lines can be used instead of the colors for black and white printing. Another possibility is to draw perspective images with iso surfaces, *i.e.* surfaces with a constant value of ELF. Fig. 10.2 shows iso surfaces with ELF = 0.8 for some molecules; from experience a value of ELF = 0.8 is well suited to reveal the distribution of electron pairs in space.

On the one hand, Fig. 10.2 exhibits iso surfaces around the fluorine atoms; on the other hand the lone electron pairs at the central atoms can be discerned quite well. The space requirement of one lone pair is larger than that of the four electron pairs at one of the more electronegative fluorine atoms. The three lone pairs at the chlorine atom of ClF_2^- add up to a rotation-symmetrical torus.

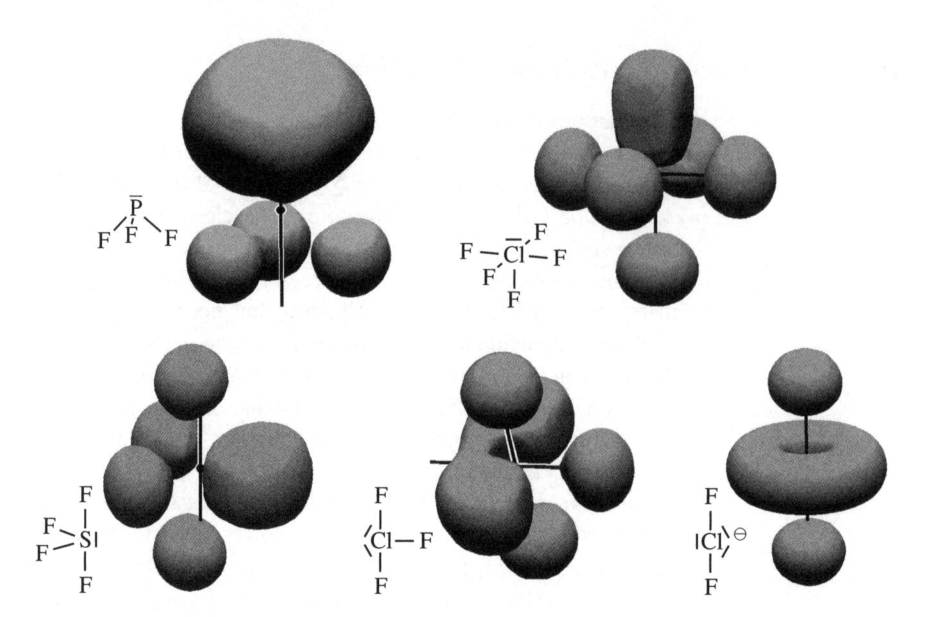

Fig. 10.2 Iso surfaces with ELF = 0.8 for some molecules having lone electron pairs (Images by T. FÄSSLER et al., Technische Universität München. Reprinted from *Angewandte Chemie* [97], with permission from Wiley-VCH)

10.4 Band Theory. The Linear Chain of Hydrogen Atoms

In a solid that cannot be interpreted on the basis of localized covalent bonds or of ions, the assessment of the bonding requires the consideration of the complete set of molecular orbitals of *all* involved atoms. This is the subject of *band theory*, which offers the most comprehensive concept of chemical bonding. Ionic bonding and localized covalent bonds result as special cases. The ideas presented in this chapter are based on the intelligible exposition by R. HOFFMANN [87], the reading of which is recommended for a deeper insight into the subject. To begin with, we consider a linear chain of $N+1$ evenly spaced hydrogen atoms. By the linear combination of their 1s functions we obtain $N+1$ wave functions $\psi_{k'}$; $k' = 0, \ldots, N$. The wave functions have some similarity to the standing waves of a vibrating chord or, better, with the vibrations of a chain of $N+1$ spheres that are connected by springs (Fig. 10.3). The chain can adopt different vibrational modes that differ in the number of vibrational nodes; we number the modes by sequential numbers k' corresponding to the number of nodes. k' cannot be larger than N, as the chain cannot adopt more nodes than spheres. We number the $N+1$ spheres from $n = 0$ to $n = N$. Every sphere vibrates with a certain amplitude:

$$A_n = A_0 \cos 2\pi \frac{k'n}{2N}$$

Each of the standing waves has a wavelength $\lambda_{k'}$:

$$\lambda_{k'} = \frac{2Na}{k'}$$

a is the distance between two spheres. Instead of numbering the vibrational modes with sequential numbers k', it is more convenient to use wave numbers k:

$$k = \frac{2\pi}{\lambda_{k'}} = \frac{\pi k'}{Na}$$

In this way one becomes independent of the number N, as the limits for k become 0 and π/a. Contrary to the numbers k' the values for k are not integral numbers.

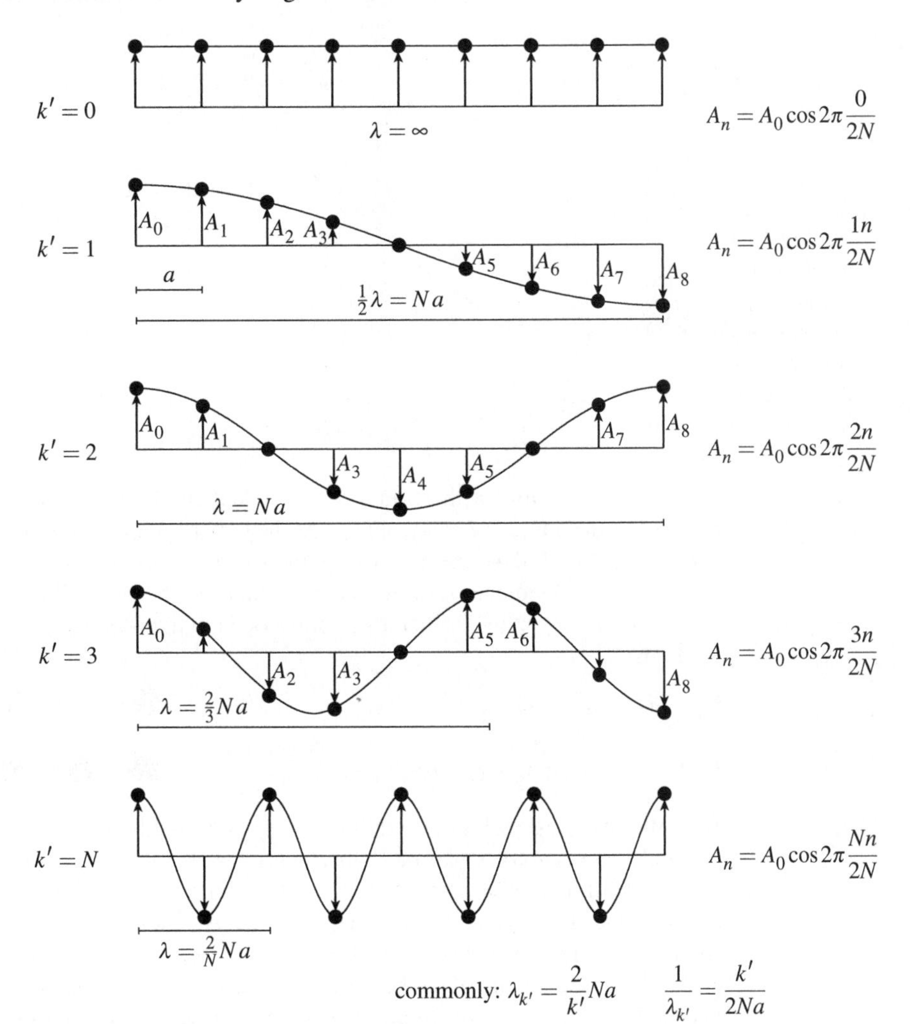

Fig. 10.3 Vibrational modes of a chain of $N+1$ spheres connected by springs

The k-th wave function of the electrons in a chain of hydrogen atoms results in a similar way. From every atom we obtain a contribution $\chi_n \cos nka$, *i.e.* the 1*s* function χ_n of the n-th atom of the chain takes the place of A_0. All atoms have the same function χ, referred to the local coordinate system of the atom, and the index n designates the position of the atom in the chain. The k-th wave function is composed of contributions of all atoms:

$$\psi_k = \sum_{n=0}^{N} \chi_n \cos nka \tag{10.4}$$

A wave function composed in this way from the contributions of single atoms is called a BLOCH function [in texts on quantum chemistry you will find this function being formulated with exponential functions $\exp(inka)$ instead of the cosine functions, since this facilitates the mathematical treatment].

The number k is more than just a simple number to designate a wave function. According to the DE BROGLIE equation, $p = h/\lambda$, every electron can be assigned a momentum p (h = PLANCK'S constant). k and the momentum are related:

$$k = \frac{2\pi}{\lambda} = \frac{2\pi p}{h} \tag{10.5}$$

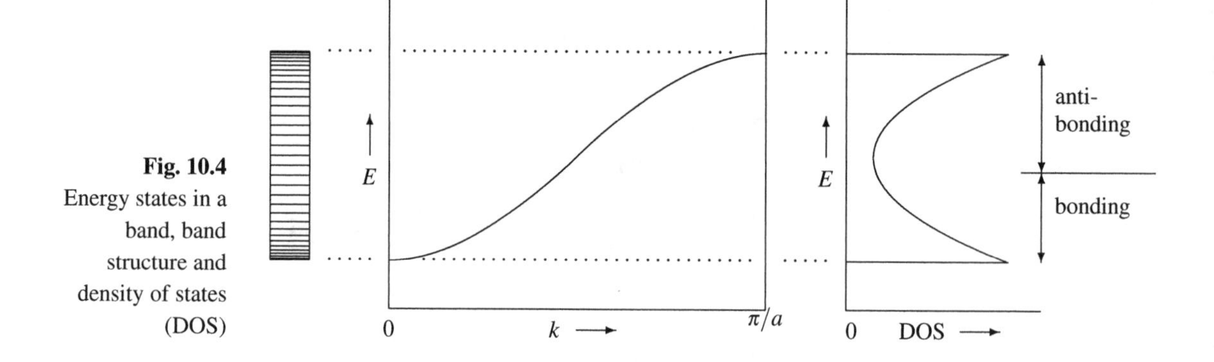

Fig. 10.4 Energy states in a band, band structure and density of states (DOS)

At the lower limit of the summation in equation (10.4) at $k = 0$ the cosine function always has the value 1, *i.e.* $\psi_0 = \Sigma \chi_n$. At the upper limit $k = \pi/a$ the cosine terms in the sum of equation (10.4) alternately have the values +1 and -1, *i.e.* $\psi_{\pi/a} = \chi_0 - \chi_1 + \chi_2 - \chi_3 + \ldots$. If we denote an H atom that contributes to the sum with $+\chi$ by ● and one that contributes with $-\chi$ by ○, then this corresponds to the following sequences in the chain of atoms:

$k = \pi/a$: $\psi_{\pi/a} = \chi_0 - \chi_1 + \chi_2 - \chi_3 + \ldots$ —●—○—●—○—●—

$k = 0$: $\psi_0 = \chi_0 + \chi_1 + \chi_2 + \chi_3 + \ldots$ —●—●—●—●—●—

ψ_0 of the chain resembles the bonding molecular orbital of the H_2 molecule. At $\psi_{\pi/a}$ there is a node between every pair of atoms, and the wave function is completely antibonding. Every wave function ψ_k is related to a definite energy state. Taking 10^6 H atoms in the chain we thus have the huge number of 10^6 energy states $E(k)$ within the limits $E(0)$ and $E(\pi/a)$.* The region between these limits is called an *energy band* or a *band* for short. The energy states are not distributed evenly in the band. Fig. 10.4 shows on the left a scheme of the band in which every line represents one energy state; only 38 instead of 10^6 lines were drawn. In the center the *band structure* is plotted, *i.e.* the energy as a function of k; the curve is not really continuous as it appears but consists of numerous tightly crowded dots, one for each energy state. The curve flattens out at the ends, showing a denser sequence of the energy levels at the band limits. The *density of states* (DOS) is shown at the right side; DOS·dE = number of energy states between E and $E + \mathrm{d}E$. The energy levels in the lower part of the band belong to bonding states, and in the upper part to antibonding states.

The *band width* or *band dispersion* is the energy difference between the highest and the lowest energy level in the band. The band width becomes larger when the interaction among the atoms increases, *i.e.* when the atomic orbitals overlap to a greater extent. A smaller interatomic distance a causes a larger band width. For the chain of hydrogen atoms a band width of 4.4 eV is calculated when adjacent atoms are separated by 200 pm, and 39 eV results when they move up to 100 pm.

According to the PAULI principle two electrons can adopt the same wave function, so that the N electrons of the N hydrogen atoms take the energy states in the lower half of the band, and the band is said to be 'half occupied'. The highest occupied energy level

* 10^6 atoms with interatomic distances of 100 pm can be accommodated in a chain of 0.1 mm length

(= HOMO = highest occupied molecular orbital) is the *Fermi limit*. Whenever the FERMI limit is inside a band, metallic electric conduction is observed. Only a very minor energy supply is needed to promote an electron from an occupied state under the FERMI limit to an unoccupied state above it; the easy switchover from one state to another is equivalent to a high electron mobility. Because of excitation by thermal energy a certain fraction of the electrons is always found above the FERMI limit.

The curve for the energy dependence as a function of k in Fig. 10.4 has a positive slope. This is not always so. When p orbitals are joined head-on to a chain, the situation is exactly the opposite. The wave function $\psi_0 = \Sigma\chi_n$ is then antibonding, whereas $\psi_{\pi/a}$ is bonding (Fig. 10.5).

Different bands can overlap each other, *i.e.* the lower limit of one band can have a lower energy level than the upper limit of another band. This applies especially to wide bands.

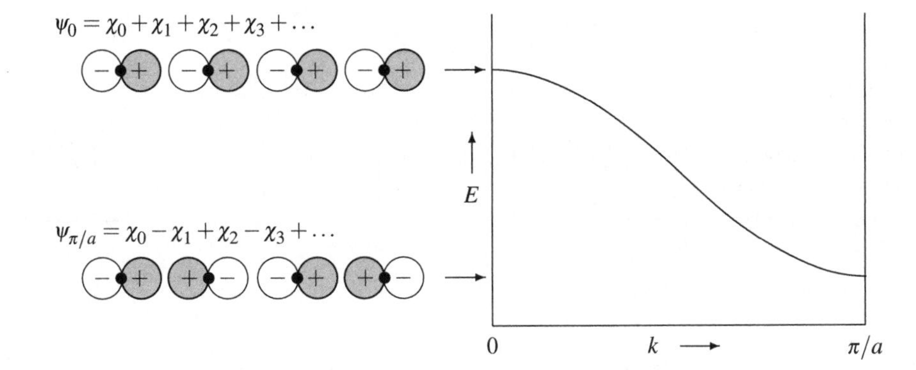

Fig. 10.5 Band structure for a chain of p orbitals oriented head-on

10.5 The Peierls Distortion

The model of the chain of hydrogen atoms with a completely delocalized (metallic) type of bonding is outlined in the preceding section. Intuitively, a chemist will find this model rather unreal, as he or she expects the atoms to combine in pairs to give H_2 molecules. In other words, the chain of equidistant H atoms is expected to be unstable, so it undergoes a distortion in such a way that the atoms approach each other in pairs. This process is called PEIERLS distortion (or strong electron–phonon coupling) in solid-state physics:

$$\cdots\vec{H}\cdots\cdots\overset{\leftarrow}{H}\cdots\cdots\vec{H}\cdots\cdots\overset{\leftarrow}{H}\cdots\cdots\vec{H}\cdots\cdots\overset{\leftarrow}{H}\cdots$$

$$\downarrow$$

$$H{-}H \qquad H{-}H \qquad H{-}H$$

The very useful chemist's intuition, however, is of no help when the question arises of how hydrogen will behave at a pressure of 500 GPa. Presumably it will be metallic then.

Let us consider once more the chain of hydrogen atoms, but this time we put it together starting from H_2 molecules. In the beginning the chain then consists of H atoms, and electron pairs occur between every other pair of atoms. Nevertheless, let us still assume equidistant H atoms. The orbitals of the H_2 molecules interact with one another to give a

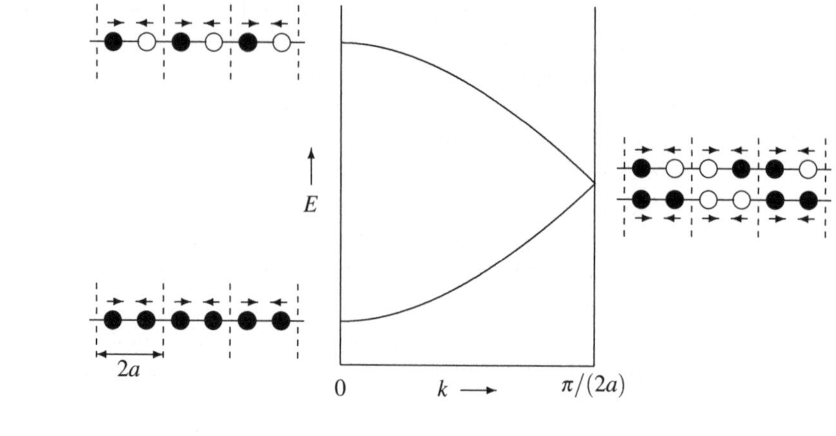

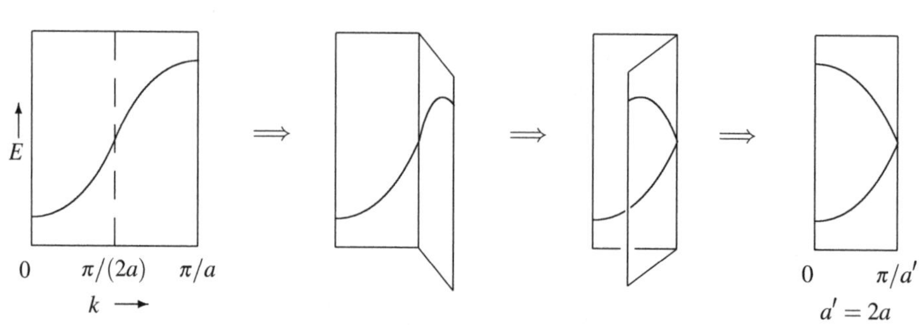

Fig. 10.6 Top: band structure for a chain of equidistant H atoms that was built up from H_2 molecules. Bottom: production of this diagram by folding the diagram of Fig. 10.4

band. As the repeating unit, *i.e.* the lattice constant in the chain, is now doubled to $2a$, the k values only run from $k = 0$ to $k = \pi/(2a)$. Instead, we have two branches in the curve for the band energy (Fig. 10.6). One branch begins at $k = 0$ and has a positive slope; it starts from the bonding molecular orbitals of the H_2, all having the same sign for their wave functions. The second branch starts at $k = 0$ with the higher energy of the antibonding H_2 orbital and has a negative slope. The two branches meet at $k = \pi/(2a)$.

As a result, the same band structure must result for the H atom chain, irrespective of whether it is based on the wave functions of N H atoms or of $N/2$ H_2 molecules. In fact, the curve of Fig. 10.4 coincides with the curve of Fig. 10.6. The apparent difference has to do with the doubling of the lattice constant from a to $a' = 2a$. As we see from equation (10.4), the same wave functions ψ_k result for $k = 0$ and for $k = 2\pi/a$, the same ones for $k = \pi/a$ and for $k = 3\pi/a$, etc. Whereas the curve in Fig. 10.4 runs steadily upwards from $k = 0$ to $k = \pi/a$, in Fig. 10.6 it only runs until $k = \pi/(2a) = \pi/a'$, then it continues upwards from right to left. We can obtain the one plot from the other by folding the diagram, as shown in the lower part of Fig. 10.6. The folding can be continued: triplication of the unit cell requires two folds, etc.

Up to now we have assumed evenly spaced H atoms. If we now allow the H atoms to approach each other in pairs, a change in the band structure takes place. The corresponding movements of the atoms are marked by arrows in Fig. 10.6. At $k = 0$ this has no consequences; at the lower (or upper) end of the band an energy gain (or loss) occurs for the atoms that approach each other; it is compensated by the energy loss (or gain) of the atoms moving apart. However, in the central part of the band, where the H atom chain has

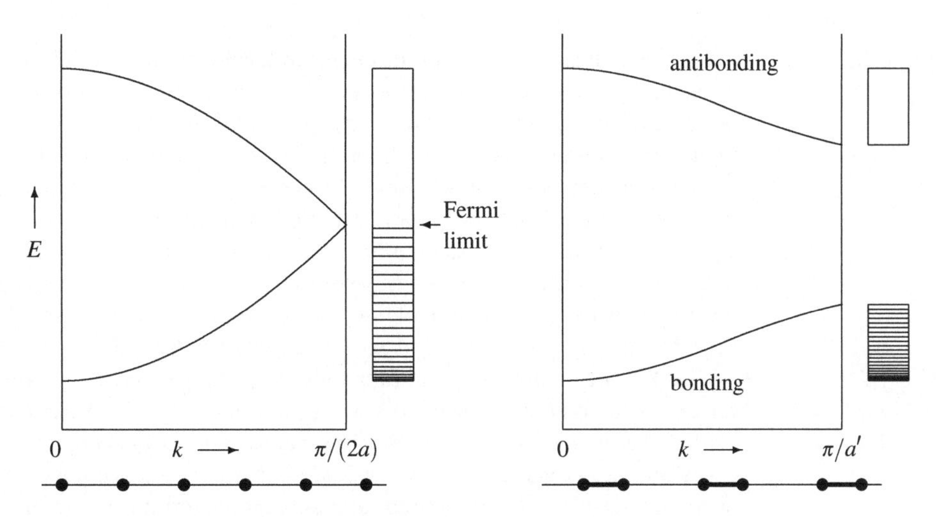

Fig. 10.7 Band structure for a chain of H atoms. Left, with equidistant atoms; right, after PEIERLS distortion to H_2 molecules. The lines in the rectangles symbolize energy states occupied by electrons

its FERMI limit, substantial changes take place. The upper branch of the curve shifts upwards, and the lower one downwards. As a result a gap opens up, and the band splits (Fig. 10.7). For the half-filled band the net result is an energy gain. Therefore, it is energetically more favorable when short and long distances between the H atoms alternate in the chain. The chain no longer is an electrical conductor, as an electron must overcome the energy gap in order to pass from one energy state to another.

The one-dimensional chain of hydrogen atoms is merely a model. However, compounds do exist to which the same kind of considerations are applicable and have been confirmed experimentally. These include polyene chains such as polyacetylene. The p orbitals of the C atoms take the place of the $1s$ functions of the H atoms; they form one bonding and one antibonding π band. Due to the PEIERLS distortion the polyacetylene chain is only stable with alternate short and long C–C bonds, that is, in the sense of the valence bond formula with alternate single and double bonds:

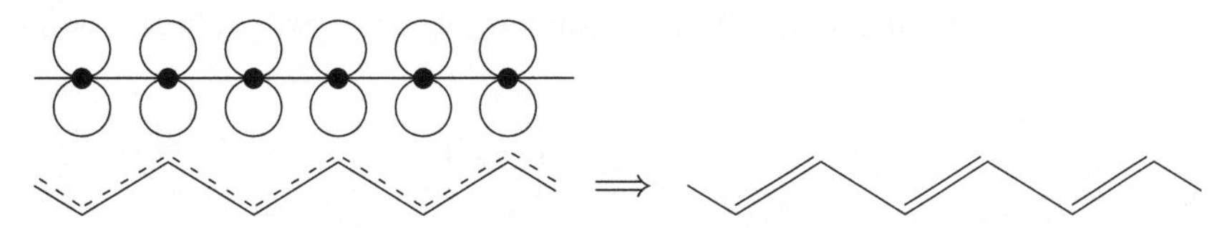

Polyacetylene is not an electrical conductor. If it is doped with an impurity that either introduces electrons into the upper band or removes electrons from the lower band, it becomes a good conductor.

The PEIERLS distortion is a substantial factor influencing which structure a solid adopts. The driving force is the tendency to maximize bonding, *i.e.* the same tendency that forces H atoms or other radicals to bond with each other. In a solid, that amounts to shifting the density of states at the FERMI level, in that bonding states are shifted towards lower and antibonding states towards higher energy values. By opening up an energy gap the bands become narrower; within a band the energy levels become more crowded. The extreme case is a band that has shrunk to a single energy value, *i.e.* all levels have the same energy. This happens, for example, when the chain of hydrogen atoms consists of widely separated H_2 molecules; then we have separate, independent H_2 molecules whose energy levels all have the same value; the bonds are localized in the molecules. Generally, the band width is a measure for the degree of localization of the bonds: a narrow band represents a high degree of localization, and with increasing band width the bonds become more delocalized. Since narrow bands can hardly overlap and are usually separated by intervening gaps, compounds with essentially localized bonds are electrical insulators.

When the atoms are forced to move closer by the exertion of pressure, their interaction increases and the bands become wider. At sufficiently high pressures the bands overlap again and the properties become metallic. The pressure-induced transition from a nonmetal to a metal has been shown experimentally in many cases, for example for iodine and other nonmetals. Under extremely high pressures even hydrogen should become metallic (metallic hydrogen is assumed to exist in the interior of Jupiter).

The PEIERLS distortion is not the only possible way to achieve the most stable state for a system. Whether it occurs is a question not only of the band structure itself, but also of the degree of occupation of the bands. For an unoccupied band or for a band occupied only at values around $k = 0$, it is of no importance how the energy levels are distributed at $k = \pi/a$. In a solid, a stabilizing distortion in one direction can cause a destabilization in another direction and may therefore not take place. The stabilizing effect of the PEIERLS distortion is small for the heavy elements (from the fifth period onward) and can be overcome by other effects. Therefore, undistorted chains and networks are observed mainly among compounds of the heavy elements.

10.6 Crystal Orbital Overlap Population (COOP)

At the end of Section 10.1 the MULLIKEN overlap population is mentioned as a quantity related to the bond order. A corresponding quantity for solids was introduced by R. HOFFMANN: the *crystal orbital overlap population* (COOP). It is a function that specifies the bond strength in a crystal, all states being taken into account by the MULLIKEN overlap populations $2c_ic_jS_{ij}$. Its calculation requires a powerful computer; however, it can be estimated in a qualitative manner by considering the interactions between neighboring atomic orbitals, such as shown in Fig. 10.8. At $k = 0$ all interatomic interactions are bonding. At $k = \pi/a$ they are antibonding for directly adjacent atoms, but they are bonding

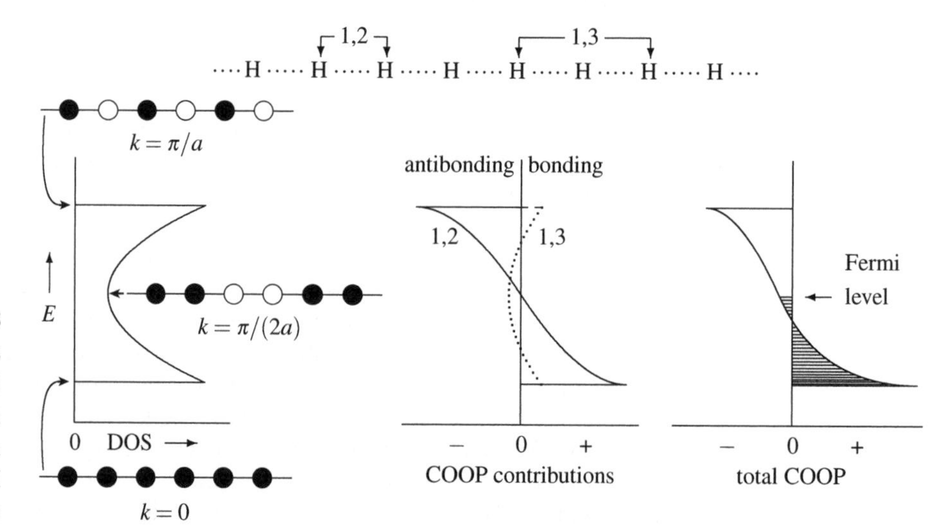

Fig. 10.8 Density of states (DOS) and crystal overlap population (COOP) for a chain of equidistant H atoms

between every other atom, albeit with reduced contributions due to the longer distance. At $k = \pi/(2a)$ the contributions between every other atom are antibonding, and those of adjacent atoms cancel each other. By also taking into account the densities of states one obtains the COOP diagram. In it net bonding overlap populations are plotted to the right and antibonding ones to the left. By marking the FERMI level it can be discerned to what extent bonding interactions predominate over antibonding interactions: they correspond to the areas enclosed by the curve below the FERMI level to the right and left sides.

Even in more complicated cases it is possible to obtain a qualitative idea. We choose the example of planar PtX_4^{2-} units that form a chain with Pt–Pt contacts. This kind of a structure is found for $K_2Pt(CN)_4$ and its partially oxidized derivatives like $K_2Pt(CN)_4Cl_{0.3}$:

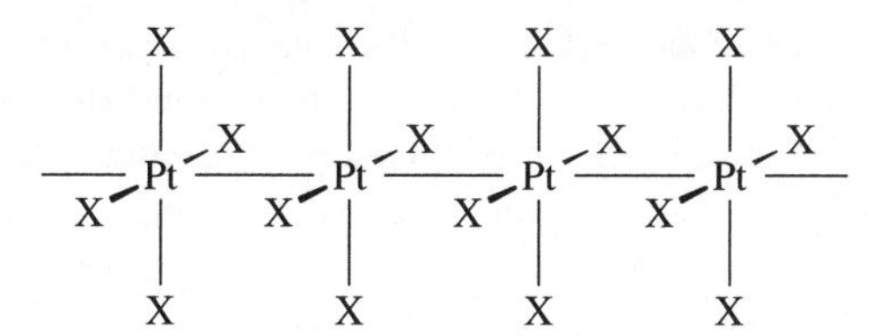

We will consider only the Pt–Pt interactions within the chain in the following. Fig. 10.9

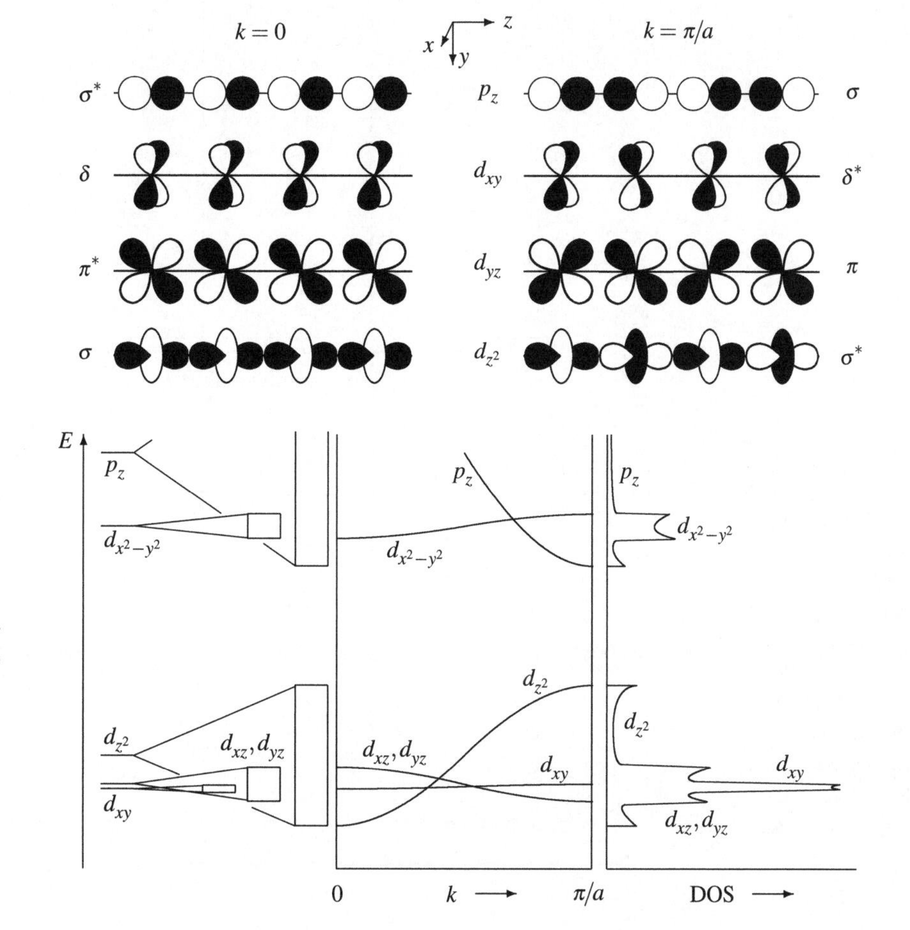

Fig. 10.9 Formation of bands by interaction of the orbitals of square PtX_4^{2-} complexes in a polymeric chain and the corresponding band structure and density of states

shows the orientations of the relevant atomic orbitals at $k = 0$ and $k = \pi/a$. Aside from d orbitals one p orbital is also taken into account. At the lower left is plotted the sequence of the energy states of the molecular orbitals of the monomeric, square complex (*cf.* Fig. 9.3, p. 77). The sketch to its right indicates how the energy levels fan out into bands when the PtX_4^{2-} ions are joined to a chain. The bands become wider the more intensely the orbitals interact with one another. With the aid of the orbital representations in the upper part of the figure the differences can be estimated: the orbitals d_{z^2} and p_z are oriented towards each other, and produce the widest bands; the interaction of the orbitals d_{xz} and d_{yz} is lower, and for d_{xy} and $d_{x^2-y^2}$ it is rather small (the band width for $d_{x^2-y^2}$ is slightly larger than for d_{xy} because of the inflation of $d_{x^2-y^2}$ due to its interaction with the ligands). The central plot shows the band structure, and the one on the right the density of states.

The DOS diagram results from the superposition of the densities of states of the different bands (Fig. 10.10). The d_{xy} band is narrow, its energy levels are crowded, and therefore it has a high density of states. For the wide d_{z^2} band the energy levels are distributed over a larger interval, and the density of states is smaller. The COOP contribution of every

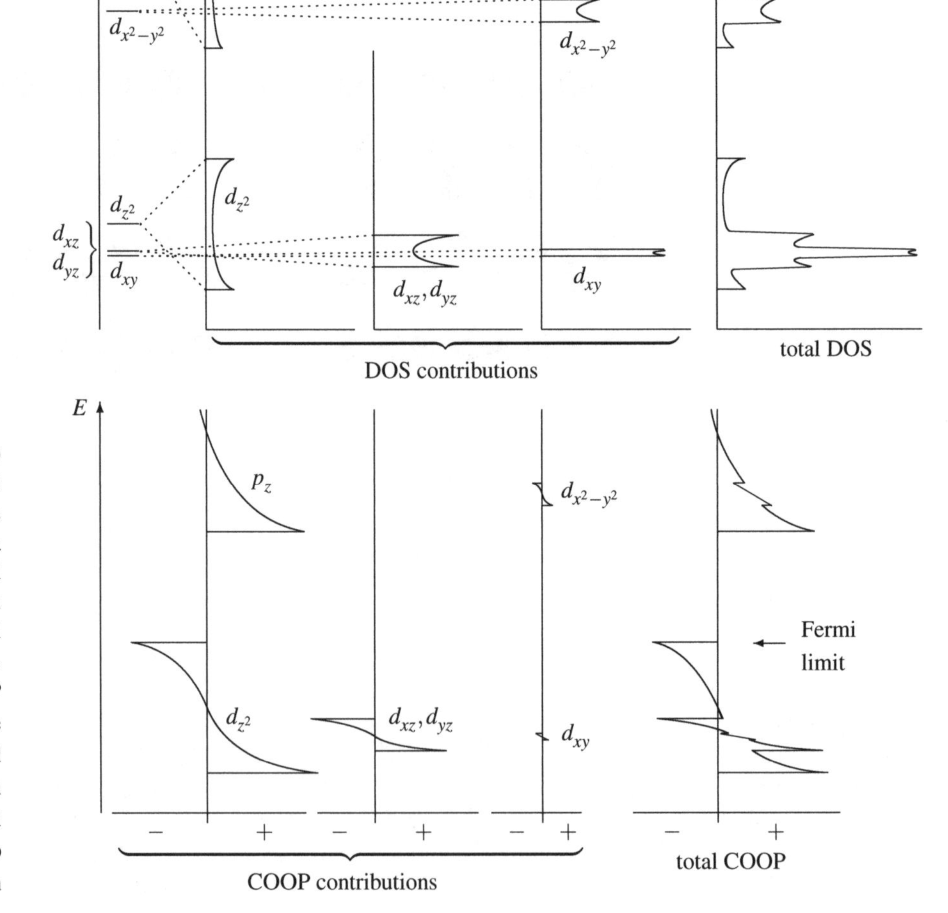

Fig. 10.10 Top: DOS contributions of the different bands of a PtX_4^{2-} chain and their superposition to give the total density of states. Bottom: COOP contributions of the different bands and their superposition to give the crystal orbital overlap population

band can be estimated. This requires consideration mainly of the bonding action (overlap population), but also of the density of states. The d_{z^2} band has a lower density of states but its bonding interaction is strong, so that its contribution to the COOP is considerable. The opposite applies to the d_{xy} band. Generally, broad bands contribute more to the crystal orbital overlap population. The diagram for the total COOP at the bottom right of Fig. 10.10 results from the superposition of the COOP contributions of the different bands; the FERMI level is also marked. Since all d orbitals except $d_{x^2-y^2}$ are occupied in the PtX_4^{2-} ion, the corresponding bands also are fully occupied, and bonding and antibonding interactions compensate each other. By oxidation, antibonding electrons are removed, the FERMI limit is lowered, and the bonding Pt–Pt interactions predominate. This agrees with observations: in $K_2Pt(CN)_4$ and similar compounds the Pt–Pt distances are about 330 pm; in the oxidized derivatives $K_2Pt(CN)_4X_x$ they are shorter (270 to 300 pm, depending on the value of x; X = Cl^- etc. Actually, in the oxidized species the ligands have a staggered arrangement along the chain, but this is of no importance for our considerations).

10.7 Bonds in Two and Three Dimensions

In principle, the calculation of bonding in two or three dimensions follows the same scheme as outlined for the chain extended in one dimension. Instead of one lattice constant a, two or three lattice constants a, b and c have to be considered, and instead of one sequential number k, two or three numbers k_x, k_y and k_z are needed. The triplet of numbers $\mathbf{k} = (k_x, k_y, k_z)$ is called *wave vector*. This term expresses the relation with the momentum of the electron. The momentum has vectorial character, its direction coincides with the direction of $\mathbf{k}$; the magnitudes of both are related by the DE BROGLIE relation [equation (10.5)]. In the directions $\mathbf{a}$, $\mathbf{b}$ and $\mathbf{c}$ the components of $\mathbf{k}$ run from 0 to π/a, π/b and π/c, respectively. As the direction of motion and the momentum of an electron can be reversed, we also allow for negative values of k_x, k_y and k_z, with values that run from 0 to $-\pi/a$ etc. However, for the calculation of the energy states the positive values are sufficient, since according to equation (10.4) the energy of a wave function is $E(\mathbf{k}) = E(-\mathbf{k})$.

The magnitude of $\mathbf{k}$ corresponds to a wave number $2\pi/\lambda$ and therefore is measured with a unit of reciprocal length. For this reason $\mathbf{k}$ is said to be a vector in a 'reciprocal space' or 'k space'.* This is a 'space' in a mathematical sense, *i.e.* it is concerned with vectors in a coordinate system, the axes of which serve to plot k_x, k_y and k_z. The directions of the axes run perpendicular to the delimiting faces of the unit cell of the crystal.

The region within which $\mathbf{k}$ is considered ($-\pi/a \leq k_x \leq \pi/a$ etc.) is the *first Brillouin zone*. In the coordinate system of k space it is a polyhedron. The faces of the first BRILLOUIN zone are oriented perpendicular to the directions from one atom to the equivalent atoms in the adjacent unit cells. The distance of a face from the origin of the k coordinate system is π/s, s being the distance between the atoms. The first BRILLOUIN zone for a cubic-primitive crystal lattice is shown in Fig. 10.11; the symbols commonly given to certain points of the BRILLOUIN zone are labeled. The BRILLOUIN zone consists of a very large number of small cells, one for each electronic state.

The pictures in Fig. 10.12 give an impression of how s orbitals interact with each other in a square lattice. Depending on the k values, *i.e.* for different points in the BRILLOUIN zone, different kinds of interactions result. Between adjacent atoms there are only bonding

*Compared to the reciprocal space commonly used in crystallography, the k space is expanded by a factor 2π, otherwise the construction for both is the same.

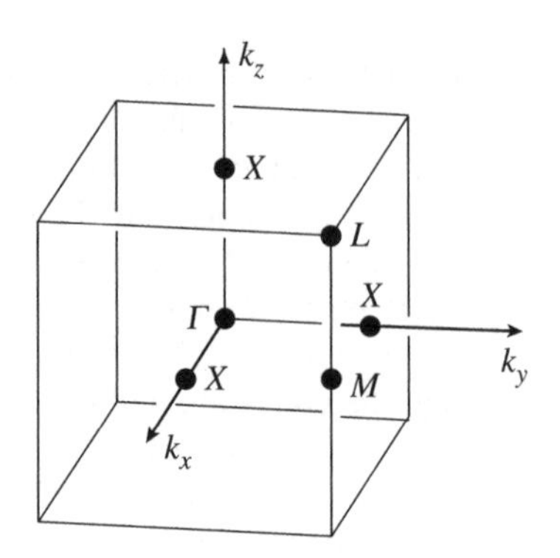

Fig. 10.11
First Brillouin zone for a cubic-primitive crystal lattice. The points X are located at $k = \pi/a$ in each case

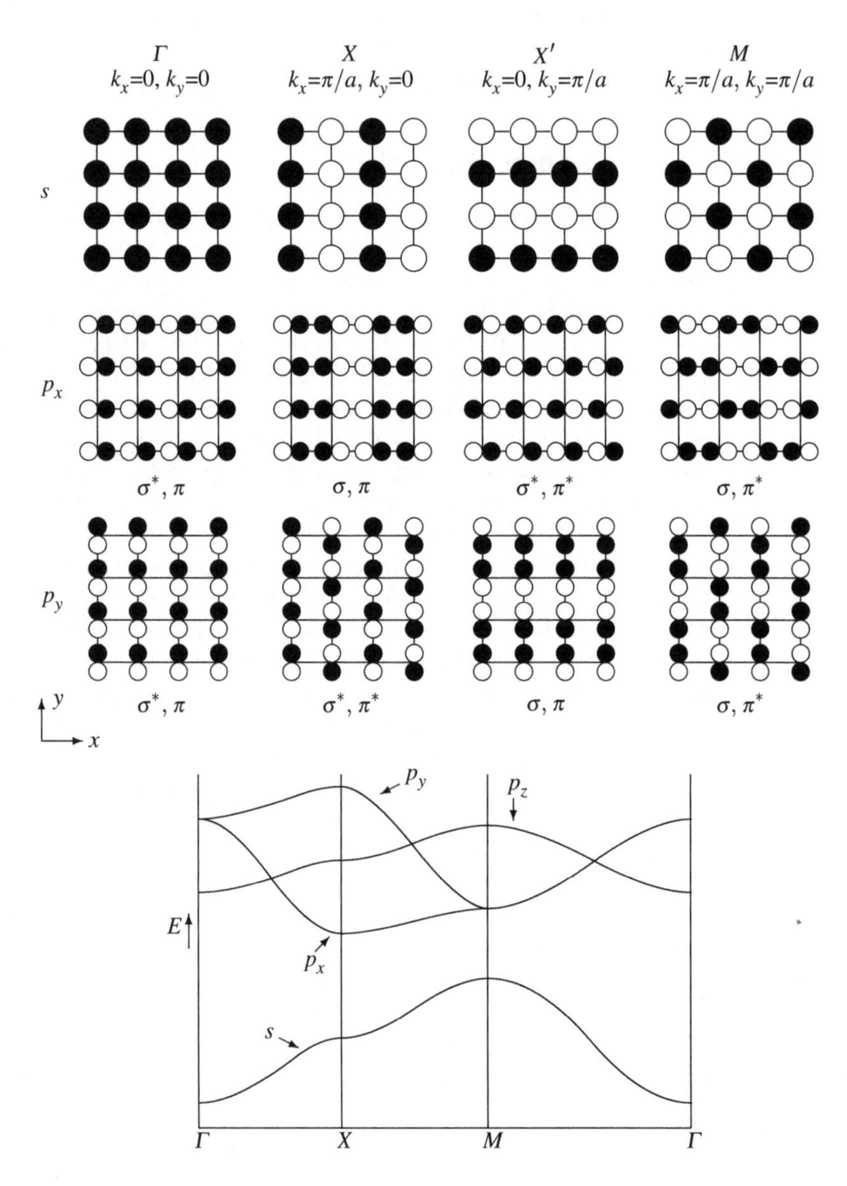

Fig. 10.12
Combination of s orbitals (top) and p orbitals in a square net and the resulting band structure (bottom)

interactions at Γ, and only antibonding interactions at M; the wave function corresponding to Γ therefore is the most favorable one energetically, and the one corresponding to M the least favorable. At X every atom has two bonding and two antibonding interactions with adjacent atoms, and its energy level is intermediate between those of Γ and M. It is hardly possible to visualize the energy levels for all of the BRILLOUIN zone, but one can plot diagrams that show how the energy values run along certain directions within the zone. This has been done in the lower part of Fig. 10.12 for three directions ($\Gamma \to X, X \to M$ and $\Gamma \to M$).

p_z orbitals that are oriented perpendicular to the square lattice interact in the same way as the s orbitals, but the π-type interactions are inferior and correspondingly the band width is smaller. For p_x and p_y orbitals the situation is somewhat more complicated, because σ and π interactions have to be considered between adjacent atoms (Fig. 10.12). For example, at Γ the p_x orbitals are σ-antibonding, but π-bonding. At X p_x and p_y differ most, one being σ and π-bonding, and the other σ and π-antibonding.

In a cubic-primitive structure (α-polonium, Fig. 2.4, p. 7) the situation is similar. By stacking square nets and considering how the orbitals interact at different points of the BRILLOUIN zone, a qualitative picture of the band structure can be obtained.

10.8 Bonding in Metals

The density of states for the elements of a long period of the periodic table can be sketched roughly as in Fig. 10.13. Due to the three-dimensional structures the more accurate consideration will no longer yield the simple DOS curves with two peaks as for a linear chain, but yields instead more or less complicated curves with numerous peaks. We will not go into the details here; in Fig. 10.13 merely a rectangle represents the DOS curve of each band. In each case the lower part of a band is bonding, and the upper part is antibonding. Correspondingly, the COOP diagram shows a contribution to the right and to the left side for every band. The p band has more antibonding than bonding contributions so that its left side predominates. In the series potassium, calcium, scandium, ... we add a valence electron from element to element, and the FERMI limit climbs; the FERMI limit is marked to the right side of the figure for some valence electron counts. As can be seen, at first

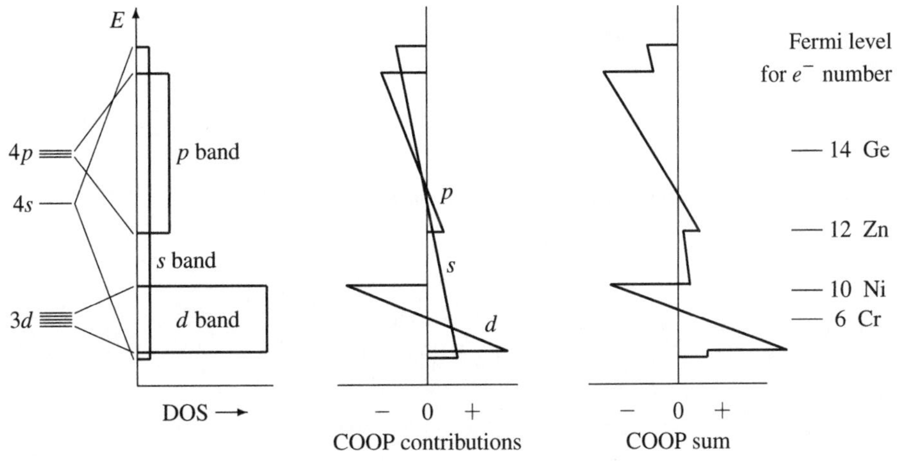

Fig. 10.13 Schematic sketch of the density of states and the crystal orbital overlap population for metals

bonding states are occupied and therefore the bond strength increases for the metals from potassium to chromium. For the seventh to tenth valence electrons only antibonding states are available, and so the bond strengths decrease from chromium to nickel. The next electrons (Cu, Zn) are weakly bonding. With more than 14 valence electrons the total overlap population for a metallic structure becomes negative; structures with lower coordination numbers become favored.

The outlined sketch is rather rough, but it correctly shows the tendencies, as can be exemplified by the melting points of the metals (values in °C):

K	Ca	Sc	Ti	V	Cr	Mn	Fe	Co	Ni	Cu	Zn
63	839	1539	1667	1915	1900	1244	1535	1495	1455	1083	420

In reality there are subtle deviations from this simple picture. The energy levels shift somewhat from element to element, and different structure types have different band structures that become more or less favorable depending on the valence electron concentration. Furthermore, in the COOP diagram of Fig. 10.13 the *s*–*p*, *s*–*d* and *p*–*d* interactions were not taken into account, although they cannot be neglected. A more exact calculation shows that only antibonding contributions are to be expected from the eleventh valence electron onwards.

10.9 Problems

10.1 What changes should occur in the band structure and the DOS diagrams (Fig. 10.4) when the chain of H atoms is compressed?

10.2 What would the band structure of a chain of *p* orbitals oriented head-on (Fig. 10.5) look like after a PEIERLS distortion?

10.3 What changes should occur in the band structure of the square net (Fig. 10.12) when it is compressed in the *x* direction?

11 The Element Structures of the Nonmetals

According to the $8-N$ rule (Chapter 8) an atom X of an element of the N-th main group of the periodic table will participate in $8-N$ covalent bonds ($N = 4$ to 7):

$$b(\mathrm{XX}) = 8 - N$$

In addition, as a rule, the *principle of maximal connectivity* holds for elements of the third and higher periods: the $8-N$ bonds usually are bonds to $8-N$ *different* atoms, and multiple bonds are avoided. For carbon, however, being an element of the second period, the less connected graphite is more stable than diamond at normal conditions. At higher pressures the importance of the principle of maximal connectivity increases; then, diamond becomes more stable.

11.1 Hydrogen and the Halogens

Hydrogen, fluorine, chlorine, bromine and iodine consist of molecules X_2, even when in the solid state. In solid hydrogen, rotating H_2 molecules take an arrangement as in a hexagonal-closet packing of spheres. In α-F_2 the F_2 molecules are packed in hexagonal layers; the molecules are oriented perpendicular to the layer, and the layers are stacked in the same way as in cubic closest-packing. Above 45.6 K up to the melting point (53.5 K) the modification β-F_2 is stable in which the molecules rotate about their centers of gravity.

The molecules in crystalline **chlorine, bromine** and **iodine** are packed in a different manner, as shown in Fig. 11.1. The rather different distances between atoms of adjacent molecules are remarkable. If we take the VAN DER WAALS distance, such as observed in organic and inorganic molecular compounds, as reference, then some of the intermolecular contacts in the *b*-*c* plane are shorter, whereas they are longer to the molecules of the next plane. We thus observe a certain degree of association of the halogen molecules within the *b*-*c* plane (dotted in Fig. 11.1, top left). This association increases from chlorine to iodine. The weaker attractive forces between the planes show up in the plate-like habit of the crystals and in their easy cleavage parallel to the layers. Similar association tendencies are also observed for the heavier elements of the fifth and sixth main groups.

The packing can be interpreted as a cubic closest-packing of halogen atoms that has been severely distorted by the covalent bonds within the molecules. By exerting pressure the distortion is reduced, *i.e.* the different lengths of the contact distances between the atoms approximate each other (Fig. 11.1). For iodine a continuous approximation is observed with increasing pressure, then at 23 GPa an abrupt phase transition takes place yielding the incommensurately modulated crystal structure of iodine-V. Such a structure cannot be described as usual with a three-dimensional space group (Section 3.6, p. 25). The four-dimensional superspace group is in this case $Fmmm(00q_3)0s0$ with $q_3 = 0.257$ at 24.6 GPa. The structure can thus be described with a three-dimensional approximant (mean structure) in the orthorhombic space group $Fmmm$, but the atoms are displaced and obey a sine wave along **c**. The wave length of the wave is $c/q_3 = c/0.257 = 3.89c$.

Inorganic Structural Chemistry, Second Edition Ulrich Müller

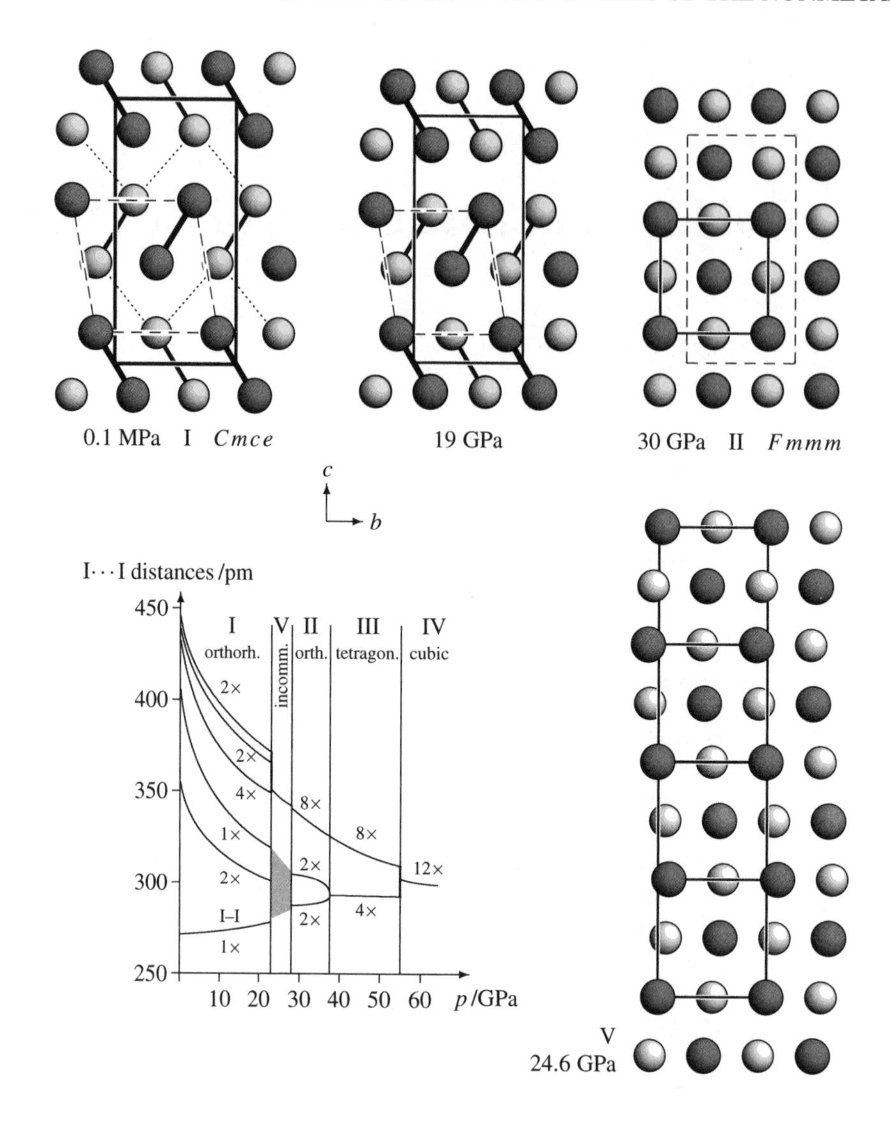

Fig. 11.1 The structure of iodine at four different pressures. The outlined face-centered unit cell in the 30-Gpa figure corresponds to that of a (distorted) cubic closest-packing of spheres. At 24.6 GPa four unit cells of the face-centered approximant structure are shown; the structure is incommensurately modulated, the atomic positions follow a sine wave with a wave length of $3.89 \times c$. The amplitude of the wave is exaggerated by a factor of two. Lower left: Dependence of the twelve interatomic contact distances on pressure

The amplitude of the wave is parallel to b and amounts to $0.053b$. The interatomic distances differ from atom to atom within an interval of 286 to 311 pm.

When pressure exceeds 28 GPa, the modulation disappears and a distorted cubic closest-packing of spheres is obtained (iodine-II). The distortion of the sphere packing decreases when the pressure is increased. A transformation to a tetragonal distorted packing of spheres (iodine-III) takes place at approximately 37.5 GPa, and finally the packing becomes undistorted cubic at 55 GPa (iodine-IV). With increasing pressure the energy gap between the fully occupied valence band and the unoccupied conduction band decreases. The energy gap disappears already at about 16 GPa, *i.e.* a transition from an insulator to a metallic conductor takes place even though molecules are still present at this pressure. Iodine thus actually becomes a metal, and at high pressures it also adopts the structure of a closest packing of spheres that is typical for metals, albeit at first in a distorted manner. A comparable transition to a metal is also expected to occur for hydrogen; the necessary pressure (not yet achieved experimentally) could amount to 450 GPa.

11.2 Chalcogens

Oxygen in the solid state consists of O_2 molecules. From 24 K to 43.6 K they are packed as in α-F_2. Under pressure (5.5 GPa) this packing is also observed at room temperature. Below 24 K the molecules are slightly tilted against the hexagonal layer. From 43.6 K up to the melting point (54.8 K) the molecules rotate in the crystal as in β-F_2. Under pressure oxygen becomes metallic at approximately 100 GPa, but it remains molecular.

No element shows as many different structures as **sulfur**. Crystal structures are known for the following forms: S_6, S_7 (four modifications), S_8 (three modifications), S_{10}, $S_6{\cdot}S_{10}$, S_{11}, S_{12}, S_{13}, S_{14}, S_{15}, S_{18} (two forms), S_{20}, S_∞ (Fig. 11.2). Many of them can be separated by chromatography from solutions that were obtained by extraction of quenched

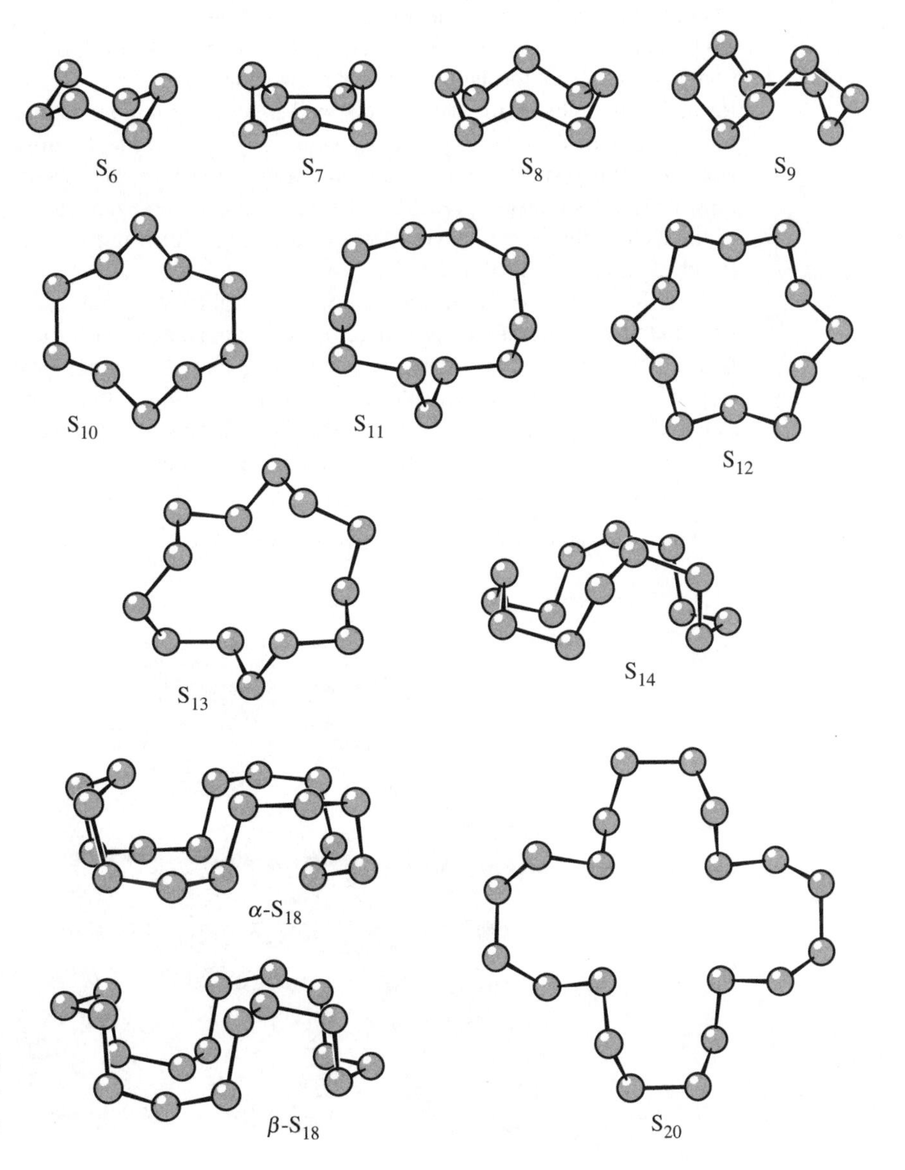

Fig. 11.2 Different molecular structures of sulfur

sulfur melts; they can also be prepared by specific chemical synthesis. Quenched sulfur melts also yield polymeric forms; the structure of one of these has been determined. All mentioned sulfur forms consist of rings or chains of S atoms, every sulfur atom being bonded with two other sulfur atoms in accordance with the $8 - N$ rule. The S–S bond lengths usually are about 206 pm, but they show a certain scatter (±10 pm). The S–S–S bond angles are between 101 and 110° and the dihedral angles between 74 and 100°.* As a consequence, a sequence of five atoms can adopt one of two arrangements:

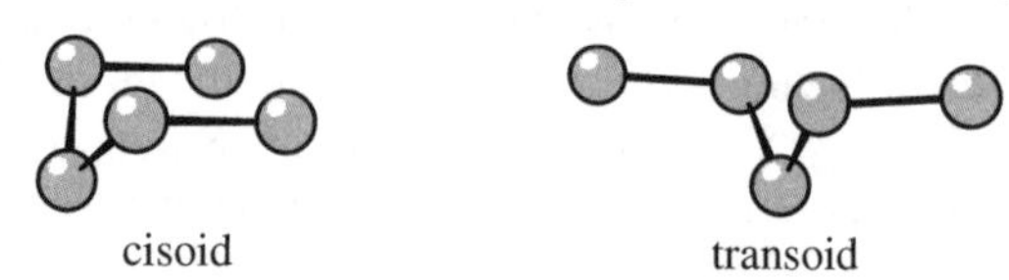

Only the cisoid arrangement occurs in the smaller rings S_6, S_7 and S_8, the dihedral angles being forced to adapt themselves (74.5° for S_6, 98° for S_8). S_6 has chair conformation, and the S_8 conformation is called crown-form (Fig. 11.2). S_7 can be imagined to be formed from S_8 by taking out one S atom. Larger rings require the presence of cisoid and transoid groups in order to be free of strain. In S_{12} cisoid and transoid groups alternate. Helical chains result when there are only transoid groups; the dihedral angle determines how many turns it takes to reach another atom directly above of an atom on a line parallel to the axis of the helix. In one form of polymer sulfur it takes ten atoms in three turns (screw axis 10_3; *cf.* Section 3.1).

In orthorhombic α-sulfur, the modification stable at normal conditions, S_8 rings are stacked to form columns. Consecutive rings are not stacked one exactly above another (as in a roll of coins), but in a staggered manner so that the column looks like a crank shaft (Fig. 11.3). This arrangement allows for a dense packing of the molecules, with columns in two mutually perpendicular directions. The columns of one direction are placed in the recesses of the perpendicular 'crank shafts'. In S_6 and in S_{12} the rings are stacked exactly one above another, and the rolls are bundled parallel to each other. In the structures a

*Dihedral angle = angle between two planes. For a chain of four atoms it is the angle between the planes through the atoms 1,2,3 and 2,3,4.

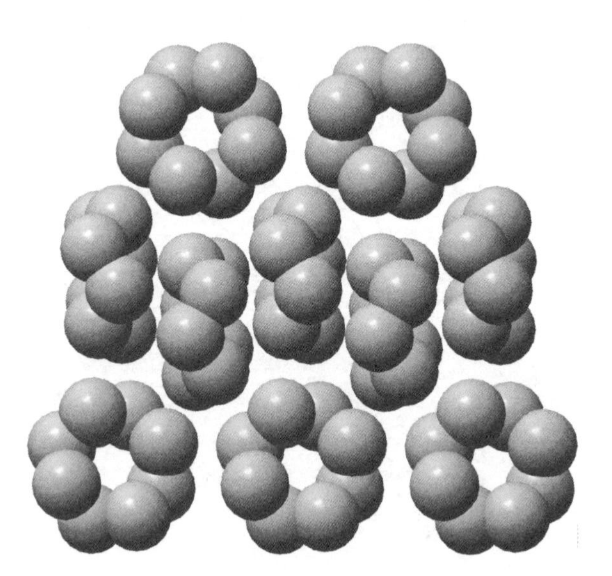

Fig. 11.3 Section of the structure of α-sulfur

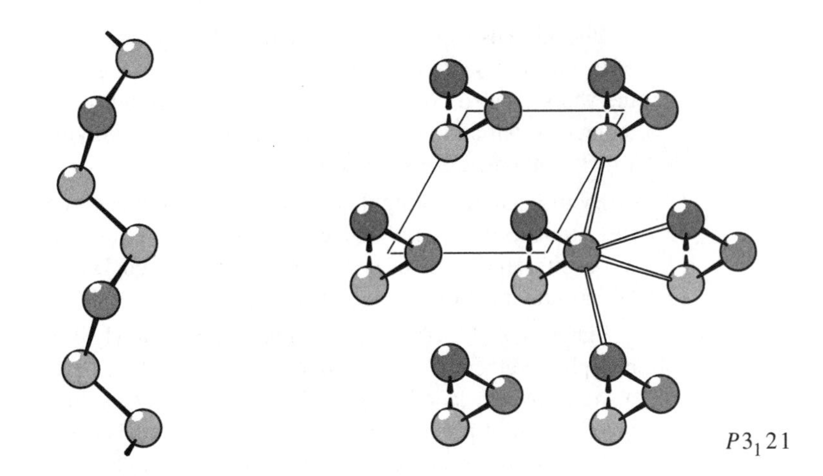

Fig. 11.4 Structure of α-selenium. Left: side view of a helix with 3_12 screw symmetry. Right: view along the helices; the unit cell and the coordination about one atom are plotted

universally valid principle can be discerned: *in the solid state molecules tend to pack as tightly as possible.*

Selenium forms three known modifications that consist of Se_8 rings. The stacking of the rings differs from that of the S_8 modifications in that they resemble coin rolls, but the rings are tilted. There also exists a modification that is isotypic with S_6, consisting of Se_6 molecules. The thermodynamically stable form of selenium, α-selenium or Se-I, consists of helical chains having three Se atoms in every turn (Fig. 11.4). The chains are bundled parallel in the crystal. Every selenium atom has four adjacent atoms from three different chains at a distance of 344 pm. Together with the two adjacent atoms within the chain at a distance of 237 pm, a strongly distorted octahedral 2+4 coordination results. The Se$\cdots$Se distance between the chains is significantly shorter than expected from the VAN DER WAALS distance. For high-pressure modifications of selenium see Section 11.4.

Tellurium crystallizes isotypic to α-selenium. As expected, the Te–Te bonds in the chain (283 pm) are longer than in selenium, but the contact distances to the atoms of the adjacent chains are nearly the same (Te$\cdots$Te 349 pm). The shortening, as compared to the VAN DER WAALS distance, is more marked and the deviation from a regular octahedral coordination of the atoms is reduced (*cf.* Table 11.1, p. 111). By exerting pressure all six distances can be made to be equal (*cf.* Section 11.4).

Two modifications are known for **polonium**. At room temperature α-polonium is stable; it has a cubic-primitive structure, every atom having an exact octahedral coordination (Fig. 2.4, p. 7). This is a rather unusual structure, but it also occurs for phosphorus and antimony at high pressures. At 54 °C α-Po is converted to β-Po. The phase transition involves a compression in the direction of one of the body diagonals of the cubic-primitive unit cell, and the result is a rhombohedral lattice. The bond angles are 98.2°.

11.3 Elements of the Fifth Main Group

For solid **nitrogen** five modifications are known that differ in the packing of the N_2 molecules. Two of them are stable at normal pressure (transition temperature 35.6 K); the others exists only under high pressure. At pressures around 100 GPa a phase transition with a marked hysteresis takes place, resulting in a non-molecular modification. It presumably corresponds to the α-arsenic type. Electrical conductivity sets in at 140 GPa.

Phosphorus vapor consists of tetrahedral P_4 molecules, and at higher temperatures also of P_2 molecules ($P{\equiv}P$ bond length 190 pm). White phosphorus forms by condensation of the vapor; it also consists of P_4 molecules. Liquid phosphorus normally consists of P_4 molecules, but at a pressure of 1 GPa and 100 °C polymeric liquid phosphorus is formed which is not miscible with liquid P_4.

By irradiation with light or by heating it to temperatures above 180 °C, white phosphorus is transformed to red phosphorus. Its tint, melting point, vapor pressure and especially its density depend on the conditions of preparation. Usually, it is amorphous or microcrystalline, and it is rather laborious to grow crystals.

Platelets of HITTORF's (violet) phosphorus slowly crystallize together with fibrous red phosphorus at temperatures around 550 °C. Single crystals of HITTORF's phosphorus were obtained by slow cooling (from 630 to 520 °C) of a solution in liquid lead. Both modifications consist of polymeric tubes with a pentagonal cross-section; the tubes are composed of cages of the same shape as in As_4S_4 and As_4S_5, being connected via P_2 dumbbells (Fig. 11.5). In fibrous phosphorus the tubes are interconnected to parallel pairs. In HITTORF's phosphorus they are connected crosswise to grids; pairs of grids are interlocked

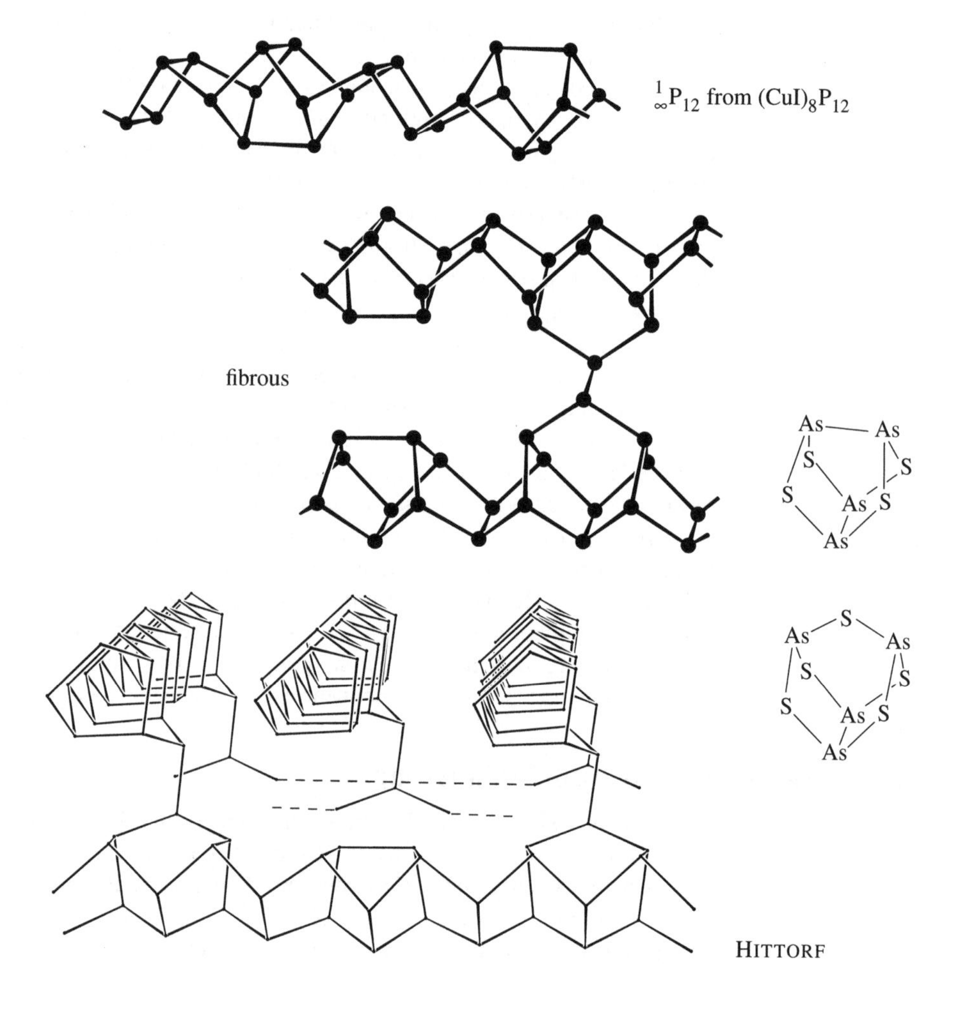

Fig. 11.5 **Top**: Repeating unit in a chain of P_8 cages and P_4 rings in $^1_\infty P_{12}$. **Middle**: Pair of parallel interconnected five-sided tubes in fibrous red phosphorus, with P_8 and P_9 cages connected by P_2 dumbbells. **Bottom**: Structure of HITTORF's phosphorus consisting of the same kind of tubes, but joined to grids; the tube shown in the center belongs to another grid than the remaining tubes. **Right**: The molecular structures of As_4S_4 and As_4S_5 are like those of the P_8 and P_9 cages

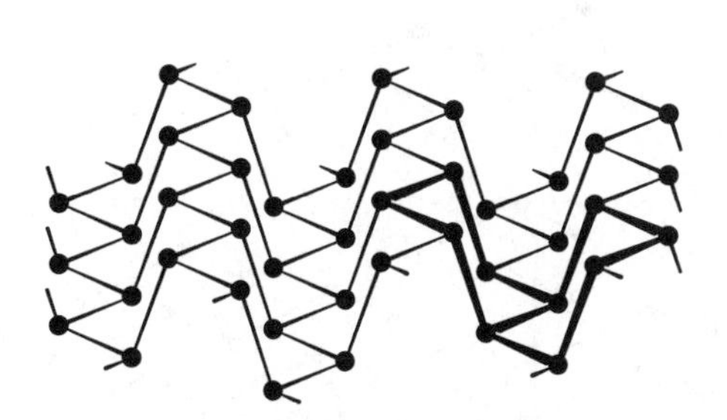
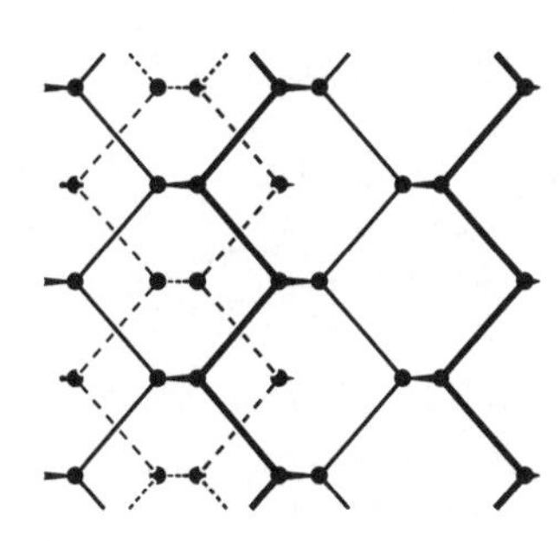

Fig. 11.6 The structure of black phosphorus. Left: section of one layer; two rings with chair conformation and relative arrangement as in *cis*-decalin are emphasized. Right: top view of a layer showing the zigzag lines; the position of the next layer is indicated

but not bonded with each other. According to the $8 - N$ rule every P atom is bonded with three other atoms. Despite its complicated structure, the linking principle of these modifications occurs frequently in the structural chemistry of phosphorus compounds. Building units that correspond to fragments of the tubes are known among the polyphosphides and polyphosphanes (*cf.* p. 133). Similar tubes also occur in $P_{15}Se$ and $P_{19}Se$.

The compound $(CuI)_8P_{12}$ can be obtained from the elements at 550 °C. It contains chains of polymeric phosphorus that can be isolated when the copper iodide is extracted with an aqueous solution of potassium cyanide. The molecules consist of P_8 cages that are joined via P_4 squares (Fig. 11.5, top). Another variety can be obtained in a similar way from $(CuI)_3P_{12}$; its chains consist of P_{10} cages and P_2 dumbbells.

Black phosphorus only forms under special conditions (high pressure, crystallization from liquid bismuth or prolonged heating in the presence of Hg); nevertheless, it is the thermodynamically stable modification at normal conditions. It consists of layers having six-membered rings in the chair conformation. Pairs of rings are connected like the rings in *cis*-decalin (Fig. 11.6). The layer can also be regarded as a system of interconnected zigzag lines that alternate in two different planes. Within the layer every P atom is bonded to three other P atoms at distances of 222 and 224 pm. The atomic distances between the layers (2×359 pm; 1×380 pm) correspond to the VAN DER WAALS distance. Certain structural features of black phosphorus are also found among the polyphosphides (*cf.* Fig. 13.2, p. 133).

Arsenic modifications with the structures of white and black phosphorus have been described. However, only gray (metallic, rhombohedral) α-arsenic is stable. It consists of layers of six-membered rings in the chair conformation that are connected with each other in the same way as in *trans*-decalin (Fig. 11.7). In the layer the atoms are situated alternately in an upper and a lower plane. The layers are stacked in a staggered manner such that over and under the center of every ring there is an As atom in an adjacent layer. In this way every As atom is in contact with three more atoms in addition to the three atoms to which it is bonded within the layer; it has a distorted octahedral $3 + 3$ coordination. The As–As bond length in the layer is 252 pm; the distance between adjacent atoms of different layers is 312 pm and thus is considerably shorter than the VAN DER WAALS distance (370 pm).

The structures of **antimony** and **bismuth** correspond to that of gray arsenic. With increasing atomic weight the distances between adjacent atoms within a layer and between layers become less different, *i.e.* the coordination polyhedra deviate less from a regular octahedron. This effect is enhanced under pressure (*cf.* next section).

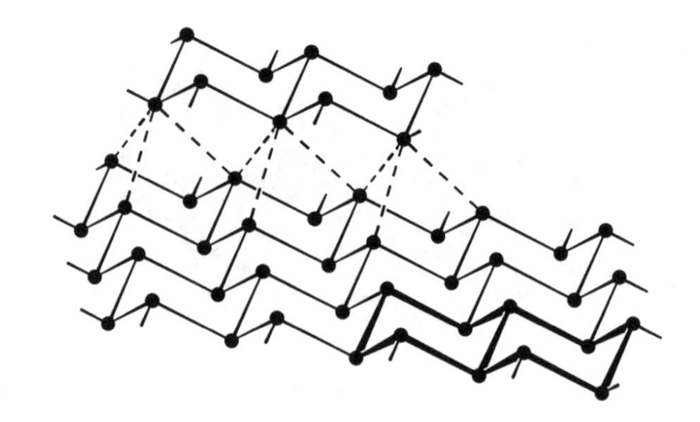

Fig. 11.7
Section of a layer in gray arsenic and the position of two rings of the next layer. Two rings with the relative arrangement as in *trans*-decalin are emphasized

The description of the structures of P, As, Sb, and Bi as layer structures and of Se and Te as chain structures neglects the presence of bonding interactions between the layers and chains. These interactions gain importance for the heavier atoms (Table 11.1). For example, the interlayer distances between adjacent atoms for Sb and Bi are only 15 % longer than the intralayer distances; the actual deviation from the α-polonium structure is rather small. Furthermore, As, Sb, and Bi show metallic conductivity. The bonding interactions can be understood using band theory: starting from the α-Po structure, a PEIERLS distortion takes place that enhances three bonds per atom (Fig. 11.8). The same applies to tellurium and selenium, two bonds per atom being enhanced.

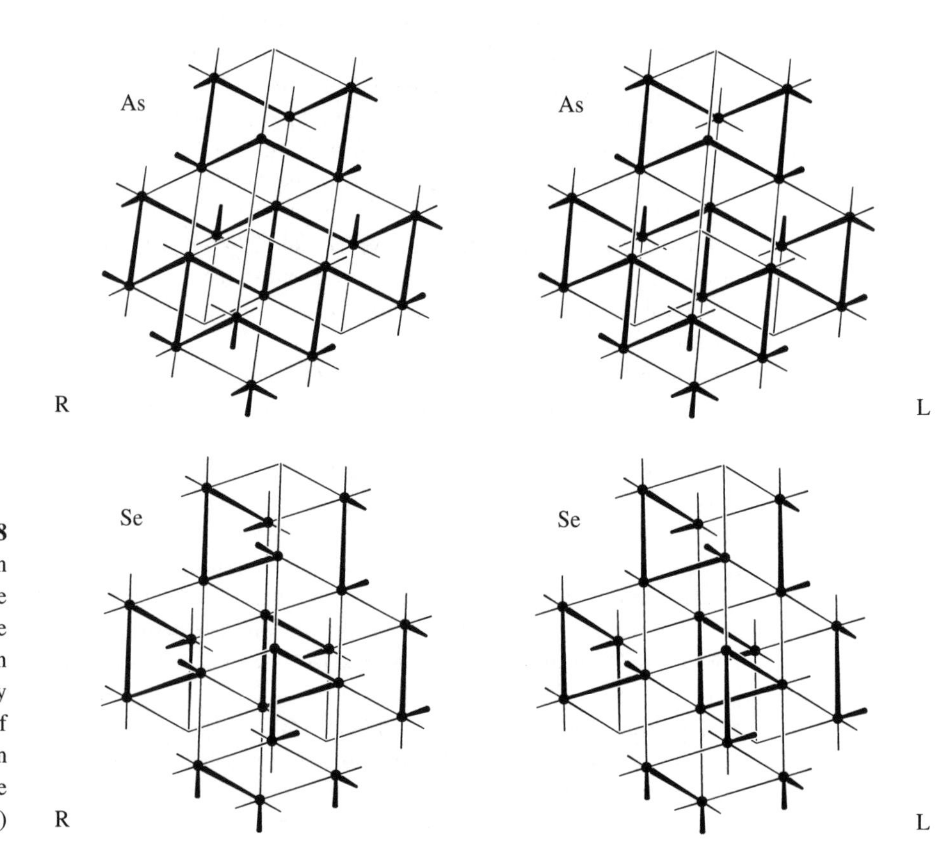

Fig. 11.8
The layer and chain structures of the elements of the fifth and sixth main groups result by contraction of certain distances in the α-Po structure (stereo images)

Table 11.1: Distances between adjacent atoms and bond angles in structures of the α-As, α-Se, α-Po and β-Po type. d_1 = bond distance, d_2 = shortest interatomic distance between layers or chains; distances in pm, angles in degrees

	structure type	d_1	d_2	d_2/d_1	angle
P (∼10 GPa)	α-As	213	327	1.54	105
P (∼12 GPa)	α-Po	238	238	1.00	90
As	α-As	252	312	1.24	96.6
As (25 GPa)	α-Po	255	255	1.00	90
Se	α-Se	237	344	1.45	103.1
Sb	α-As	291	336	1.15	95.6
Te	α-Se	283	349	1.23	103.2
Te-IV (11.5 GPa)	β-Po	295	295	1.00	102.7
Bi	α-As	307	353	1.15	95.5
Po	α-Po	337	337	1.00	90
Po	β-Po	337	337	1.00	98.2

11.4 Elements of the Fifth and Sixth Main Groups under Pressure

Crystal structure determinations from very small samples have become possible due to the high intensities of the X-rays from a synchrotron. Very high pressures can be exerted on a small sample situated between two anvils made from diamond. In this way, our knowledge of the behavior of matter under high pressures has been widened considerably. Under pressure the elements of the fifth and sixth main groups exhibit rather unusual structures. A synopsis of the structures that occur is given in Fig. 11.9.

Under normal conditions an atom in elemental tellurium has coordination number $2+4$. It has been known for a long time that pressure causes the interatomic distances to approximate each other until finally every tellurium atom has six equidistant neighboring atoms at 297 pm; the structure (now called Te-IV) corresponds to β-polonium. However, before this is attained, two other modifications (Te-II and Te-III) that are out of the ordinary appear at 4 GPa and 7 GPa. Te-II contains parallel, linear chains that are mutually shifted in such a way that each Te atom has, in addition to its two neighboring atoms within the chain

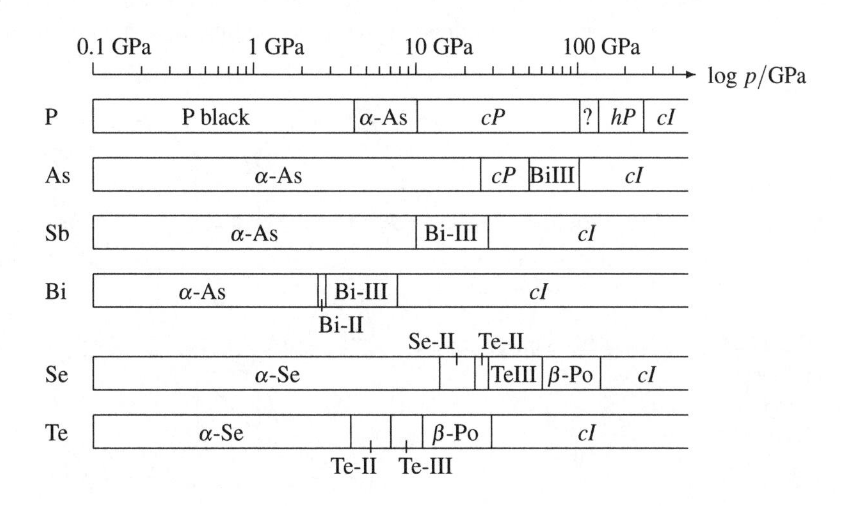

Fig. 11.9 Ranges of stabilities of the elements of the fifth and sixth main groups in dependence on pressure at room temperature. cP = cubic primitive (α-Po); hP = hexagonal primitive; cI = body-centered cubic packing of spheres

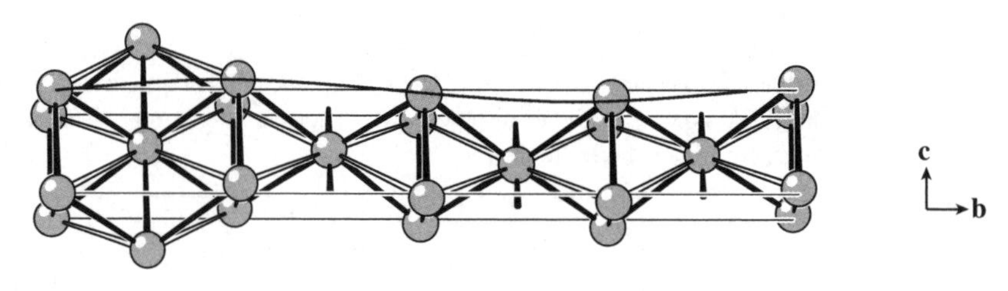

Fig. 11.10 Four body-centered unit cells of the incommensurately modulated structure of tellurium-III. Short bonds are drawn black, long bonds open; two more long contacts along $a = 392$ pm are not shown. In the direction **b** the atoms follow a sine wave having a wave length of $3.742 \times b$

(310 pm), two closer (286 – 299 pm) and four more distant neighboring atoms (331 – 364 pm). Te-III has an incommensurately modulated structure in which every tellurium atom has six neighboring atoms at distances of 297 to 316 pm and six more distant ones at 368 to 392 pm; these distances vary slightly from atom to atom (Fig. 11.10). Finally, at 27 GPa tellurium transforms to a body-centered cubic arrangement, which is a typically metallic structure (Te-V). Sulfur forms at least five high-pressure modifications; one of them (> 80 GPa) has the β-polonium structure.

Under pressure black phosphorus transforms first to a modification that corresponds to gray arsenic. At an even higher pressure this is converted to the α-polonium structure. Then follows a hexagonal-primitive structure, which has also been observed for silicon under pressure (p. 122), but that hardly ever occurs otherwise. Above 262 GPa phosphorus is body-centered cubic; this modification becomes superconducting below 22 K..

Arsenic also adopts the α-polonium structure at 25 GPa and becomes body-centered cubic at the highest pressures. The rather unusual Bi-III structure appears intermediately between these two modifications.

This bismuth-III structure is also observed for antimony from 10 to 28 GPa and for bismuth from 2.8 to 8 GPa. At even higher pressures antimony and bismuth adopt the body-centered cubic packing of spheres which is typical for metals. Bi-III has a peculiar *incommensurate composite crystal* structure. It can be described by two intergrown partial structures that are not compatible metrically with one another (Fig. 11.11). The partial structure 1 consists of square antiprisms which share faces along c and which are connected by tetrahedral building blocks. The partial structure 2 forms linear chains of atoms that run along c in the midst of the square antiprisms. In addition, to compensate for the

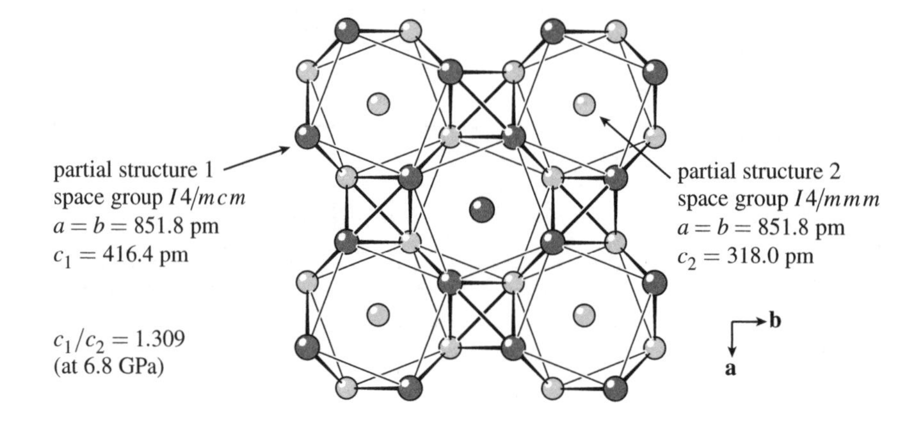

Fig. 11.11 The incommensurate composite structure of bismuth-III

differing distances between the atoms of the chains and the surrounding antiprisms, both partial structures are incommensurately modulated. The atoms of the chains are slightly displaced along *c*, those of the antiprisms perpendicular to *c*.

As a summary, one can state the following tendencies: the larger the atomic number, the lower is the pressure needed to attain a typically metallic structure. Intermediate between the non-metallic and the metallic structures, peculiar structures appear that cannot be integrated in the common chemical models.

11.5 Carbon

Graphite is the modification of carbon which is stable under normal conditions. It has a structure consisting of planar layers (Fig. 11.12). Within the layer each C atom is bonded covalently with three other C atoms. Every atom contributes one *p* orbital and one electron to the delocalized π bond system of the layer. This constitutes a half-filled band, so we have a metallic state with two-dimensional electrical conductivity. Between the layers weak VAN DER WAALS forces are the essential attractive forces. The bonds within the layers have a length of 142 pm and the distance from layer to layer is 335 pm. The high electric conductivity therefore only exists parallel to the layers, and not perpendicular to them. The layers are stacked in a staggered manner; half of the atoms of one layer are situated exactly above atoms of the layer below, and the other half are situated over the ring centers (Fig. 11.12). Three layer positions are possible, *A, B* and *C*. The stacking sequence in normal (hexagonal) graphite is *ABAB*..., but frequently a more or less statistical layer sequence is found, in which regions of the predominating sequence *ABAB*... are separated by regions with the sequence *ABC*. This is called a *one-dimensional disorder*, *i.e.* within the layers the atoms are ordered, but in the direction of stacking the periodic order is missing.

Graphite forms *intercalation compounds* with alkali metals. They have compositions such as LiC_6, LiC_{12}, LiC_{18} or KC_8, KC_{24}, KC_{36}, KC_{48}. Depending on the metal content they have colors extending from a golden luster to black. They are better electric conductors than graphite. The alkali metal ions are intercalated between every pair of graphite layers in KC_8 (first stage intercalation), between every other pair in KC_{24} (second stage) etc. (Fig. 11.13). The metal atoms turn over their valence electrons to the valence band of the graphite. The possibility of the reversible electrochemical intercalation of Li^+ ions into graphite in varying amounts is taken advantage of in the electrodes of lithium ion batter-

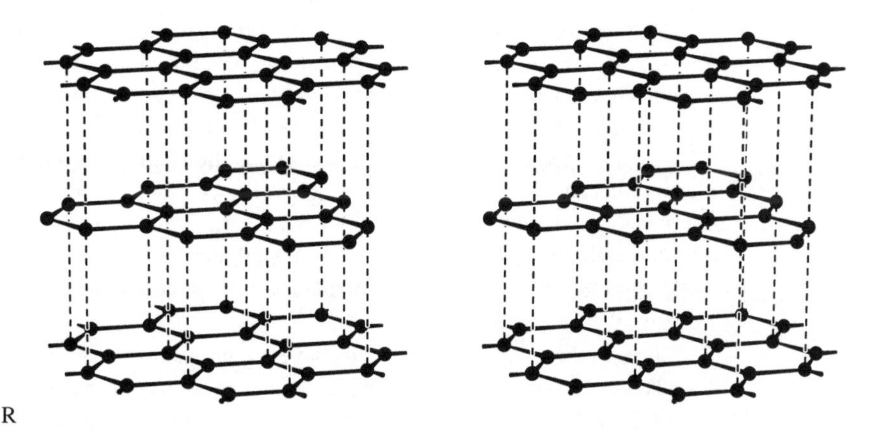

Fig. 11.12 Structure of graphite (stereo image)

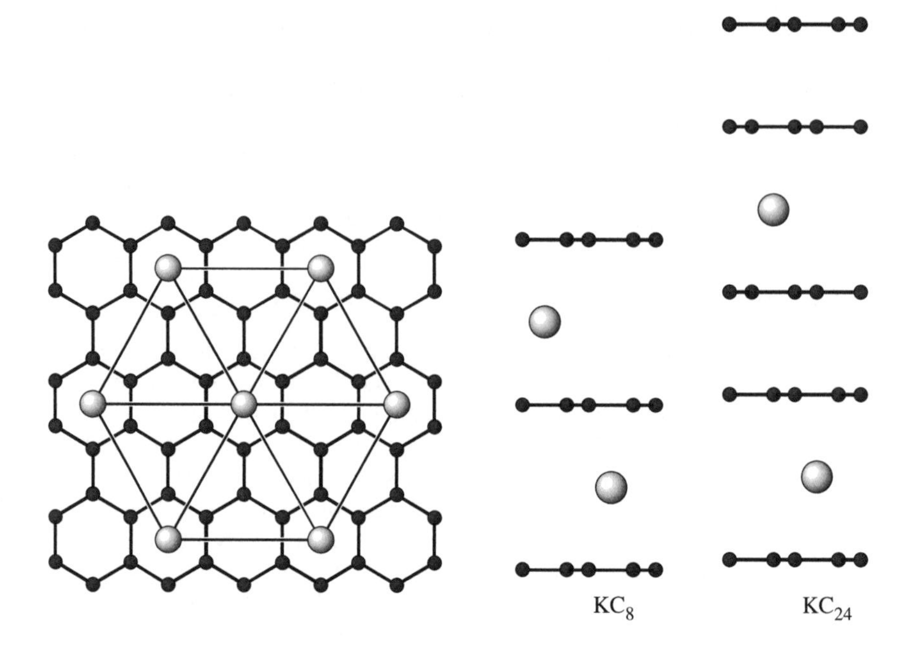

Fig. 11.13 Left: arrangement of the K^+ ions relative to an adjacent graphite layer in KC_8; in KC_{24} a K^+ ion layer only contains two thirds as many ions, they are disordered and highly mobile. Right: stacking sequence of graphite layers and K^+ ions in KC_8 and KC_{24}

ies. A different kind of intercalation compound is those with metal chlorides MCl_n (M = nearly all metals; n = 2 to 6) and some fluorides and bromides. The intercalated halide layers have structures that essentially correspond to the structures in the pure compounds; for example, intercalated $FeCl_3$ layers have the same structure as in pure $FeCl_3$, as shown on the front cover.

Carbon in its different forms such as pit-coal, coke, charcoal, soot etc., is in principle graphite-like, but with a low degree of ordering. It can be microcrystalline or amorphous; OH groups and possibly other atom groups are bonded at the edges of the graphite layer fragments. Many species of carbon have numerous pores and therefore have a large inner surface; for this reason they can adsorb large quantities of other substances and act as catalysts. In this respect crystalline graphite is less active. Carbon fibers that can be made, for example, by pyrolysis of polyacrylonitrile fibers, consist of graphite layers that are oriented parallel to the fiber direction.

Fullerenes are modifications of carbon that consist of cage-like molecules. They can be obtained by setting up an electric arc between two graphite electrodes in a controlled atmosphere of helium and condensing the evaporated carbon, and then recrystallizing it from magenta-colored benzene solution. The main product is the fullerene C_{60}, called buckminsterfullerene. The C_{60} molecule has the shape of a soccer ball, consisting of 12 pentagons and 20 benzene-like hexagons (Fig. 11.14). Second in yield from this preparation is C_{70}, which has 12 pentagons and 25 hexagons and a shape reminiscent of a peanut. Cages with other sizes can also be produced, but they are less stable (they may have any even number of C atoms, beginning at C_{32}). Independent of its size, a fullerene molecule always has twelve pentagons.

In crystalline C_{60} the molecules have a face-centered cubic arrangement, *i.e.* they are packed as in a cubic closest-packing of spheres; as they are nearly spherical, the molecules spin in the crystal. The crystals are as soft as graphite. Similar to the intercalation com-

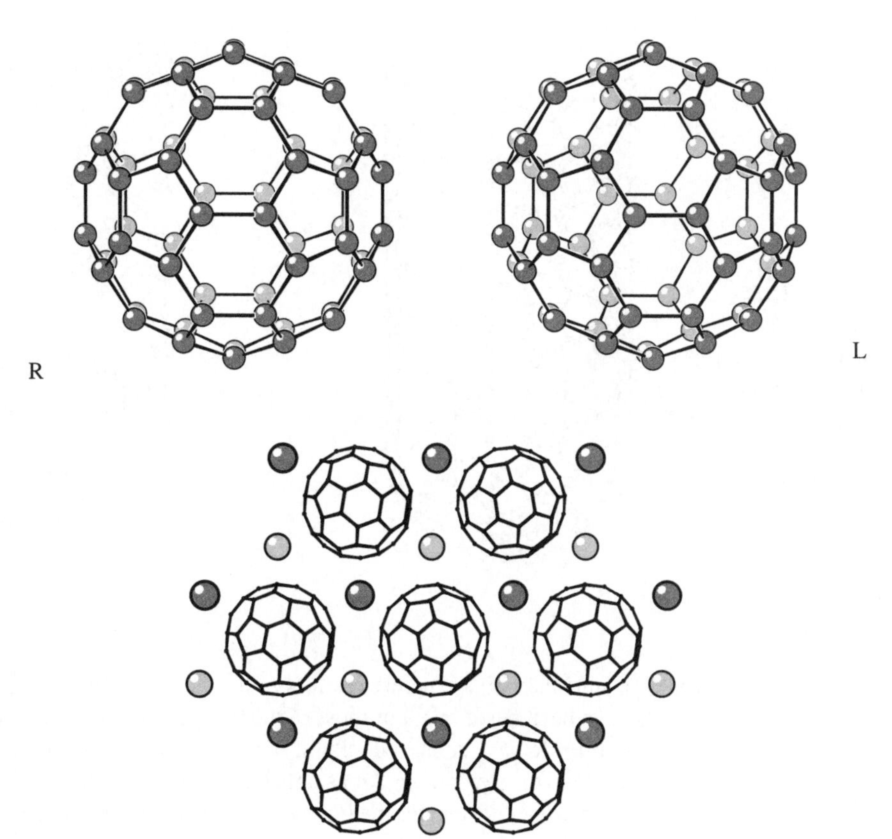

Fig. 11.14 Top: molecular structure of the C_{60} molecule (stereo view). Bottom: packing of C_{60} molecules and K^+ ions in K_3C_{60}

pounds of graphite, potassium atoms can be enclosed; they occupy the cavities between the C_{60} balls. With all cavities occupied (tetrahedral and octahedral interstices if the C_{60} balls are taken as closest-packed spheres), the composition is K_3C_{60}. This compound has metallic properties and becomes superconducting when cooled below 18 K. Even more potassium can be intercalated; in K_6C_{60} the C_{60} molecules have a body-centered cubic packing.

Carbon nanotubes can be made from graphite in an electric arc or by laser ablation. These tubes are tangled. Catalyzed pyrolysis of gaseous hydrocarbons at temperatures of 700 to 1100 °C is a method to produce ordered arrays of carbon nanotubes on appropriate carrier materials. For example, single-walled nanotubes with a diameter of 1.4 nm are obtained by pyrolysis of acetylene in the presence of ferrocene at 1100 °C; pyrolysis of benzene in the presence of $Fe(CO)_5$ yields multi-walled nanotubes. Nanotubes consist of bent graphite-like layers; they can be as long as 0.1 mm (Fig. 11.15). As a rule, the tubes are closed at their ends by half a fullerene sphere; these caps can be removed with ultrasound in a suspension of a strong acid. The six-membered rings can have different orientations relative to the tube axis (Fig. 11.15); this has influence on the electric conductivity. Multilayered nanotubes consist of tubes arranged concentrically around each other.

The structures of diamond, silicon, germanium and tin are discussed in Chapter 12.

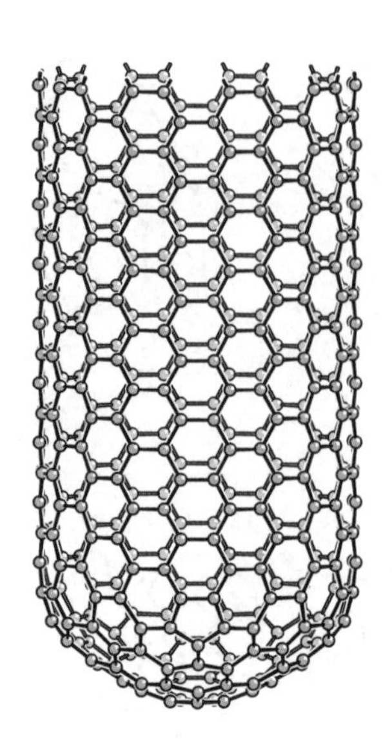
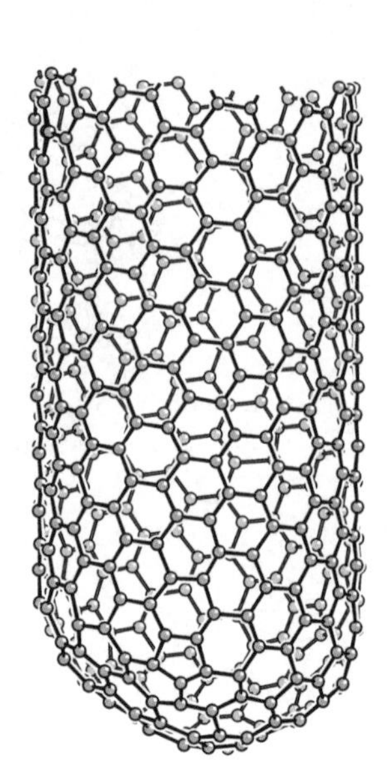

Fig. 11.15 Structures of two kinds of mono-layered carbon nanotubes. The left one shows metallic conduction, the right one is semiconducting

11.6 Boron

Boron is as unusual in its structures as it is in its chemical behavior. Sixteen boron modifications have been described, but most of them have not been well characterized. Many samples assumed to have consisted only of boron were possibly boron-rich borides (many of which are known, *e.g.* YB_{66}). An established structure is that of rhombohedral α-B_{12} (the subscript number designates the number of atoms per unit cell). The crystal structures of three further forms are known, tetragonal α-B_{50}, rhombohedral β-B_{105} and rhombohedral β-$B_{\sim 320}$, but probably boron-rich borides were studied. α-B_{50} should be formulated $B_{48}X_2$. It consists of B_{12} icosahedra that are linked by tetrahedrally coordinated X atoms. These atoms are presumably C or N atoms (B, C and N can hardly be distinguished by X-ray diffraction).

The outstanding building unit in all modifications of boron that have been described is the B_{12} icosahedron, which also is present in the anionic *closo*-borane $B_{12}H_{12}^{2-}$. The twelve atoms of an icosahedron are held together by multicenter bonds; according to MO theory, 13 bonding orbitals occupied by 26 electrons should be present; 10 valence electrons are left over. In the $B_{12}H_{12}^{2-}$ ion 14 additional electrons are present (12 from the H atoms, 2 from the ionic charge), which amounts to a total of 24 electrons or 12 electron pairs; these are used for the 12 covalent B–H bonds that are oriented radially outwards from the icosahedron. In elemental boron the B_{12} icosahedra are linked with one another by such radial bonds, but for 12 bonds only 10 valence electrons are available; therefore, not all of them can be normal two-center two-electron bonds.

In α-B_{12} the icosahedra are arranged as in a cubic closest-packing of spheres (Fig. 11.16). In one layer of icosahedra every icosahedron is surrounded by six other icosahedra that are linked by three-center two-electron bonds. Every boron atom involved contributes an average of $\frac{2}{3}$ electrons to these bonds, which amounts to $\frac{2}{3} \cdot 6 = 4$ electrons per icosahedron. Every icosahedron is surrounded additionally by six icosahedra of the two adjacent layers, to which it is bonded by normal B–B bonds; this requires 6 electrons per icosahedron. In total, this adds up exactly to the above-mentioned 10 electrons for the inter-icosahedron bonds.

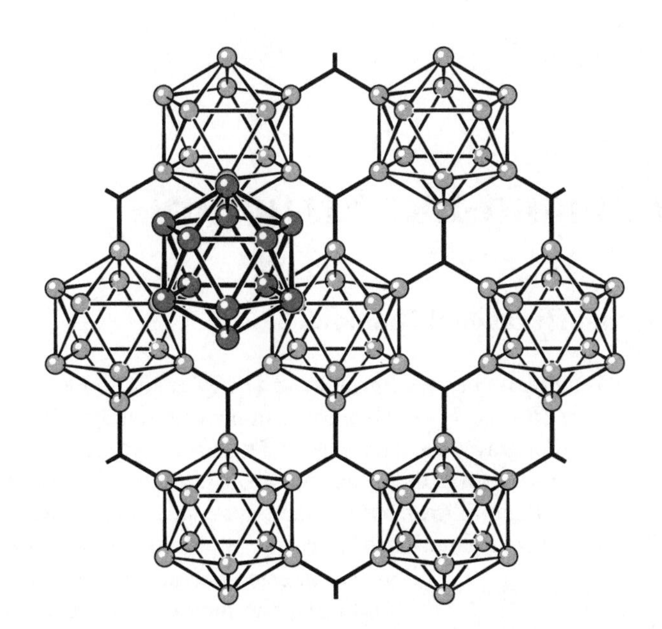

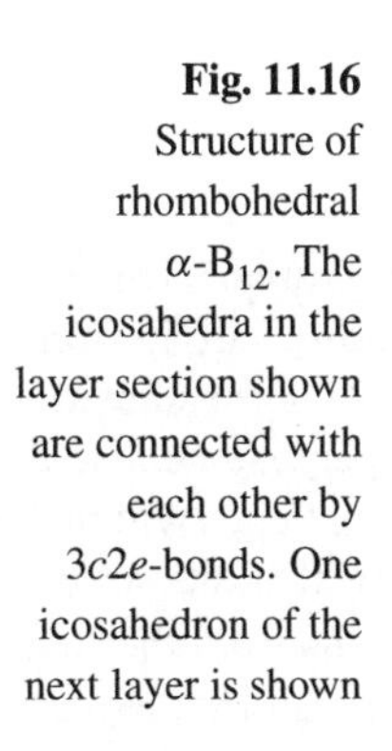

Fig. 11.16 Structure of rhombohedral α-B_{12}. The icosahedra in the layer section shown are connected with each other by 3*c*2*e*-bonds. One icosahedron of the next layer is shown

12 Diamond-like Structures

12.1 Cubic and Hexagonal Diamond

Diamond, silicon, germanium and (gray) α-tin (stable below 13 °C) are isotypic. Diamond consists of a network of carbon atoms with four covalent bonds per atom. Starting from a layer of gray arsenic (*cf.* Fig. 11.7), all As atoms can be thought of as being substituted by C atoms; each of these can participate in a fourth bond that is oriented perpendicular to the layer. Relative to any one of the chair conformation rings of the layer the bonds within the layer take equatorial positions; the remaining bonds correspond to axial positions that are directed alternately upwards and downwards from the layer. In graphite fluoride $(CF)_x$ every axial position is occupied by a fluorine atom. In diamond the axial bonds serve to link the layers with each other (Fig. 12.1). Thereby new six-membered rings are formed that can have either a chair or a boat conformation, depending on how the joined layers are positioned relative to each other. If in projection the layers are staggered, then all resulting rings have a chair conformation; this is the arrangement in normal, cubic diamond. In hexagonal diamond the layers in projection are eclipsed, and the new rings have a boat conformation. Hexagonal diamond occurs very seldom as the mineral lonsdaleite; it has been found in meteorites.

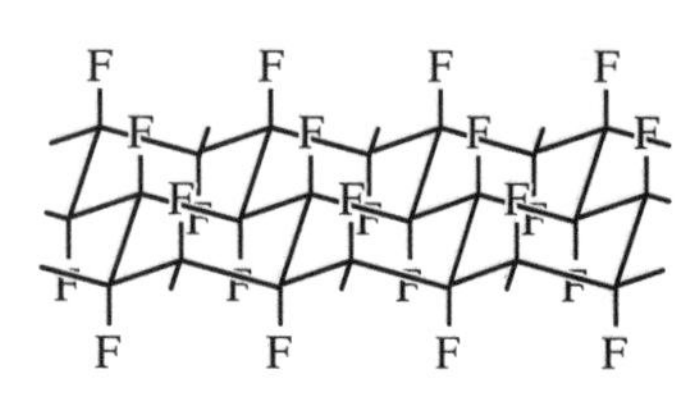

graphite fluoride

The unit cell of cubic diamond corresponds to a face-centered packing of carbon atoms. Aside from the four C atoms in the vertices and face centers, four more atoms are present in the centers of four of the eight octants of the unit cell. Since every octant is a cube having four of its eight vertices occupied by C atoms, an exact tetrahedral coordination results for the atom in the center of the octant. The same also applies to all other atoms — they are all symmetry-equivalent. In the center of every C–C bond there is an inversion center. As in alkanes the C–C bonds have a length of 154 pm and the bond angles are 109.47°.

12.2 Binary Diamond-like Compounds

By substituting alternately the carbon atoms in cubic diamond by zinc and sulfur atoms, one obtains the structure of zinc blende (sphalerite). By the corresponding substitution in hexagonal diamond, the wurtzite structure results. As long as atoms of one element are allowed to be bonded only to atoms of the other element, binary compounds can only have a 1 : 1 composition. For the four bonds per atom an average of four electrons per atom are needed; this condition is fulfilled if the total number of valence electrons is four times the number of atoms. Possible element combinations and examples are given in Table 12.1.

Inorganic Structural Chemistry, Second Edition Ulrich Müller

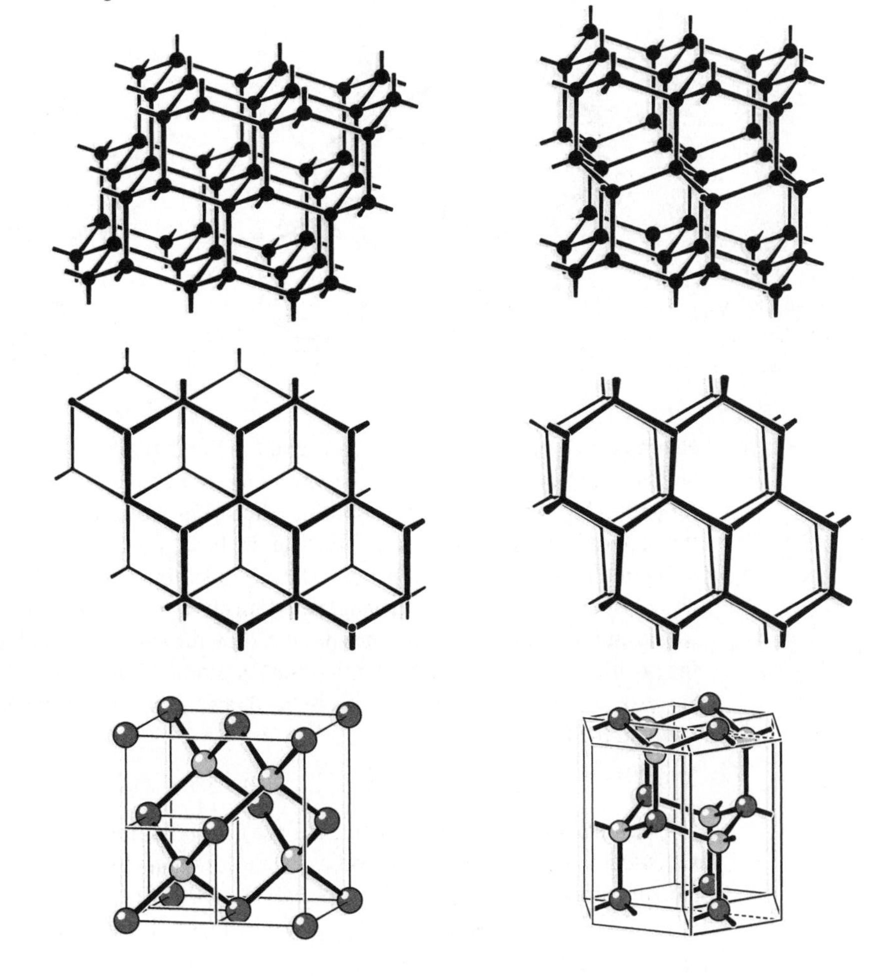

Fig. 12.1 Structure of cubic (left) and hexagonal (right) diamond. Top row: connected layers as in α-As. Central row: the same layers in projection perpendicular to the layers. Bottom: unit cells; when the light and dark atoms are different, this corresponds to the structures of zinc blende (sphalerite) and wurtzite, respectively

Table 12.1: Possible element combinations for the ZnS structure types

combination*		examples, zinc blende type	examples, wurtzite type
IV	IV	β-SiC	SiC
III	V	BP, GaAs, InSb	AlN, GaN
II	VI	BeS, CdS, ZnSe	BeO, ZnO, CdS (high temp.)
I	VII	CuCl, CuBr, AgI	CuCl (high temp.), β-AgI

* group numbers in the periodic table

The GRIMM–SOMMERFELD rule is valid for the bond lengths: if the sum of the atomic numbers is the same, the interatomic distances are the same. For example:

MX	Z(M) + Z(X)	d(M–X)
GeGe	$32+32=64$	245.0 pm
GaAs	$31+33=64$	244.8 pm
ZnSe	$30+34=64$	244.7 pm
CuBr	$29+35=64$	246.0 pm

The sections of the structures of zinc blende and wurtzite shown in Fig. 12.2 correspond to the central row of Fig. 12.1 (projections perpendicular to the arsenic-like layers).

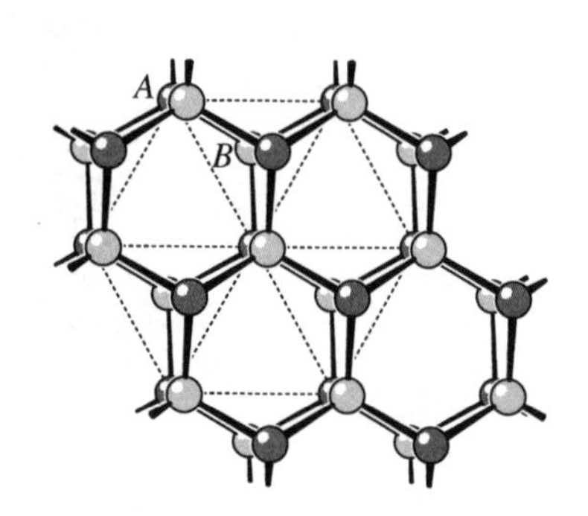

Fig. 12.2 Positions of the Zn and S atoms in zinc blende (left) and wurtzite

Behind every sulfur atom there is a zinc atom bonded to it the direction of view. The zinc atoms within one of the arsenic-like layers are in one plane and form a hexagonal pattern (dotted in Fig. 12.2); the same applies to the sulfur atoms on top of them. The position of the pattern is marked by an A. In wurtzite the hexagonal pattern of the following atoms is staggered relative to the first pattern; the atoms of this position B are placed over the centers of one half of the dotted triangles. Atoms over the centers of the remaining triangles (position C) do not occur in wurtzite, but they do occur in zinc blende. If we designate the positions of the planes containing the Zn atoms by A, B, and C, respectively, and the corresponding planes of the S atoms by α, β, and γ, then the following stacking sequences apply to the planes:

zinc blende: $A\alpha B\beta C\gamma\ldots$ wurtzite: $A\alpha B\beta\ldots$

Other stacking sequences than these are also possible, for example $A\alpha B\beta A\alpha C\gamma\ldots$ or statistical sequences without periodic order. More than 70 stacking varieties are known for silicon carbide, and together they are called α-SiC. Structures that can be considered as stacking variants are called polytypes. We deal with them further in the context of closest-sphere packings (Chapter 14).

Several of the binary diamond-like compounds have industrial applications because of their physical properties. They include silicon carbide and cubic boron nitride (obtainable from graphite-like BN under pressure at 1800 °C); they are almost as hard as diamond and serve as abrasives. SiC is also used to make heating devices for high-temperature furnaces as it is a semiconductor with a sufficiently high conductivity at high temperatures, but also is highly corrosion resistant and has a low thermal expansion. Yellow CdS and red CdSe are excellent color pigments, and ZnS is used as a luminophore in cathode ray displays. The III–V compounds are semiconductors with electric properties that can be adapted by variation of the composition and by doping; light-emitting diodes and photovoltaic cells are made on the basis of GaAs.

12.3 Diamond-like Compounds under Pressure

The diamond-type structure of α-tin is stable at ambient pressure only up to 13 °C; above 13 °C it transforms to β-tin (white tin). The transition α-Sn $\rightarrow$ β-Sn can also by achieved below 13 °C by exerting pressure. Silicon and germanium also adopt the structure of β-Sn at higher pressures. The transformation involves a considerable increase in density (for Sn +21%). The β-Sn structure evolves from the α-Sn structure by a drastic compression

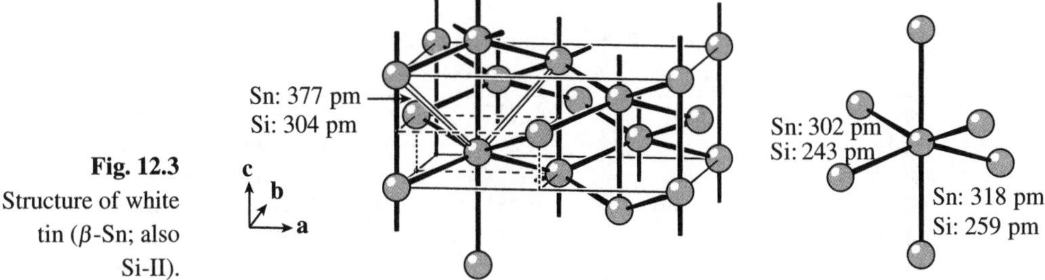

Fig. 12.3 Structure of white tin (β-Sn; also Si-II).
The drawn cell corresponds to a unit cell of diamond (α-Sn; Si-I) that has been strongly compressed in one direction. Right: coordination about a tin (or Si) atom with bond lengths; *cf.* the atom in the dashed octant

in the direction of one of the edges of the unit cell (Fig. 12.3). In this way two atoms that previously were further away in the direction of the compression become neighbors to an atom; together with the four atoms that were already adjacent in α-Sn, a coordination number of 6 results. The regular coordination tetrahedron of α-Sn is converted to a flattened tetrahedron with Sn–Sn distances of 302 pm; the two atoms above and below the flattened tetrahedron are at a distance of 318 pm. These distances are *longer* than in α-Sn (281 pm). Although β-Sn forms from α-Sn by the action of pressure and has a higher density, the transformation involves an increase of the interatomic distances.

Generally, the following rules apply for pressure-induced phase transitions:

Pressure–coordination rule by A. NEUHAUS: *with increasing pressure an increase of the coordination number takes place.*

'Pressure–distance paradox' by W. KLEBER: *When the coordination number increases according to the previous rule, the interatomic distances also increase.*

Further examples where these rules are observed are as follows. Under pressure, some compounds with zinc blende structure, such as AlSb and GaSb, transform to modifications that correspond to the β-Sn structure. Others, such as InAs, CdS, and CdSe, adopt the NaCl structure when compressed, and their atoms thus also attain coordination number 6. Graphite (c.n. 3, C–C distance 141.5 pm, density 2.26 g cm^{-3}) $\stackrel{\text{pressure}}{\longrightarrow}$ diamond (c.n. 4, C–C 154 pm, 3.51 g cm^{-3}).

The rules are also reflected in the behavior of silicon and germanium at even higher pressures. Fig. 12.4 shows which other structure types are observed. Silicon adopts a complicated variety of structures at high pressures. However, in general, the higher the

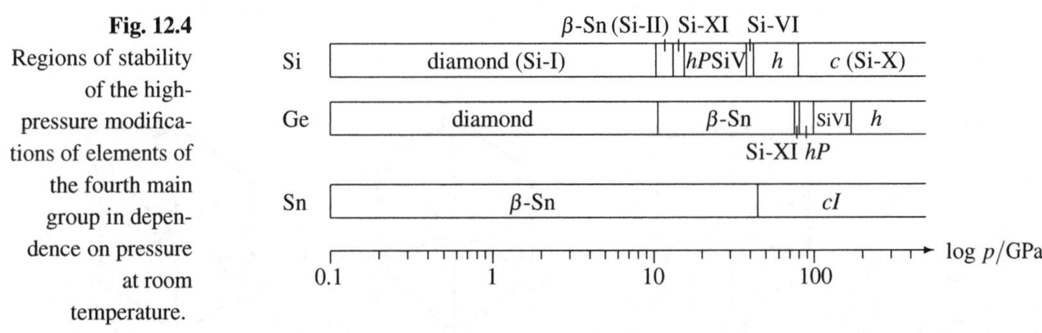

Fig. 12.4 Regions of stability of the high-pressure modifications of elements of the fourth main group in dependence on pressure at room temperature.
hP = hexagonal-primitive (Si-V); *cI* = body-centered cubic packing of spheres; *h* = hexagonal closest-packing of spheres; *c* = cubic closest-packing of spheres

Table 12.2: High-pressure modifications of silicon

	structure type	c.n.: d/ pm*	range of stability	space group
Si-I	diamond	4: 235	$<$ 10.3 GPa	$Fd\bar{3}m$
Si-II	β-Sn	6: 248	10.3 – 13.2 GPa	$I4_1/amd$
Si-XI		6 + 2: 253	13.2 – 15.6 GPa	$Imma$
Si-V	hexagonal-primitive	8: 251	15.6 – 38 GPa	$P6/mmm$
Si-VI		10: 248; 11: 249	38 – 42 GPa	$Cmce$
Si-VII	Mg**	12: 248	42 – 79 GPa	$P6_3/mmc$
Si-X	Cu†	12: 248	$>$ 79 GPa	$Fm\bar{3}m$

* c.n. = coordination number, d = mean value of the bond lengths
** hexagonal closest-packing of spheres
† cubic closest-packing of spheres

pressure, the higher is the coordination number of the atoms (Table 12.2). At very high pressures the pressure–distance paradox becomes nearly imperceptible.

Si-XI, which is the next modification formed under pressure after Si-II, can be described as a compressed variant of the β-tin type. The structure is compressed in the direction of one of the diagonals of the a–b plane. As a consequence, the two bonds drawn as open bonds in Fig. 12.3 are shortened to 275 pm, whereas the other six bonds approximately keep their lengths. The coordination polyhedron is a distorted hexagonal bipyramid (Si-XI, Fig. 12.5). From this, with further increase in pressure, an undistorted hexagonal bipyramid results in a simple, hexagonal-primitive structure (Si-V, Fig. 12.6). Then, a major rearrangement follows at 38 GPa. In the Si-VI structure obtained one can discern two kinds of alternating layers. One kind of layer has a slightly corrugated square pattern with atoms of coordination number 10; the other kind of layer consists of squares and rhombs and has atoms of coordination number 11 (layers at $x = \frac{1}{2}$ and $x = 1$, Fig. 12.6). Finally, at the highest pressures, silicon forms closest packings of spheres with atoms of coordination number 12.

In addition, silicon adopts a number of metastable structures that can be obtained, depending on pressure, by rapid release of the pressure: from Si-II, Si-XII is formed, and from this Si-III; upon heating, Si-III transforms to the hexagonal diamond structure (Si-IV). Si-III has a peculiar structure with a distorted tetrahedral coordination of its atoms. The atoms are arranged to interconnected right- and left-handed helices (Fig. 12.7). The structure being cubic, the helices run in the directions a, b as well as c. Si-VIII and Si-IX

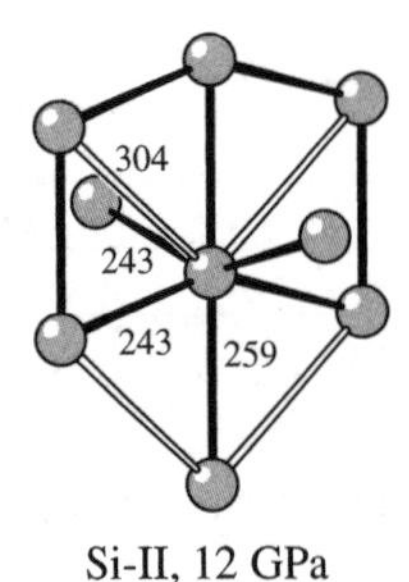

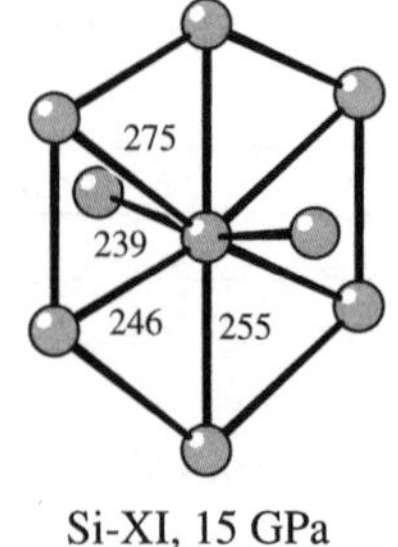

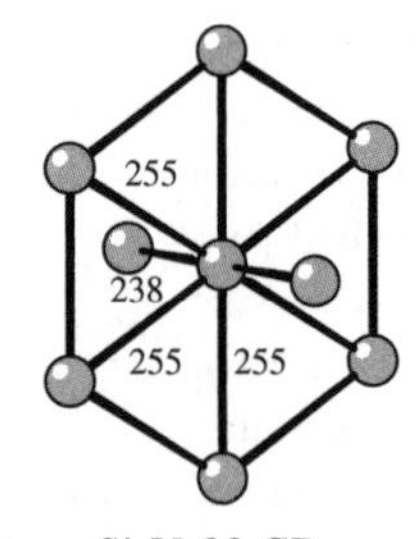

Fig. 12.5 Change of the coordination polyhedron of a silicon atom at increasing pressures; same perspective as for the atom in the dashed octant of Fig. 12.3. Si–Si distances in pm

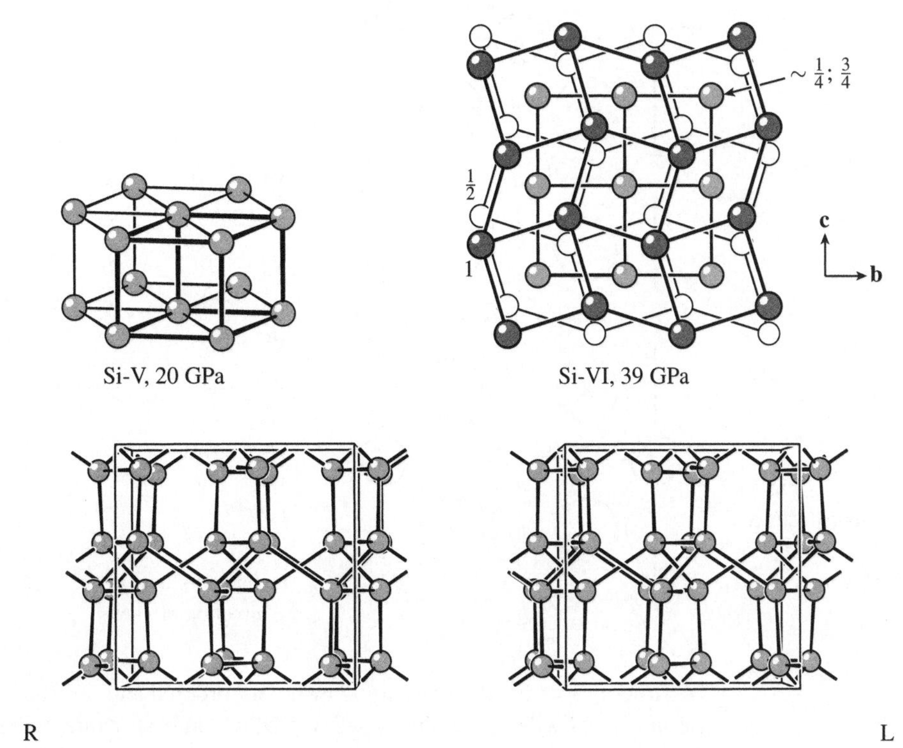

Fig. 12.6 Hexagonal-primitive packing of Si-V; one unit cell is outlined. Si-VI; only atomic contacts within the layers parallel to the *b*–*c* plane have been drawn; Numbers: *x* coordinates

Si-V, 20 GPa

Si-VI, 39 GPa

Fig. 12.7 The metastable cubic structures of Si-III and Ge-IV (stereo image)

are obtained from Si-XI upon sudden release of pressure. All high-pressure modifications of silicon are metallic.

Germanium forms the same kinds of modifications as silicon at similar conditions (Fig. 12.4). Tin, however, does not exhibit this diversity; β-tin transforms to a body-centered cubic packing of spheres at 45 GPa. Lead already adopts a cubic closest-packing of spheres at ambient pressure.

12.4 Polynary Diamond-like Compounds

Of the numerous ternary and polynary diamond-like compounds we deal only with those that can be considered as superstructures of zinc blende. A superstructure is a structure that, while having the same structural principle, has an enlarged unit cell. When the unit cell of zinc blende is doubled in one direction (*c* axis), different kinds of atoms can occupy the doubled number of atomic positions. All the structure types listed in Fig. 12.8 have the tetrahedral coordination of all atoms in common, except for the variants with certain vacant positions.

$CuFeS_2$ (chalcopyrite) is one of the most important copper minerals. Red β-Cu_2HgI_4 and yellow β-Ag_2HgI_4 ($CdGa_2S_4$ type) are thermochromic: they transform at 70 °C and 51 °C, respectively, to modifications having different colors (black and orange); in these the atoms and the vacancies have a disordered distribution.

Aside from the superstructures mentioned, other superstructures with other enlargement factors for the unit cell are known, as well as superstructures of wurtzite. Defect structures,

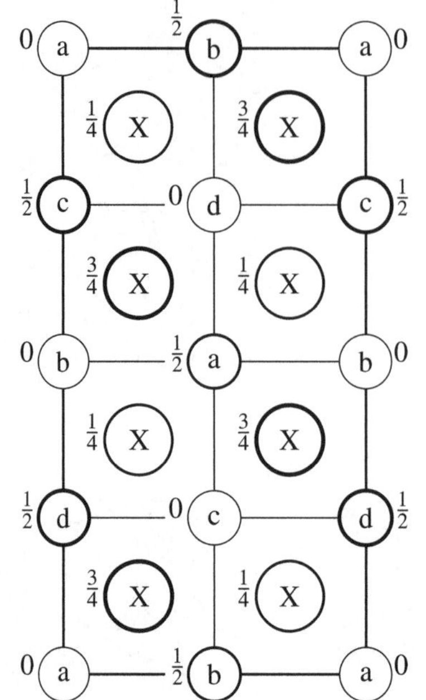

Fig. 12.8 Superstructures of the zinc blende type with doubled *c* axis

	Atomic position				
structure type	X	a	b	c	d
$CuFeS_2$ *	S	Fe	Cu	Fe	Cu
Cu_3SbS_4 †	S	Sb	Cu	Cu	Cu
Cu_2FeSnS_4 ‡	S	Fe	Sn	Cu	Cu
$CdGa_2S_4$	S	Cd	Ga	Ga	□
β-Cu_2HgI_4	I	Hg	□	Cu	Cu

□ = vacancy
* chalcopyrite
† famatinite
‡ stannite

The numbers next to the circles designate the height in the direction of view

i.e. structures with vacancies, are known with ordered and disordered distributions of the vacancies. γ-Ga_2S_3, for example, has the zinc blende structure with statistically only two thirds of the metal positions occupied by Ga atoms.

12.5 Widened Diamond Lattices. SiO_2 Structures

Take elemental silicon (diamond structure) and insert an oxygen atom between every pair of silicon atoms; in this way, every Si–Si bond is replaced by an Si–O–Si group and every Si atom is surrounded tetrahedrally by four O atoms. The result is the structure of cristobalite. The SiO_4 tetrahedra are all linked by common vertices. As there are twice as many Si–Si bonds than Si atoms in silicon, the composition is SiO_2. Cristobalite is one of the polymorphic forms of SiO_2; it is stable between 1470 and 1713 °C and is metastable at lower temperatures. It occurs as a mineral. The oxygen atoms are situated to the side of the Si$\cdots$Si connecting lines, so that the Si–O–Si bond angle is 147°. The structure model shown in Fig. 12.9, left side, however, is only a snapshot. Above 250 °C the tetrahedra perform coupled tilting vibrations that on average result in a higher symmetry, with O atoms exactly on the Si–Si connecting lines (Fig. 12.9, right); the large ellipsoids show the vibration. When cooled below ~240 °C the vibrations 'freeze' ($\rightarrow$ α-cristobalite; the $\alpha \rightleftharpoons \beta$ transition temperature depends on the purity of the sample).

The insertion of the oxygen atoms widens the silicon lattice considerably. A relatively large void remains in each of the four vacant octants of the unit cell. In natural cristobalite they usually contain foreign ions (mainly alkali and alkaline earth metal ions) that probably stabilize the structure and allow the crystallization of this modification at temperatures far below the stability range of pure cristobalite. To conserve electrical neutrality, probably one Si atom per alkali metal ion is substituted by an Al atom.* The substitution of Si

* Al and Si can hardly be distinguished by X-ray structure analysis owing to the nearly equal number of electrons

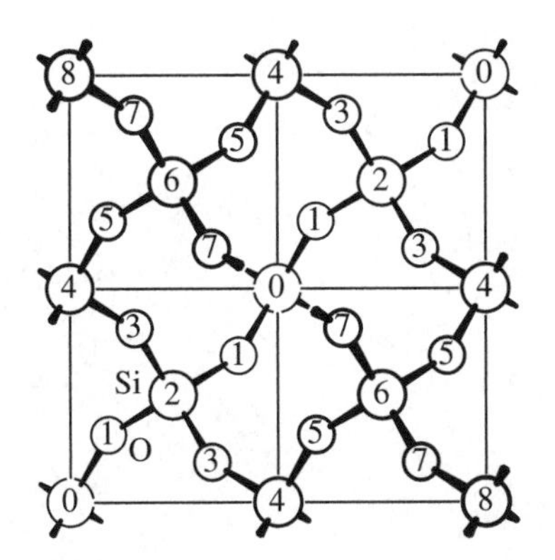

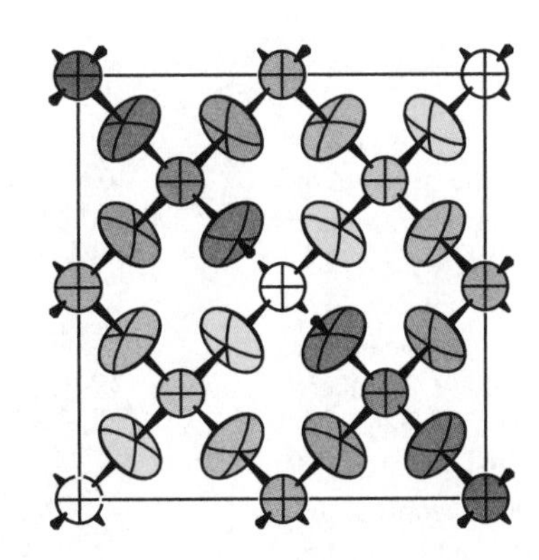

Fig. 12.9 Unit cell of β-cristobalite. Left: snapshot; the numbers indicate the height of the atoms in the direction of view as multiples of $\frac{1}{8}$. Right: with ellipsoids of thermal motion at 300 °C

by Al atoms in an SiO_2 framework with simultaneous inclusion of cations in voids is a very common phenomenon; silicates of this kind are called aluminosilicates. The mineral carnegieite, $Na[AlSiO_4]$, has a cristobalite structure in which half of the Si atoms have been substituted by Al atoms and all voids have been occupied by Na^+ ions. The LOEWENSTEIN rule has been stated for aluminosilicates: AlO_4 tetrahedra tend not to be linked directly with each other; the group Al–O–Al is avoided.

Tridymite is another form of SiO_2 which is stable between 870 and 1470 °C, but it can also be maintained in a metastable state at lower temperatures and occurs as a mineral. Its structure can be derived from that of hexagonal diamond in the same way as that of cristobalite from cubic diamond. In this case the oxygen atoms are also situated to the side of the Si···Si connecting lines and the Si–O–Si bond angles are approximately 150°. At temperatures below 380 °C several variants occur that differ in the kind of mutual tilting of the SiO_4 tetrahedra. Tridymite also encloses larger voids that can be occupied by alkali or alkaline earth ions. The anionic framework of some aluminosilicates corresponds to the tridymite structure, for example in nepheline, $Na_3K[AlSiO_4]_4$.

Quartz is the modification of SiO_2 that is stable up to 870 °C, with two slightly different forms, α-quartz occurring below and β-quartz above 573 °C. We discuss the quartz structure here, although it cannot be derived from one of the forms of diamond. Nevertheless, quartz also consists of a network of SiO_4 tetrahedra sharing vertices, but with smaller voids than in cristobalite and tridymite (this is manifested in the densities: quartz 2.65, cristobalite 2.33, tridymite 2.27 g cm^{-3}). As shown in Fig 12.10, the tetrahedra form helices, and in a given crystal these are all either right-handed or left-handed. Right-handed and left-handed quartz can also be intergrown in a well-defined manner, forming twinned crystals ('Brazilian twins'). Due to the helical structure quartz crystals are optically active and have piezoelectric properties (Section 19.2). Quartz crystals are produced industrially by hydrothermal synthesis. For this purpose quartz powder is placed in one end of a closed vial at 400 °C, while a seed crystal is placed at the opposite end at 380 °C; the vial is filled with an aqueous alkaline solution that is maintained liquid by a pressure of 100 MPa. The quartz powder slowly dissolves while the seed crystal grows.

The phase diagram for SiO_2 is shown in Fig. 12.11. The transition between α- and β-quartz only requires minor rotations of the SiO_4 tetrahedra, the linkage pattern remaining unaltered (Fig. 18.9, p. 224); this transition takes place rapidly. The other transitions, on the other hand, require a reconstruction of the structure, Si–O bonds being untied and rejoined; they proceed slowly and thus render possible the existence of the metastable mod-

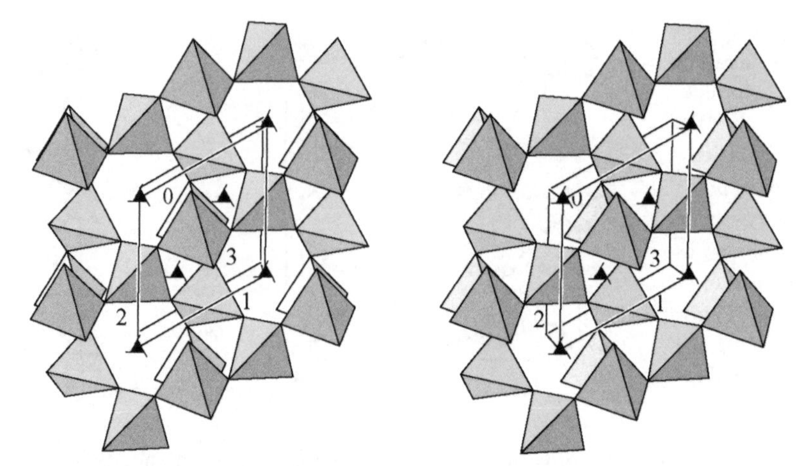

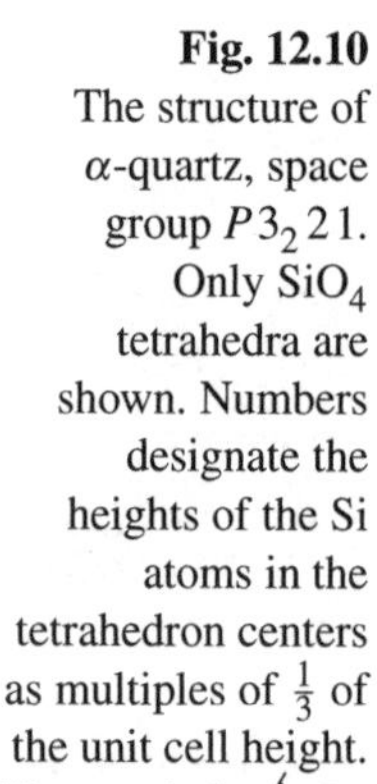

Fig. 12.10 The structure of α-quartz, space group $P3_221$. Only SiO_4 tetrahedra are shown. Numbers designate the heights of the Si atoms in the tetrahedron centers as multiples of $\frac{1}{3}$ of the unit cell height. The symbols ▲ for 3_2 screw axes mark the axes of the helical chains. The slight tilting of the tetrahedra relative to the direction of view (c axis) vanishes in β-quartz (stereo image)

ifications. Coesite and stishovite are stable only at high pressures, but they are metastable at ambient temperature and pressure. Coesite also consists of a framework of SiO_4 tetrahedra sharing vertices. Stishovite, however, has the rutile structure, *i.e.* silicon atoms having coordination number 6. Further metastable modifications are quartz glass (supercooled melt), moganite, keatite and fibrous SiO_2 with the SiS_2 structure (Fig. 16.27, p. 189).

Further compounds that occur with the structure types of SiO_2 are H_2O and BeF_2. Ice normally crystallizes in the hexagonal tridymite type (ice I_h), the oxygen atoms occupying the Si positions of tridymite while the hydrogen atoms are placed between two oxygen atoms each. An H atom is shifted towards one of the O atoms so that it belongs to one H_2O molecule and participates in a hydrogen bridge to another H_2O molecule. Metastable ice I_c crystallizes from the gas phase at temperatures below $-140\,°C$; it has a cubic structure like cristobalite. Eleven further modifications can be obtained under pressure, some of

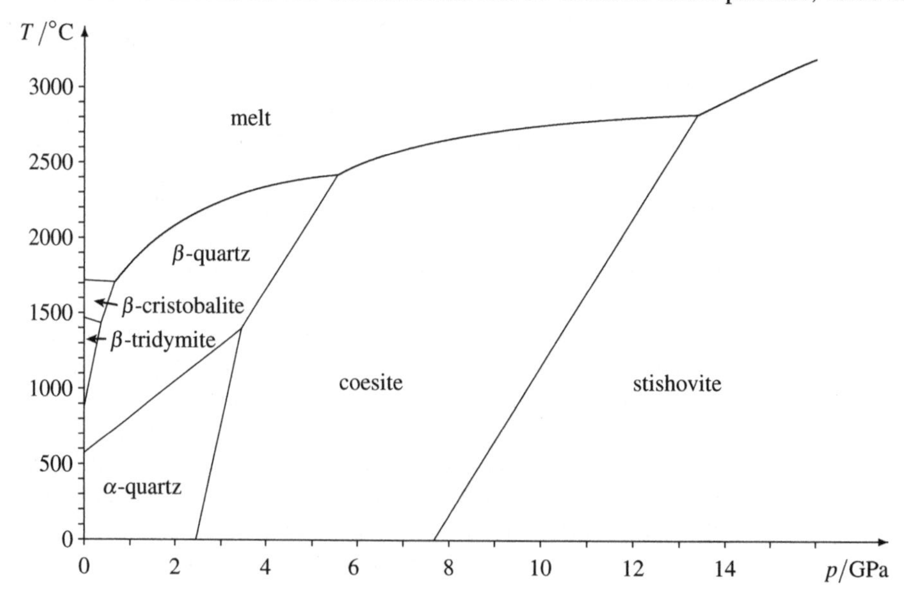

Fig. 12.11 Phase diagram for SiO_2. In addition, modifications of the α-PbO_2 and $CaCl_2$ type occur at pressures above 35 MPa

which correspond to other SiO_2 modifications (*e.g.* keatite; see the phase diagram of H_2O, Fig. 4.3, p. 35).

In the same way as the zinc blende structure is derived from diamond by the alternating substitution of C atoms by Zn and S atoms, the Si atoms in the SiO_2 structures can be substituted alternately by two different kinds of atoms. Examples include $AlPO_4$, $MnPO_4$, and $ZnSO_4$. The cristobalite and tridymite structures with filled voids also are frequently encountered. Examples in addition to the above-mentioned aluminosilicates $Na[AlSiO_4]$ and $Na_3K[AlSiO_4]_4$ are $K[FeO_2]$ and MILLON's base, $[NHg_2]^+OH^-{\cdot}H_2O$.

The large voids in the network of cristobalite can also be filled in another way, namely by a second network of the same kind that interpenetrates the first. Cuprite, Cu_2O, has this structure. Take a cristobalite structure in which the Si positions are occupied by O atoms that are linked via Cu atoms having coordination number 2. As the bond angle at a Cu atom is 180°, the packing density is even less than in cristobalite itself. Two exactly equal networks of this kind interpenetrate each other, one being shifted against the other (Fig. 12.12). The two networks 'float' one within the other; there are no direct bonds between them. This kind of a structure is possible when tetrahedrally coordinated atoms are held at a distance from each other by linear linking groups like –Cu– or –Ag– (in isotypic Ag_2O). Cyanide groups between tetrahedrally coordinated zinc atoms, Zn–C≡N–Zn, act in the same way as spacers in $Zn(CN)_2$, which has the same structure as Cu_2O (with metal atom positions interchanged with the anion positions).

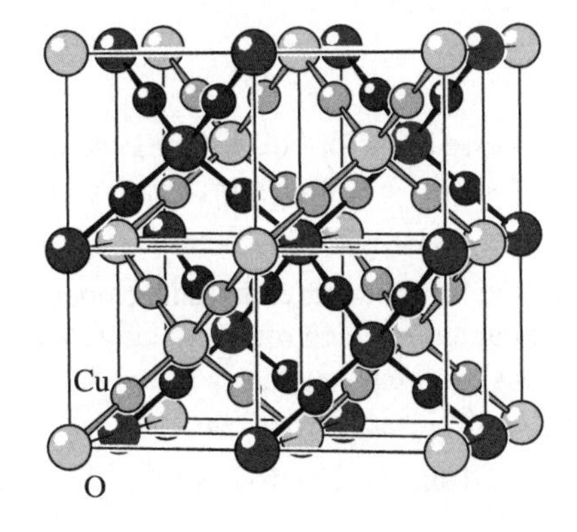

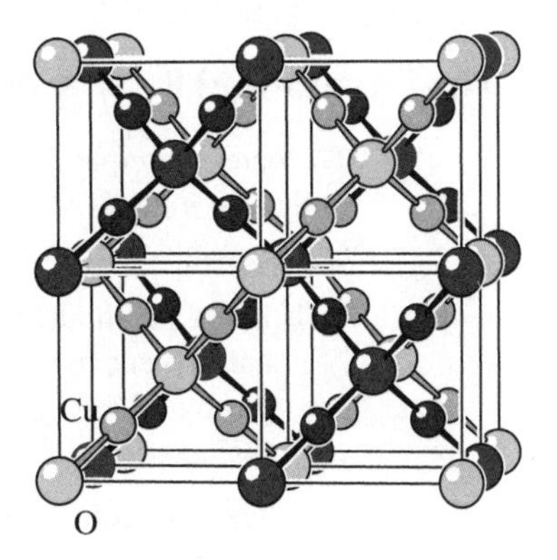

Fig. 12.12 The structure of Cu_2O (cuprite). Eight unit cells are shown; they correspond to one unit cell of cristobalite if only one of the two networks is present. The gray network has no direct bonds to the black network (stereo image)

12.6 Problems

12.1 The bond length in β-SiC is 188 pm. For which of the following compounds would you expect longer, shorter or the same bond lengths?
BeO, BeS, BN, BP, AlN, AlP.

12.2 Stishovite is a high-pressure modification of SiO_2 having the rutile structure. Should it have longer or shorter Si–O bond lengths than quartz?

12.3 Whereas AgCl has the NaCl structure, AgI has the zinc blende structure. Could you imagine conditions under which both compounds would have the same structure?

12.4 What is the coordination number of the iodine atoms in β-Cu_2HgI_4?

12.5 If well-crystallized Hg_2C could be made, what structure should it have?

13 Polyanionic and Polycationic Compounds. Zintl Phases

The compounds dealt with first in this chapter belong to the *normal valence compounds*; these are compounds that fulfill the classical valence concept of stable eight-electron shells. They include not only the numerous molecular compounds of nonmetals, but also compounds made up of elements from the left side with elements from the right side of the ZINTL line. The ZINTL line is a delimiting line that runs in the periodic table of the elements between the third and the fourth main groups. According to the classical concepts, such compounds consist of ions, for example NaCl, K_2S, Mg_2Sn, Ba_3Bi_2. Judging by the composition, however, in many cases the octet rule seems to be violated as, for example, in $CaSi_2$ or NaP. This impression is erroneous: the octet rule is still being fulfilled; the formation of covalent bonds renders it possible. In $CaSi_2$ the Si atoms are joined in layers as in gray arsenic (Si^- and As are isoelectronic), and in NaP the phosphorus atoms form helical chains analogous to polymeric sulfur (P^- and S are isoelectronic). Whether a compound fulfills the octet rule can only be decided when its structure is known.

13.1 The Generalized 8−*N* Rule

The octet principle can be expressed as a formula by the *generalized 8−N rule* according to E. MOOSER & W. B. PEARSON. We restrict our considerations to binary compounds, and presuppose the following:

1. Let X be an element of the fourth to seventh main groups of the periodic table, *i.e.* an element that tends to attain the electronic configuration of the following noble gas by taking up electrons (the heavy elements of the third main group may also be included). An X atom has e(X) valence electrons.
2. The electrons needed to fill up the electron octet at X are supplied by the more electropositive element M. An M atom has e(M) valence electrons.

The composition being M_mX_x, $8x$ electrons are required in order to achieve the octet shells for the x X atoms:

$$m \cdot e(\mathrm{M}) + x \cdot e(\mathrm{X}) = 8x \quad (13.1)$$

If covalent bonds exist between M atoms, then not all of the e(M) electrons of M can be turned over to X, and the number e(M) in equation (13.1) must be reduced by the number b(MM) of covalent bonds per M atom. If the M atoms retain nonbonding electrons (lone electron pairs as for Tl^+), then e(M) must also be reduced by the number E of these electrons. On the other hand, the X atoms require fewer electrons if they take part in covalent bonds with each other; the number e(X) can be increased by the number b(XX) of covalent bonds per X atom:

$$m[e(\mathrm{M}) - b(\mathrm{MM}) - E] + x[e(\mathrm{X}) + b(\mathrm{XX})] = 8x \quad (13.2)$$

By rearrangement of this equation we obtain:

$$\frac{m \cdot e(\mathrm{M}) + x \cdot e(\mathrm{X})}{x} = 8 + \frac{m[b(\mathrm{MM}) + E] - x \cdot b(\mathrm{XX})}{x} \qquad (13.3)$$

We define the *valence electron concentration per anion*, VEC(X), as the total number of *all* valence electrons in relation to the number of anionic atoms:

$$\mathrm{VEC(X)} = \frac{m \cdot e(\mathrm{M}) + x \cdot e(\mathrm{X})}{x} \qquad (13.4)$$

By substituting equation (13.3) into equation (13.4) and solving for b(XX) we obtain:

$$b(\mathrm{XX}) = 8 - \mathrm{VEC(X)} + \frac{m}{x}[b(\mathrm{MM}) + E] \qquad (13.5)$$

Equation (13.5) represents the generalized $8-N$ rule. Compared to the simple $8-N$ rule (p. 62), it is enlarged by the term $\frac{m}{x}[b(\mathrm{MM}) + E]$, and VEC(X) has taken the place of the main group number N. The following specialized cases are of importance:

1. Elements. For pure elements that belong to the right side of the ZINTL line, we have $m = 0$, VEC(X) = e(X) = N, and equation (13.5) becomes:

$$b(\mathrm{XX}) = 8 - \mathrm{VEC(X)} = 8 - N \qquad (13.6)$$

This is none other than the simple $8-N$ rule. For example, in sulfur ($N = 6$) the number of covalent bonds per S atom is $b(\mathrm{SS}) = 8 - N = 2$.

2. Polyanionic compounds. Frequently, the M atoms lose all their valence electrons to the X atoms, *i.e.* no cation–cation bonds occur and no nonbonding electrons remain at the cations, $b(\mathrm{MM}) = 0$ and $E = 0$. Equation (13.5) then becomes:

$$b(\mathrm{XX}) = 8 - \mathrm{VEC(X)} \qquad (13.7)$$

This once again is the $8-N$ rule, but only for the anionic component of the compound. For example: Na_2O_2; VEC(O) = 7; $b(\mathrm{OO}) = 8 - 7 = 1$, there is one covalent bond per O atom.

By comparing equations (13.6) and (13.7) we can deduce:

The geometric arrangement of the atoms in a polyanionic compound corresponds to the arrangement in the structures of the elements of the fourth to seventh main groups when the number of covalent bonds per atom b(XX) is equal. According to this concept, put forward by E. ZINTL and further developed by W. KLEMM and E. BUSMANN, the more electronegative partner in a compound is treated like that element which has the same number of electrons. This statement is therefore a specialized case of the general rule according to which isoelectronic atom groups adopt the same kind of structures.

3. Polycationic compounds. Provided that no covalent bonds occur between the anionic atoms, $b(\mathrm{XX})=0$, equation (13.5) becomes:

$$b(\mathrm{MM}) + E = \frac{x}{m}[\mathrm{VEC(X)} - 8] \qquad (13.8)$$

When applying this equation, note that for the calculation of VEC(X) according to equation (13.4) *all* valence electrons have to be considered, including those that take part in M–M bonds.

For example: Hg_2Cl_2; e(Hg) = 2; VEC(Cl) = 9; b(HgHg) = 1 [when the 10 d electrons of an Hg atom are also considered as being valence electrons, then VEC(Cl) = 19, E = 10, b(HgHg) = 1].

4. Simple ionic compounds, *i.e.* compounds having no covalent bonds, $b(\text{MM}) = b(\text{XX}) = E = 0$. Equation (13.5) becomes:

$$\text{VEC(X)} = 8$$

which is the octet rule.

We now can classify compounds according to the values of VEC(X). Since b(MM), E and b(XX) cannot adopt negative values, VEC(X) in equation (13.7) must be smaller than 8, and in equation (13.8) it must be greater than 8. We thus deduce the criterion:

$\text{VEC(X)} < 8$	polyanionic
$\text{VEC(X)} = 8$	simple ionic
$\text{VEC(X)} > 8$	polycationic

As VEC(X) is easy to calculate according to equation (13.4), we can quickly estimate the kind of bonding in a compound, for example:

polycationic	VEC(X)	polyanionic	VEC(X)	simple ionic	VEC(X)
Ti_2S	14	Ca_5Si_3	$7\frac{1}{3}$	Mg_2Sn	8
$MoCl_2$	10	Sr_2Sb_3	$6\frac{1}{3}$	Na_3P	8
$Cs_{11}O_3$	$9\frac{2}{3}$	CaSi	6	wrong:	
GaSe	9	KGe	5	InBi	8

As we can see from the last entry in this table, we have deduced only a *rule*. In InBi there are Bi–Bi contacts and it has metallic properties. Further examples that do not fulfill the rule are LiPb (Pb atoms surrounded only by Li) and K_8Ge_{46}. In the latter, all Ge atoms have four covalent bonds; they form a wide-meshed framework that encloses the K^+ ions (Fig. 16.26, p. 188); the electrons donated by the potassium atoms are not taken over by the germanium, and instead they form a band. In a way, this is a kind of a solid solution, with germanium as 'solvent' for K^+ and 'solvated' electrons. K_8Ge_{46} has metallic properties. In the sense of the $8-N$ rule the metallic electrons can be 'captured': in $K_8Ga_8Ge_{38}$, which has the same structure, all the electrons of the potassium are required for the framework, and it is a semiconductor. In spite of the exceptions, the concept has turned out to be very fruitful, especially in the context of understanding the ZINTL phases.

13.2 Polyanionic Compounds, Zintl Phases

Table 13.1 lists some binary polyanionic compounds, arranged according to the valence electron concentration per anion atom. Only compounds with integral values for VEC(X) are listed. In agreement with the above-mentioned rule, in fact structures like those of pure elements with the corresponding numbers of valence electrons occur for the anionic components. However, the variety of structures is considerably larger than for the pure elements. For example, three-bonded atoms occur not only in the layer structures as in phosphorus and arsenic, but also in several other connection patterns (Fig. 13.1). This seems reasonable, since the anionic grid has to make allowance for the space requirements of the cations. $CaSi_2$, for example, has layers $(Si^-)_\infty$ as in arsenic; the Ca^{2+} ions are located between the layers. $SrSi_2$, however, has a network structure in which helical chains with fourfold screw symmetry are interconnected; each Si atom has three bonds. Under pressure, both $CaSi_2$ and $SrSi_2$ are transformed to the α-$ThSi_2$ type, with yet another kind of network of three-bonded Si atoms. Contrary to expectations based on the $8-N$ rule, the

Table 13.1: Examples of polyanionic compounds which have integral valence electron concentrations per anion atom

Example	VEC(X)	b(XX)	structure of the anion part
Li_2S_2	7	1	S_2^{2-} pairs as in Cl_2
FeS_2 / FeAsS	7	1	S_2^{2-} and AsS^{3-} pairs
NiP	7	1	P_2^{4-} pairs
CaSi	6	2	zigzag chains
LiAs	6	2	helical chains
$CoAs_3$	6	2	four-membered As_4^{4-} rings
InP_3	6	2	P_6^{6-} rings (chair) as in S_6
$CaSi_2$	5	3	undulated layers as in α-As
$SrSi_2$	5	3	interconnected helical chains
K_4Ge_4	5	3	Ge_4^{4-} tetrahedra as in P_4
CaC_2	5	3	C_2^{2-} pairs as in N_2
NaTl	4	4	diamond-like
$SrGa_2$	4	4	graphite-like

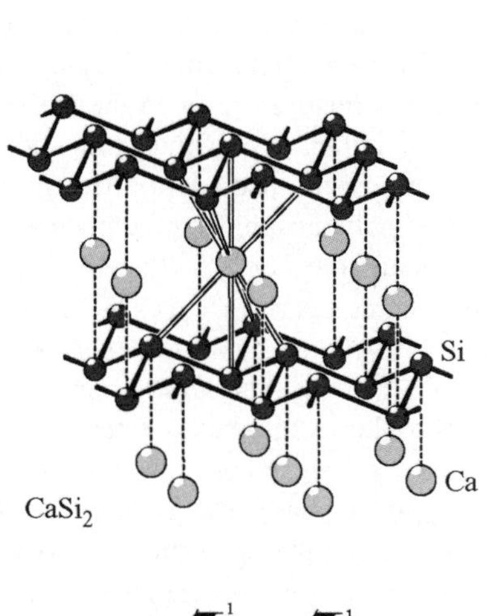

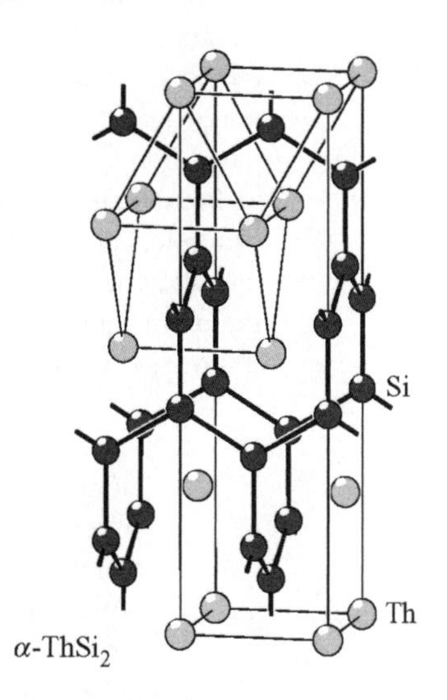

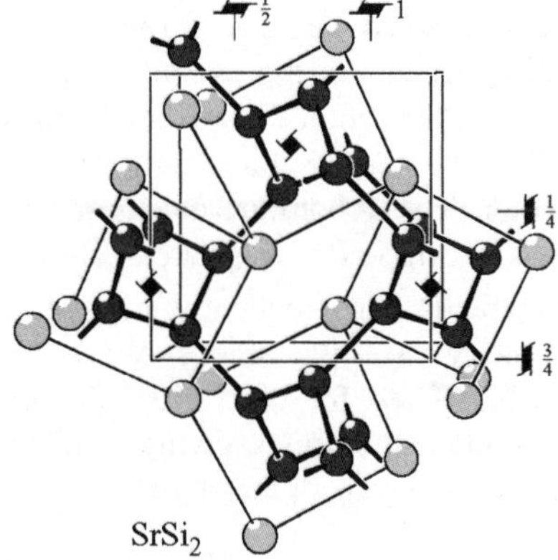

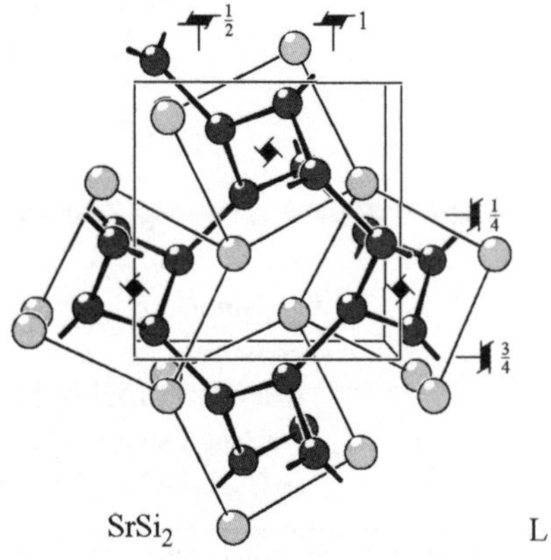

Fig. 13.1 Sections of the structures of some polysilicides with three-bonded Si atoms. In the stereo image for $SrSi_2$, the positions of the 4_3 screw axes of the cubic space group $P4_332$ are shown

Si atoms in α-$ThSi_2$ do not have pyramidal coordination, but planar coordination; in $SrSi_2$ the coordination is nearly planar. The Si atoms in the α-$ThSi_2$ type are located in the centers of trigonal prisms formed by the cations.

The calculation of VEC(X) for many compounds results in non-integral numbers. According to equation (13.7) fractional numbers then also result for the number b(XX) of covalent bonds. This happens when structurally different atoms occur in the anion. The following examples help to illustrate this:

Na_2S_3: with VEC(X) $= \frac{20}{3}$ we obtain b(XX) $= \frac{4}{3}$. This is due to the chain structure of the S_3^{2-} ion. For the two terminal atoms we have b(XX) = 1, and for the central one b(XX) = 2; the average is $(2 \times 1 + 2)/3 = \frac{4}{3}$. For unbranched chains with specific lengths as in polysulfides S_n^{2-}, 6 < VEC(X) < 7 holds as long as no multiple bonds occur. When there are multiple bonds, VEC(X) < 6 is possible, *e.g.* VEC(N) = 5.33 for the azide ion, ⟨N=N=N⟩ $^-$.

Ba_3Si_4: VEC(X) $= \frac{11}{2}$, b(XX) $= \frac{5}{2}$. An average value of $2\frac{1}{2}$ covalent bonds per Si atom results when half of the Si atoms are bonded with two covalent bonds, and the other half with three covalent bonds. This corresponds to the real structure.

The number of negative charges of the anion can also be counted in the following way: an atom of the N-th main group that participates in $8-N$ covalent bonds obtains a formal charge of zero; for every bond less than $8-N$ it obtains a negative formal charge. A four-bonded silicon atom thus obtains a formal charge of 0, a three-bonded one obtains 1⊖ and a two-bonded one obtains 2⊖. The sum of all formal charges is equal to the ionic charge.*

Sometimes rather complicated structures occur in the anionic part of a structure. For example, approximately 50 different binary polyphosphides are known only for the alkali and alkaline earth metals, which, in part, also adopt different modifications. In addition, there are more than 120 binary polyphosphides of other metals. Fig. 13.2 conveys an impression of how manifold the structures are.

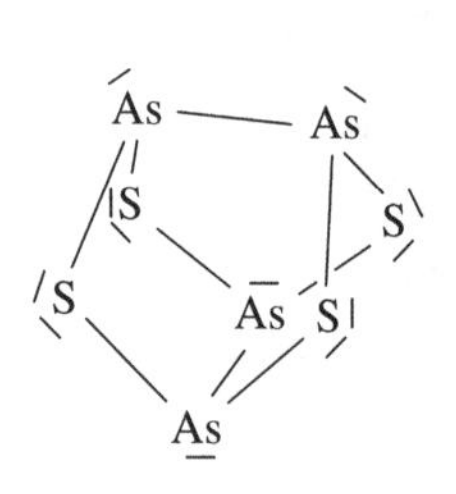

Apart from simple chains and rings, cages like those in sulfides such as As_4S_4 and P_4S_3 (and others) have been observed; every P atom that substitutes an S atom is to be taken as a $P^{\ominus}$. Layer structures can be regarded as sections of the structures of black phosphorus or arsenic. Other structures correspond to fragments of the structure of fibrous red phosphorus. The diversity in polyarsenides, polyantimonides and polysilicides is just as complicated. In addition, several different kinds of anions can be present simultaneously. For example, Ca_2As_3 or rather $Ca_8[As_4][As_8]$ contains unbranched chain-like As_4^{6-} and As_8^{10-} ions .

Atoms which are assigned negative formal charges in fact bear negative charges, as can be recognized in the overall structures: these atoms are those which are coordinated to the cations. For example, in NaP_5 there are four neutral P atoms for each $P^{\ominus}$. The neutral atoms form ribbons of connected rings having chair conformation. The ribbons are interconnected by individual $P^{\ominus}$ atoms (Fig. 13.2, bottom). Only these $P^{\ominus}$ atoms are in close contact with the Na^+ ions.

Binary polyanionic compounds can frequently be synthesized directly from the elements. In some cases, intact cage-like anions can be extracted from the solids when a complexing ligand is offered for the cation. For example, the Na^+ ions of Na_2Sn_5 can be captured by cryptand molecules, giving $[NaCrypt^+]_2Sn_5^{2-}$. Cryptands like $N(C_2H_4OC_2H_4OC_2H_4)_3N$ enclose the alkali metal ion.

*The signs of formal charges should always be encircled, ⊕ and ⊖, and the signs of ionic charges should never be encircled.

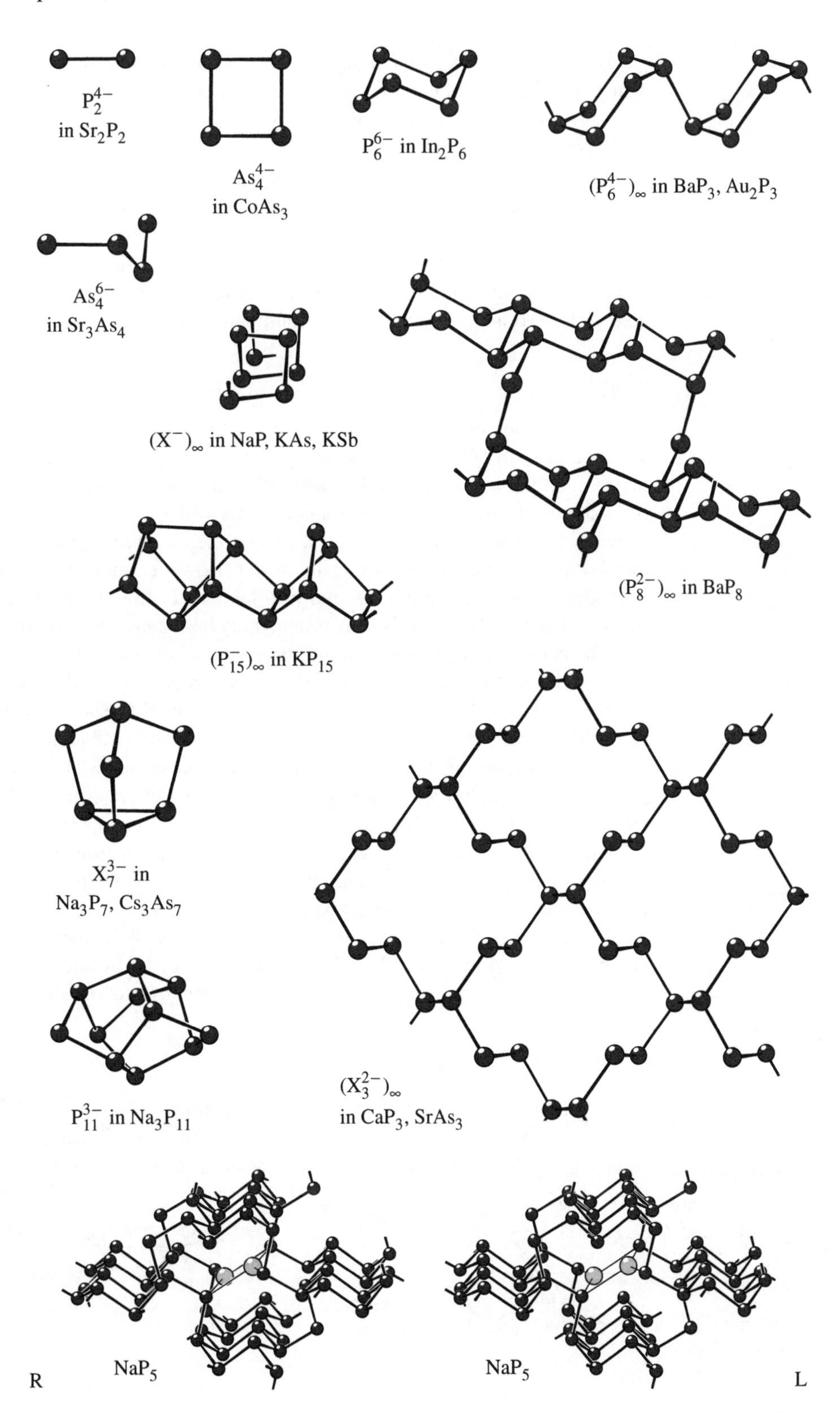

Fig. 13.2 Examples of the anionic structures in polyphosphides, polyarsenides and polyantimonides. For comparison, recall the structures of red and black phosphorus and of arsenic (pp. 108, 109 and 110). Stereo image for NaP_5

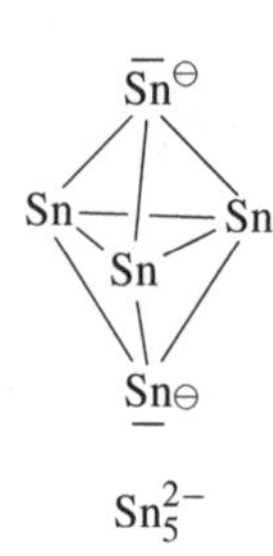

For some of the cage-like anions the kind of bonding is consistent with the preceding statements, but for some others they do not seem to apply. The ionic charges of As_7^{3-} or P_{11}^{3-} correspond exactly to the numbers of two-bonded $P^{\ominus}$ or $As^{\ominus}$ atoms (Fig. 13.2). For Sn_5^{2-} this is not so clear. The 22 valence electrons of the Sn_5^{2-} ion could be accommodated in exact agreement with the octet rule according to the formula given in the margin. However, calculations with the electron localization function show that lone electron pairs are also present at the equatorial atoms; therefore, only six electron pairs remain for the bonds. This corresponds to the number expected according to the WADE rules, as for boranes ($n+1$ multicenter bonds in a *closo* cluster with $n = 5$ vertices, *cf.* p. 144). We will deal with the bonding in such cluster compounds in Section 13.4.

Zintl Phases

Many of the compounds presented in the preceding paragraphs belong to the **Zintl phases**. This is a class of compounds consisting of an electropositive, cationic component (alkali metal, alkaline earth metal, lanthanoid) and an anionic component of main group elements of moderate electronegativity. The anionic part of the structure fulfills the simple concept of normal valence compounds. Nevertheless, the compounds are not salt-like, but have metallic properties, especially metallic luster. However, they are not 'full-value' metals; instead of being metallic-ductile, many of them are brittle. As far as the electrical properties have been studied, mostly semiconductivity has been found. There are many analogies with the half-metallic elements: in the structures of germanium, α-tin, arsenic, antimony, bismuth, selenium and tellurium the $8-N$ rule can be discerned; although these elements can be considered to be normal valence compounds, they show metallic luster, but they are brittle and are semiconductors or moderate metallic conductors.

The classic example of a ZINTL phase is the compound NaTl which can be interpreted as Na^+Tl^-; its thallium partial structure has the diamond structure (Fig. 13.3). In NaTl the Tl–Tl bonds are significantly shorter than the contact distances in metallic thallium (324 instead of 343 pm, albeit with a reduced coordination number). Although the valence electron concentration is the same, the Ga^- particles in $SrGa_2$ do not form a diamond-like structure, but layers as in graphite (AlB_2 type; AlB_2 itself does not fulfill the octet rule). MgB_2, which becomes superconducting below 39 K, has the same structure. All compounds listed in Table 13.1 with the exception of Li_2S_2 and CaC_2 are ZINTL phases (recall the golden luster of pyrite, FeS_2). The number of known ZINTL phases is enormous.

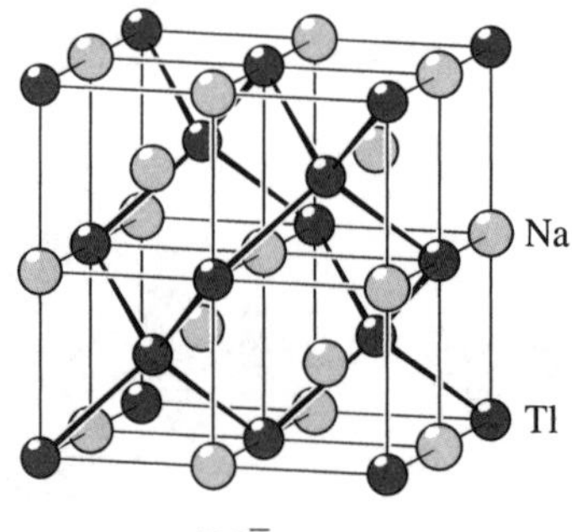

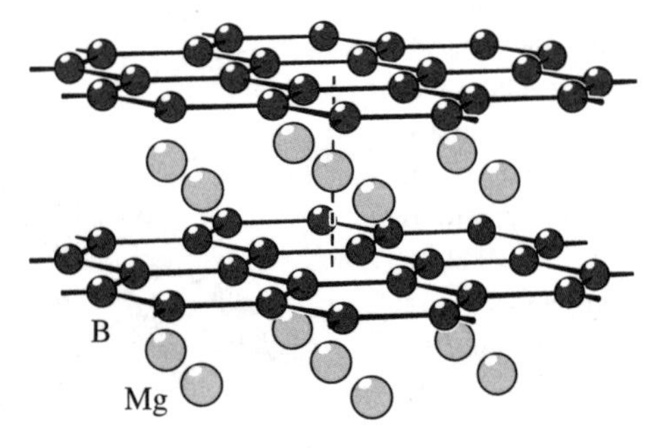

Fig. 13.3 Left: unit cell of NaTl. The plotted bonds of the thallium partial structure correspond to the C–C bonds in diamond. Right: section of the structure of $SrGa_2$ and MgB_2 (AlB_2 type)

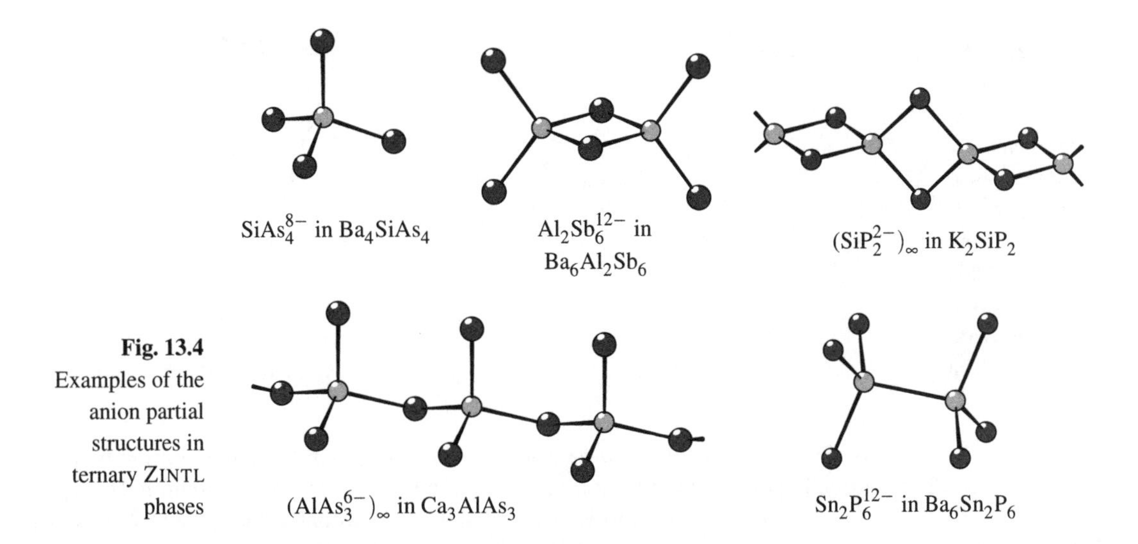

Fig. 13.4 Examples of the anion partial structures in ternary ZINTL phases

The structural principles of the elements can also be found in a number of ternary ZINTL phases. For example, KSnSb contains $(SnSb^-)_\infty$ layers as in α-arsenic. In other ternary ZINTL phases the anionic part of the structures resembles halo or oxo anions or molecular halides. For example, in Ba_4SiAs_4 there are tetrahedral $SiAs_4^{8-}$ particles that are isostructural to $SiBr_4$ molecules. In Ba_3AlSb_3 dimeric groups $Al_2Sb_6^{12-}$ are present, with a structure as in Al_2Cl_6 molecules (Fig. 13.4). Ca_3AlAs_3 contains polymeric chains of linked tetrahedra $(AlAs_3^{6-})_\infty$ as in chain silicates $(SiO_3^{2-})_\infty$. Instead of polymeric chains as in $(SiO_3^{2-})_\infty$, monomeric ions can occur that correspond to the carbonate ion CO_3^{2-}, such as SiP_3^{5-} ions in $Na_3K_2SiP_3$. The compound $Ca_{14}AlSb_{11}$ = $[Ca^{2+}]_{14}[Sb^{3-}]_4[Sb_3^{7-}][AlSb_4^{9-}]$ contains three kinds of anions, namely single ions Sb^{3-}, ions Sb_3^{7-} that are isostructural with I_3^-, and tetrahedral $AlSb_4^{9-}$ ions. $Ba_6Sn_2P_6$ has $Sn_2P_6^{12-}$ particles with an Sn–Sn bond; their structure is like that of ethane. Also, complicated chains and frameworks are known that are reminiscent of the manifold structures of the silicates; however, the possible varieties are far greater than for silicates because the anionic component is not restricted to the linking of SiO_4 tetrahedra.

The octet principle, primitive as it may appear, has not only been applied very successfully to the half-metallic ZINTL phases, but it is also theoretically well founded (requiring a lot of computational expenditure). Evading the purely metallic state with delocalized electrons in favor of electrons more localized in the anionic partial structure can be understood as the PEIERLS distortion (*cf.* Section 10.5).

Polyanionic Compounds that do not Fulfill the Octet Rule

The generalized $8-N$ rule can hold only as long as the atoms of the more electronegative element fulfill the octet principle. Especially for the heavier non-metals it is quite common for this principle not to be fulfilled. The corresponding atoms are termed *hypervalent*. The polyhalides offer an example. Among these the polyiodides show the largest variety. They can be regarded as association products of I_2 molecules and I^- ions, with a weakened bond in the I_2 molecules and a relatively weak bond between the I_2 and I^- (Fig. 13.5). The structures fulfill the GILLESPIE–NYHOLM rules.

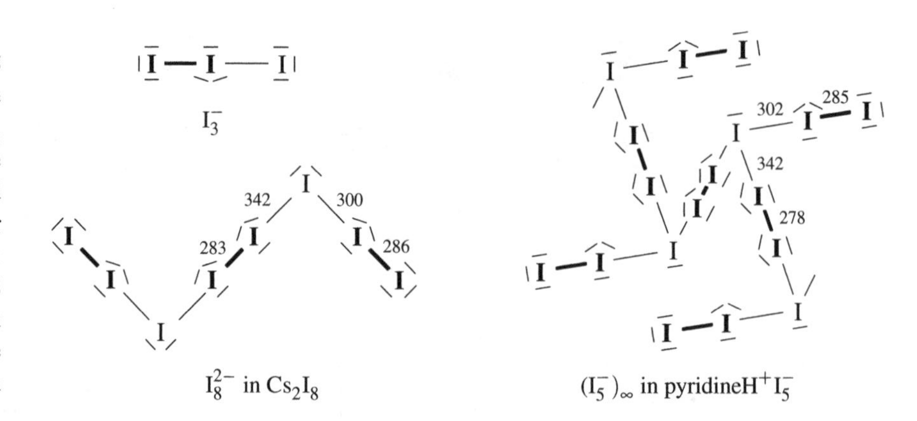

Fig. 13.5 Structures of some polyiodides. The I_2 building units are in bold face. Bond lengths in pm. For comparison: molecule I—I 268 pm, VAN DER WAALS distance I···I 396 pm

According to MO theory, a *three-center four-electron bond* accounts for the bonding. The central, hypervalent iodine atom in the I_3^- ion has an *s* orbital, two *p* orbitals perpendicular to the molecular axis and one *p* orbital in the molecular axis. The last-mentioned *p* orbital interacts with the corresponding *p* orbitals of the neighboring atoms. The situation is just as in the chain of atoms having *p* orbitals joined head-on (Fig. 10.5, p. 93), but the chain consists of only thee atoms. The result is one bonding, one nonbonding and one antibonding molecular orbital. Two electron pairs have to be placed in these three orbitals. The bonding orbital causes a bond between all three atoms, but it is relatively weak, since it must join three atoms. The two electron pairs correspond to the two bond lines in the valence-bond formula (Fig. 13.5). The valence-bond formula does not show that the bonds are weaker than normal single bonds (bond order $\frac{1}{2}$), but with the aid of the GILLESPIE–NYHOLM rules it yields the correct (linear) structure.

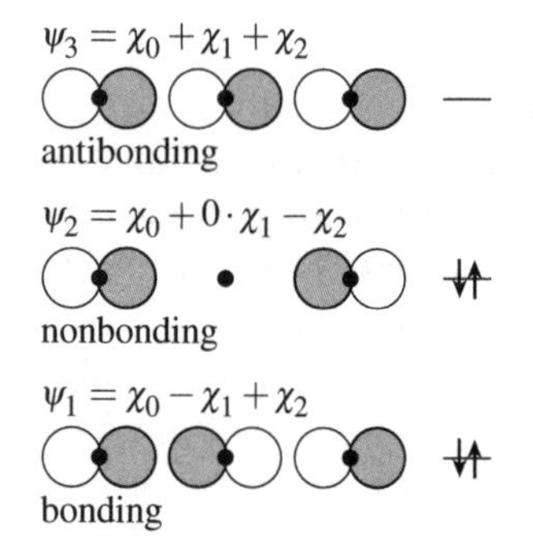

The GILLESPIE–NYHOLM rules usually can also be applied to other polyanionic compounds with hypervalent atoms. As an example, some polytellurides are depicted in Fig. 13.6. The Te_5^{6-} ion is square like the BrF_4^- ion.

In Li_2Sb we can assume Sb^{2-} particles with seven valence electrons. Therefore, we expect Sb_2^{4-} dumbbells (isoelectronic with I_2) and observance of the octet rule. In fact, such dumbbells are present in the structure (Sb–Sb bond length 297 pm); however, this applies only to half of the Sb atoms. The other half form linear chains of Sb atoms (Sb–Sb distance 326 pm). For the bonds in the chain we assume a band according to Fig. 10.5 (p. 93); every Sb atom contributes to this band with one *p* orbital and one electron. With one electron per Sb atom the band is half-occupied, and therefore it is bonding. The

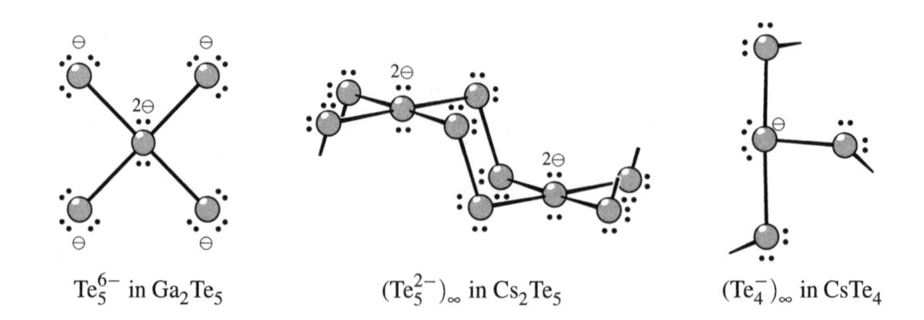

Fig. 13.6 Structures of some polytellurides. Lone electron pairs are marked by double dots. Te_5^{2-} also forms simple chain structures like S_5^{2-}

remaining six electrons occupy the *s* orbital and the other two *p* orbitals of the Sb atom and act as nonbonding lone electron pairs. In the mean we have one bonding electron per Sb–Sb bond, which corresponds to a bond order of $\frac{1}{2}$, just as in an I_3^- ion. We draw the conclusion: *Seven valence electrons per atom are needed for a linear chain of main-group atoms.* To express this with a valence-bond formula, we can use bonding dots instead of bonding lines (this does not mean that there are unpaired electrons). The GILLESPIE–NYHOLM rules can be applied with the aid of this formulation. The occurrence of both kinds of building blocks in Li_2Sb, chains and dumbbells, shows that in this case the PEIERLS distortion contributes only a minor stabilization and is partially overridden by other effects. The PEIERLS distortion cannot be suppressed that easily with lighter elements.

The formation of linear chains can be extended to two dimensions. Parallel $^1_\infty Sb^{2-}$ chains lying side by side can be joined to a square net. One more singly occupied *p* orbital per Sb atom is needed. Formally, an oxidation, $^1_\infty Sb^{2-} \xrightarrow{-e^-} {}^2_\infty Sb^-$, has to take place. Six valence electrons per atom are needed for the square net. Nets of this kind occur, for example, in $YbSb_2$ (with Yb^{2+}). Starting from the square nets, another formal oxidation, $^2_\infty Sb^- \xrightarrow{-e^-} {}^3_\infty Sb$, yields the primitive-cubic polonium-type structure, which is known as a high-pressure modification of arsenic. Therefore, five electrons per atom are needed for this structure. Remarkably, polonium itself has one electron per atom too many for its structure.

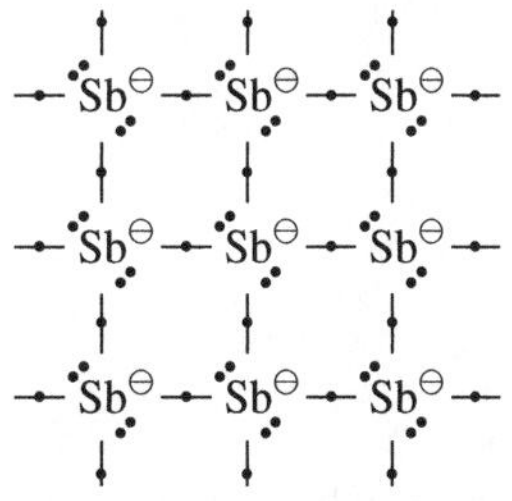

13.3 Polycationic Compounds

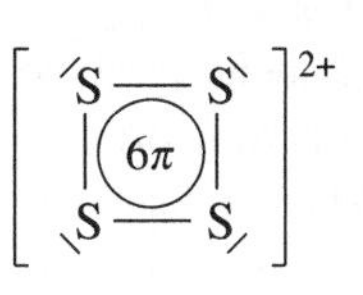

The number of known polycationic compounds of main group elements is far less than that of polyanionic compounds. Examples include the chalcogen cations S_4^{2+}, S_8^{2+}, Se_{10}^{2+} and Te_6^{4+} that are obtained when the elements react with Lewis acids under oxidizing conditions. The ions S_4^{2+}, Se_4^{2+} and Te_4^{2+} have a square structure that can be assumed to have a 6π electron system.

The structures of S_8^{2+} and Se_8^{2+} can be interpreted with the $8-N$ rule: a bond is generated across an S_8 ring, resulting in two atoms having three bonds and one positive formal charge each (Fig. 13.7). The new bond is remarkably long (289 pm as compared to 203 pm for the other bonds), but the occurrence of abnormally long S–S bonds is also known for some other sulfur compounds. Several varieties are known of Te_8^{2+} ions, in which triply bonded $Te^{\oplus}$ and uncharged Te atoms are discernible. The structure of the $Te_3S_3^{2+}$ and Te_6^{4+} ions can also be understood in terms of the $8-N$ rule. Te_6^{4+} can be described as a trigonal-pyramidal structure in which one prism edge has been strongly elongated; according to the $8-N$ rule, this edge would not be considered to be a bond. However, some weak bonding interaction must still be present, otherwise the structure would not be as it is. The $8-N$ rule is somewhat too simple, just as it is too simple to understand elemental tellurium. For the trigonal-prismatic Bi_5^{3+} ion, one could formulate a simple valence bond formula as given on page 134 for the isoelectronic Sn_5^{2-} ion (without lone electron pairs at the equatorial atoms). However, for the square-antiprismatic Bi_8^{2+} ion this is not possible. In this case multicenter bonds are needed, such as are used to describe the bonding in cluster compounds. In a broader sense numerous cluster compounds can be considered to be polycationic compounds; due to their variety we deal with them next in a section of their own.

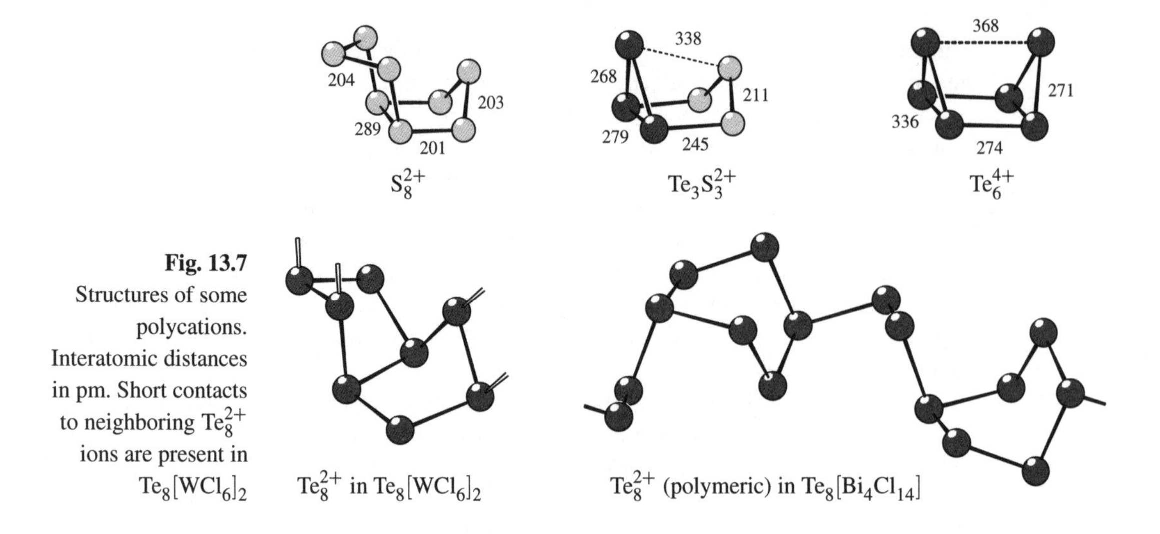

Fig. 13.7 Structures of some polycations. Interatomic distances in pm. Short contacts to neighboring Te_8^{2+} ions are present in $Te_8[WCl_6]_2$

13.4 Cluster Compounds

Links between atoms serve to compensate for the lack of the electrons which are necessary to attain the electron configuration of the next noble gas in the periodic table. With a common electron pair between two atoms each of them gains one electron in its valence shell. As the two electrons link two 'centers',* this is called a two-center two-electron bond or, for short, $2c2e$ bond. If, for an element, the number of available partner atoms of a different element is not sufficient to fill the valence shell, atoms of the same element combine with each other, as is the case for polyanionic compounds and for the numerous organic compounds. For the majority of polyanionic compounds a sufficient number of electrons is available to satisfy the demand for electrons with the aid of $2c2e$ bonds. Therefore, the generalized $8-N$ rule is usually fulfilled for polyanionic compounds.

For more electropositive elements, which have an inferior number of valence electrons in the first place, and which in addition have to supply electrons to a more electronegative partner, the number of available electrons is rather small. They can gain electrons in two ways: first, as far as possible, by complexation, *i.e.* by the acquisition of ligands; and second, by combining their own atoms with each other. This can result in the formation of clusters. A *cluster* is an accumulation of three or more atoms of the same element or of similar elements that are directly linked with each other. If the accumulation of atoms yields a sufficient number of electrons to allow for one electron pair for every connecting line between two adjacent atoms, then each of these lines can be taken to be a $2c2e$ bond just as in a common valence bond (LEWIS) formula. Clusters of this kind have been called *electron precise*.

For low values of the valence electron concentration (VEC$<$ 4 for main group elements), covalent $2c2e$ bonds are not sufficient to overcome the electron deficiency. We have the case of 'electron-deficient compounds'. For these, relief comes from *multicenter bonds*. In a three-center two-electron bond ($3c2e$) three atoms share an electron pair. An even larger number of atoms can share one electron pair. With increasing numbers of

*In recent times, in chemistry it has become a fatuous habit to call atoms 'centers'. See comments on page 247.

atoms sharing the same electron pair, each atom is less tightly bonded. The electron pair in a $3c2e$ bond essentially is located in the center of the triangle defined by the three atoms:

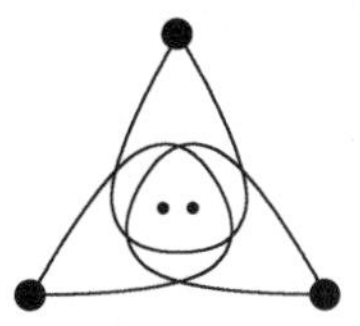

The location of electrons linking more than three atoms cannot be illustrated as easily. The simple, descriptive models must give way to the theoretical treatment by molecular orbital theory. With its aid, however, certain electron counting rules have been deduced for cluster compounds that set up relations between the structure and the number of valence electrons. A bridge between molecular-orbital theory and vividness is offered by the electron-localization function (*cf.* p. 89).

Completely closed, convex, single-shell clusters are called *closo* clusters; their atoms form a polyhedron. If the polyhedron has only triangular faces, it is also called a *deltahedron*. Depending on the number of available electrons, we can distinguish four general bonding types for *closo* clusters:

1. Electron precise clusters with exactly one electron pair per polyhedron edge;
2. Clusters with one $3c2e$ bond for every triangular face;
3. Clusters that satisfy the WADE rules discussed on page 144;
4. Clusters not matching any of these patterns.

Electron Precise Clusters

Molecules such as P_4 and the polyanionic clusters such as Si_4^{4-} or As_7^{3-} that are discussed in Section 13.2 are representatives of electron precise closo clusters. Organic cage molecules like tetrahedrane (C_4R_4), prismane (C_6H_6), cubane (C_8H_8), and dodecahedrane ($C_{20}H_{20}$) also belong to this kind of cluster.

Numerous clusters with electron numbers that account for exactly one electron pair per polyhedron edge are also known for the more electron-rich transition group elements (beginning with group six). In addition, every cluster atom obtains electrons from coordinated ligands, with a tendency to attain a total of 18 valence electrons per atom. The easiest way to count the number of electrons is to start from uncharged metal atoms and uncharged ligands. Ligands such as NH_3, PR_3, and CO supply two electrons. Nonbridging halogen atoms, H atoms and groups such as SiR_3 supply one electron (for halogen atoms this amounts to the same as assuming a Hal^- ligand that makes available two electrons, but that had previously obtained an electron from a metal atom). A μ_2-bridging halogen atom supplies three electrons (one as before plus one of its lone electron pairs), and a μ_3-bridging halogen atom five. Table 13.2 lists how many electrons are to be taken into account for some ligands.

The electrons supplied by the ligands and the valence electrons of the n metal atoms of an M_n cluster are added to a total electron number g. The number of M–M bonds (polyhedron edges) then is:

$$\text{main group element clusters:} \quad b = \tfrac{1}{2}(8n-g) \tag{13.9}$$

$$\text{transition element clusters:} \quad b = \tfrac{1}{2}(18n-g) \tag{13.10}$$

Table 13.2: Number of electrons supplied by ligands to metal atoms in complexes when the metal atoms are considered to be uncharged.
μ_1 = terminal ligand, μ_2 = ligand bridging two atoms, μ_3 = ligand bridging three atoms; *int* = interstitial atom inside a cluster

ligand		electrons	ligand		electrons
H	μ_1	1	NR_3	μ_1	2
H	μ_2	1	NCR	μ_1	2
H	μ_3	1	NO	μ_1	3
CO	μ_1	2	PR_3	μ_1	2
CO	μ_2	2	OR	μ_1	1
CS	μ_1	2	OR	μ_2	3
CR_2	μ_1	2	OR_2	μ_1	2
η^2-C_2R_4	μ_1	2	O, S, Se, Te	μ_1	0
η^2-C_2R_2	μ_1	2	O, S, Se, Te	μ_2	2
η^5-C_5R_5	μ_1	5	O, S, Se, Te	μ_3	4
η^6-C_6R_6	μ_1	6	O, S	*int*	6
C	*int*	4	F, Cl, Br, I	μ_1	1
SiR_3	μ_1	2	F, Cl, Br, I	μ_2	3
N, P	*int*	5	Cl, Br, I	μ_3	5

This mode of calculation has been called the 'EAN rule' (effective atomic number rule). It is valid for arbitrary metal clusters (*closo* and others) if the number of electrons is sufficient to assign one electron pair for every M–M connecting line between adjacent atoms, and if the octet rule or the 18-electron rule is fulfilled for main group elements or for transition group elements, respectively. The number of bonds b calculated in this way is a limiting value: the number of polyhedron edges in the cluster can be greater than or equal to b, but never smaller. If it is equal, the cluster is electron precise.

Since an M atom gains one electron per M–M bond, the calculation can also be performed in the following way: the total number g of valence electrons of the cluster must be equal to:

$$\text{main group element clusters:} \quad g = 7n_1 + 6n_2 + 5n_3 + 4n_4 \qquad (13.11)$$

$$\text{transition element clusters:} \quad g = 17n_1 + 16n_2 + 15n_3 + 14n_4 \qquad (13.12)$$

n_1, n_2, n_3, and n_4 are the numbers of polyhedron vertices at which 1, 2, 3 or 4 polyhedron edges (M–M bonds) meet, respectively. Polyhedra with five or more edges per vertex are generally not electron precise (for this reason no numbers $n_5, n_6, \ldots$ occur in the equations). Therefore, the expected valence electron numbers for some simple polyhedra are:

	main group elements	transition group elements
triangle	18	48
tetrahedron	20	60
trigonal bipyramid	22	72
octahedron	–	84
trigonal prism	30	90
cube	40	120

No value is given for the octahedron in the list for the main group elements, because their octahedral clusters do not fit into the scheme of electron precise clusters. This is explained below for Tl_6^{6-} (p. 146). As an exercise, one could calculate the numbers for some of the polyanionic compounds in Section 13.2. Further examples include:

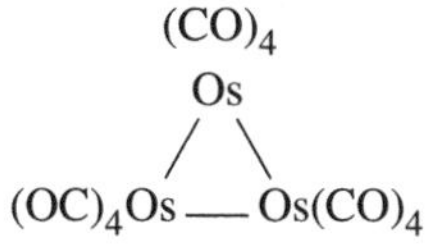

$(CO)_4Os$ / $(OC)_4Os$—$Os(CO)_4$

$Os_3(CO)_{12}$	3 Os	$3 \times 8 = 24$	
	12 CO	$12 \times 2 = 24$	
	$g =$	48	$= 16n_2$

$b = \frac{1}{2}(18 \times 3 - 48) = 3$

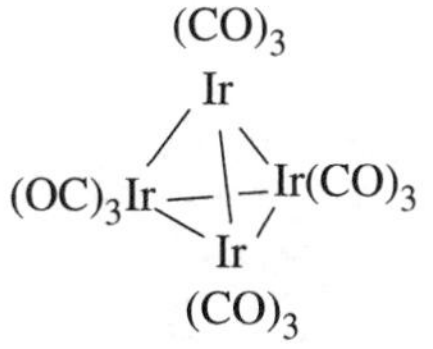

$(CO)_3Ir$ / $(OC)_3Ir$—$Ir(CO)_3$ / $Ir(CO)_3$

$Ir_4(CO)_{12}$	4 Ir	$4 \times 9 = 36$	
	12 CO	$12 \times 2 = 24$	
	$g =$	60	$= 15n_3$

$b = \frac{1}{2}(18 \times 4 - 60) = 6$

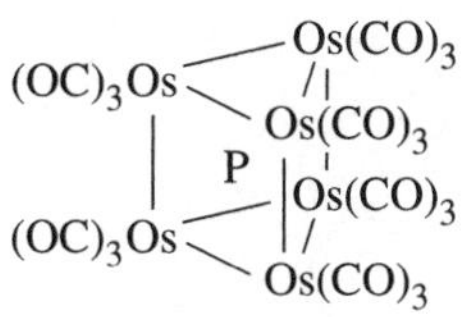

$(OC)_3Os$, $Os(CO)_3$, $Os(CO)_3$, P, $Os(CO)_3$, $(OC)_3Os$, $Os(CO)_3$

$[Os_6(CO)_{18}P]^-$	6 Os	$6 \times 8 = 48$	
	18 CO	$18 \times 2 = 36$	
	P	5	
	charge	1	
	$g =$	90	$= 15n_3$

$b = \frac{1}{2}(18 \times 6 - 90) = 9$

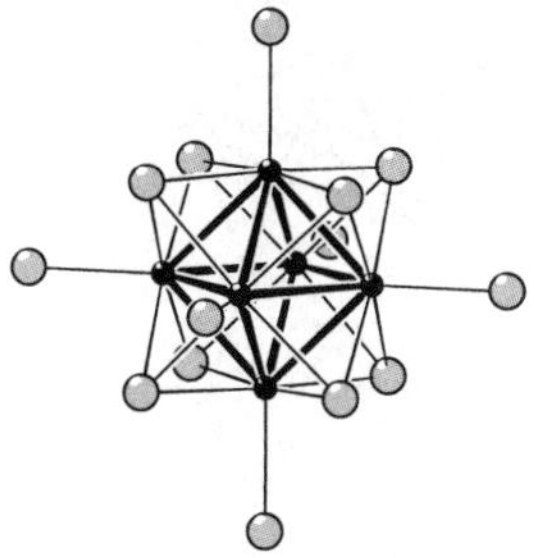

$[Mo_6Cl_{14}]^{2-}$	6 Mo	$6 \times 6 = 36$	
	8 μ_3-Cl	$8 \times 5 = 40$	
	6 μ_1-Cl	$6 \times 1 = 6$	
	charge	2	
	$g =$	84	$= 14n_4$

$b = \frac{1}{2}(18 \times 6 - 84) = 12$

The cluster mentioned last, $[Mo_6Cl_{14}]^{2-}$, also occurs in $MoCl_2$. It consists of an Mo_6 octahedron inscribed in a Cl_8 cube; each of the eight Cl atoms of the cube is situated on top of one of the octahedron faces and is coordinated to three molybdenum atoms (Fig. 13.8). The formula $[Mo_6Cl_8]^{4+}$ applies to this unit; in it, every Mo atom is still short of two electrons it needs to attain 18 valence electrons. They are supplied by the six Cl^- ions bonded at each octahedron vertex. This also applies to $MoCl_2$, but there are only four Cl^- per cluster; however, two of them act as bridging ligands between clusters, corresponding to the formula $[Mo_6Cl_8]Cl_{2/1}Cl_{4/2}$ (Fig. 13.8).

The situation is very similar in the CHEVREL phases. These are ternary molybdenum chalcogenides $A_x[Mo_6X_8]$ (A = metal, X = S, Se) that have attracted much attention because of their physical properties, especially as superconductors. The 'parent compound' is $PbMo_6S_8$; it contains Mo_6S_8 clusters that are linked with each other in such a way that the free coordination sites of one cluster are occupied by sulfur atoms of adjacent clusters (Fig. 13.9). The electric properties of CHEVREL phases depend on the number of valence electrons. With 24 electrons per cluster (one electron pair for each edge of the

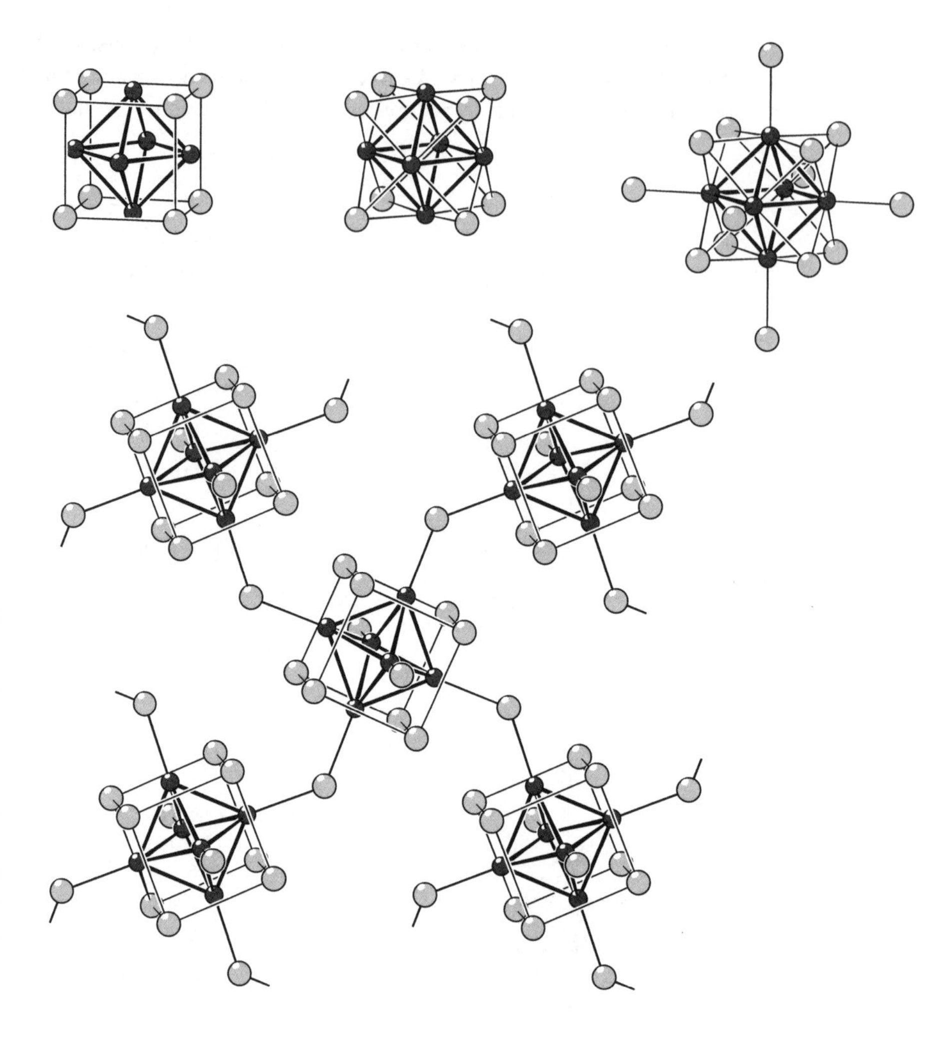

Fig. 13.8 Top: two representations of the $[Mo_6Cl_8]^{4+}$ cluster and the structure of the $[Mo_6Cl_{14}]^{2-}$ ion. Bottom: array of Mo_6Cl_8 clusters linked via chlorine atoms to a layer in Mo_6Cl_{12}

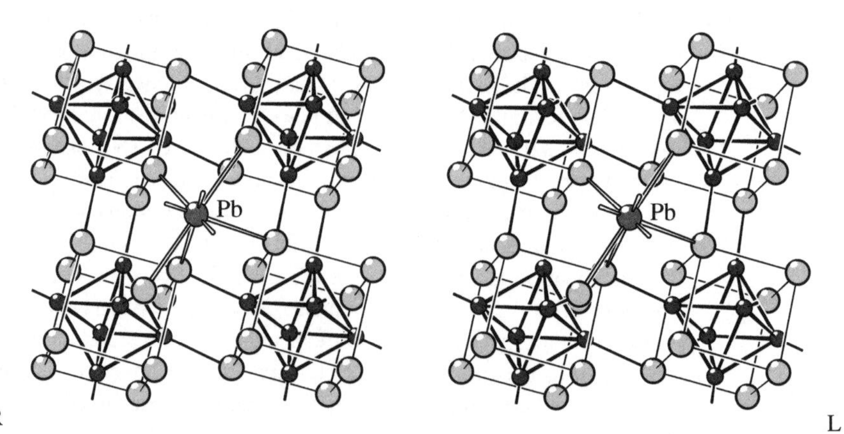

Fig. 13.9 Association of Mo_6S_8 clusters in the CHEVREL phase $PbMo_6S_8$ (stereo image)

Mo_6 octahedron) the cluster is electron-precise, the valence band is fully occupied and the compounds are semiconductors, as, for example, $(Mo_4Ru_2)Se_8$ (it has two Mo atoms substituted by Ru atoms in the cluster). In $PbMo_6S_8$ there are only 22 electrons per cluster; the 'electron holes' facilitate a better electrical conductivity; below 14 K it becomes a superconductor. By incorporating other elements in the cluster and by the choice of the electron-donating element A, the number of electrons in the cluster can be varied within certain limits (19 to 24 electrons for the octahedral skeleton). With the lower electron numbers the weakened cluster bonds show up in trigonally elongated octahedra.

If electrons are added to an electron precise cluster, cleavage of bonds is to be expected according to equation (13.9) or (13.10); for every additional electron pair g increases by 2 and b decreases by 1. The Si_4^{6-} ion presented on p. 132 is an example; it can be thought of having been formed from a tetrahedral Si_4^{4-} by the addition of two electrons. Another example is $Os_3(CO)_{12}(SiCl_3)_2$ with a linear Os–Os–Os group; by attaching two $SiCl_3$ groups to triangular $Os_3(CO)_{12}$, two more electrons are supplied, and one Os–Os bond has to be cleaved.

However, certain polyhedra allow the inclusion of another electron pair without cleavage of any bond. This applies especially to octahedral clusters which should have 84 valence electrons according to equation (13.12), but they frequently have 86 electrons. The additional electron pair assumes a bonding action as a six-center bond inside the octahedron. An octahedral cluster with 86 valence electrons fulfills the WADE rule discussed below.

Clusters with 3*c*2*e* Bonds

If there are not enough electrons for all of the polyhedron edges, 3*c*2*e* bonds on the triangular polyhedron faces can be the next best solution to compensate for the lack of electrons. This solution is only possible for deltahedra that have no more than four edges (and faces) meeting at any vertex. These include especially the tetrahedron, trigonal bipyramid and octahedron.

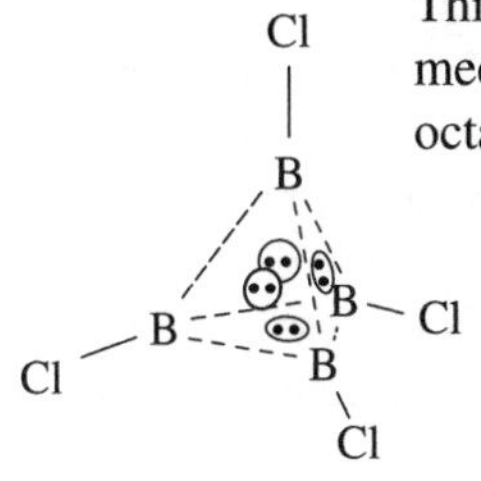

For example, the bonding in B_4Cl_4 can be interpreted in the following way: every boron atom takes part in four bonds, one 2*c*2*e* B–Cl bond and three 3*c*2*e* bonds on the faces of the B_4 tetrahedron. In this way every boron atom attains an electron octet. Eight of the valence electrons take part in the multicenter bonds; the other eight are needed for the B–Cl bonds.

In the $Nb_6Cl_{18}^{4-}$ ion the octahedral Nb_6 cluster can be assumed to have eight 3*c*2*e* bonds on its eight octahedron faces. A chlorine atom bonded with two Nb atoms is situated next to each octahedron edge. This makes twelve Cl atoms in an $Nb_6Cl_{12}^{2+}$ unit. The remaining six Cl^- ions are terminally bonded to the octahedron vertices (Fig. 13.10). The number of valence electrons is:

6 Nb	$6 \times 5 =$	30
12 μ_2-Cl	$12 \times 3 =$	36
6 μ_1-Cl	$6 \times 1 =$	6
charge		4
		76

12 of these 76 electrons are needed for the bonds with the μ_1-Cl atoms. Four electrons are needed for every Cl atom on top of an octahedron edge, altogether $4 \times 12 = 48$. $76 - 12 - 48 = 16$ electrons remain for the Nb_6 skeleton, *i.e.* exactly one electron pair per octahedron face.

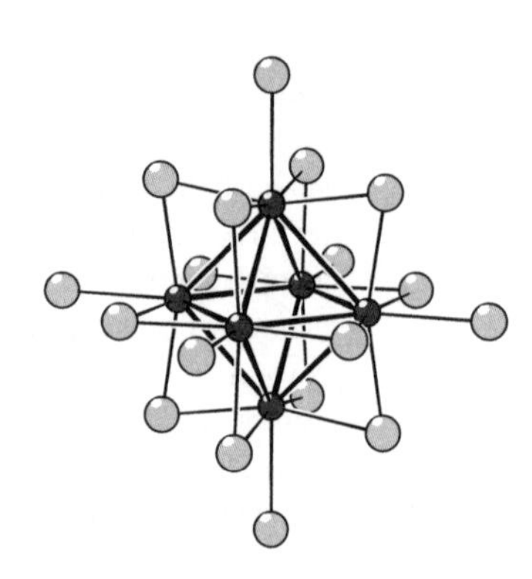

Fig. 13.10 Structures of the $Nb_6Cl_{12}^{2+}$ cluster and the $Nb_6Cl_{18}^{4-}$ ion

For each Nb atom the situation is the same as in the $Mo_6Cl_{14}^{2-}$ ion: the metal atom is surrounded by five Cl atoms and is involved in four metal–metal bonds in the cluster. However, the MCl_5 unit is rotated with respect to the octahedron: Cl atoms on top of the Mo_6 octahedron faces become Cl atoms on top of the Nb_6 octahedron edges, and the bonding electron pairs switch over from the edges to the faces. In both cases the valence electrons for a metal atom add up to 18. In Nb_6Cl_{14} the $Nb_6Cl_{12}^{2+}$ clusters are associated via intervening chlorine atoms, similar to Mo_6Cl_{12}.

Just as the Mo_6X_8 units in the CHEVREL phases tolerate a certain lack of electrons (*e.g.* 20 instead of 24 skeleton electrons), clusters with M_6X_{12} units which have fewer than 16 skeleton electrons are also possible. For example, in Zr_6I_{12} there are only 12 skeleton electrons, and $Sc_7Cl_{12} = Sc^{3+}[Sc_6Cl_{12}]^{3-}$ has only nine.

Wade Clusters

K. WADE has put forward some rules that relate the composition of a cluster to the number of its valence electrons. The rules were first derived for boranes. To calculate the wave functions of a *closo* cluster with n atoms, the coordinate systems of all n atoms are oriented with their z axes radially to the center of the polyhedron. The contribution of the s orbitals can be estimated best by combining them with the p_z orbitals to form sp hybrid orbitals. One of the two sp orbitals of an atom points radially to the center of the cluster, the other one radially outwards. The latter is used for bonding with external atoms (*e.g.* with the H atoms of the $B_6H_6^{2-}$ ion). The n sp orbitals pointing inwards combine to give one bonding and $n-1$ nonbonding or antibonding orbitals. The orbitals p_x and p_y of every atom are oriented tangentially to the cluster and combine to give n bonding and n antibonding orbitals (Fig. 13.11). Altogether, we obtain $n+1$ bonding orbitals for the cluster skeleton. From this follows the WADE rule: *a stable closo cluster requires* $2n+2$ *skeleton electrons.* This is a smaller number of electrons than that required for an electron precise cluster or for a cluster with $3c2e$ bonds, with one exception: a tetrahedral cluster with $3c2e$ bonds on its four faces requires only 8 electrons, whereas it should have 10 electrons according to the WADE rule; the WADE rule does not apply to tetrahedra. In fact, *closo*-boranes with the composition $B_nH_n^{2-}$ are known only for $n \geq 5$. For a trigonal bipyramid it makes no difference whether one assumes $3c2e$ bonds on the six faces or $n+1 = 6$ electron pairs according to the WADE rule.

According to calculations with the electron localization function (ELF) the electron pairs of the $B_6H_6^{2-}$ cluster are essentially concentrated on top of the octahedron edges and faces (Fig. 13.12).

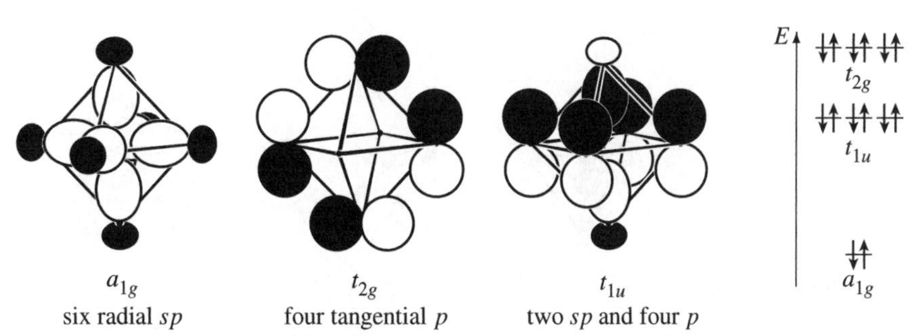

Fig. 13.11 Combinations of atomic orbitals that result in bonding molecular orbitals in an octahedral cluster such as $B_6H_6^{2-}$. For the triply degenerate orbitals t_{2g} and t_{1u} only one of each is plotted; for each of them, two further, equal orbitals exist which are oriented along the other two octahedron axes. Right: energy sequence of the seven occupied bonding orbitals

The *closo*-boranes $B_nH_n^{2-}$ ($5 \leq n \leq 12$) and the carboranes $B_nC_2H_{n+2}$ are showpieces for the mentioned WADE rule. Further examples include the B_{12} icosahedra in elemental boron (Fig. 11.16) and certain borides such as CaB_6. In CaB_6, B_6 octahedra are linked with each other via normal 2*c*2*e* bonds (Fig. 13.13). Six electrons per octahedron are required for these bonds; together with the $2n + 2 = 14$ electrons for the octahedron skeleton this adds up to a total of 20 valence electrons. The boron atoms supply $3 \times 6 = 18$ of them, and calcium the remaining two.

WADE stated some further rules for open clusters that are interpreted as deltahedra with missing vertices. They are of special importance for boranes:

nido cluster: one missing polyhedron vertex, $n + 2$ bonding skeleton orbitals;
arachno cluster: two missing vertices, $n + 3$ bonding skeleton orbitals;
hypho cluster: three missing vertices, $n + 4$ bonding skeleton orbitals.

The WADE rules can be applied to ligand-free cluster compounds of main-group elements. If we postulate one lone electron pair pointing outwards on each of the n atoms, then $g - 2n$ electrons remain for the polyhedron skeleton (g = total number of valence elec-

Fig. 13.12
Electron localization function for $B_6H_6^{2-}$ (only valence electrons, without regions around the H atoms), shown as iso-surface with ELF = 0.80. (Reprinted from *Angewandte Chemie* [97] with permission from Wiley-VCH)

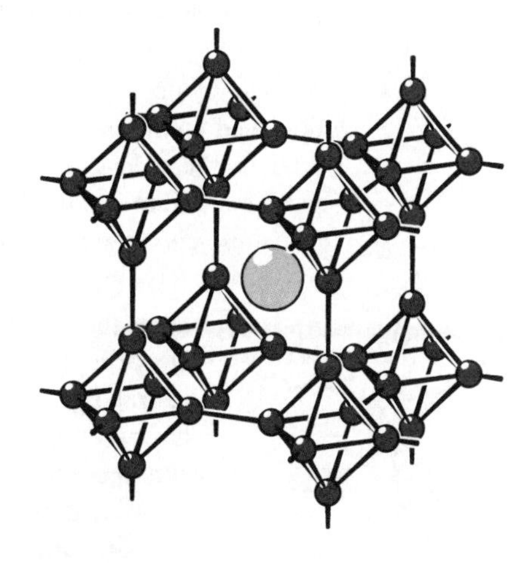

Fig. 13.13
The structure of CaB_6

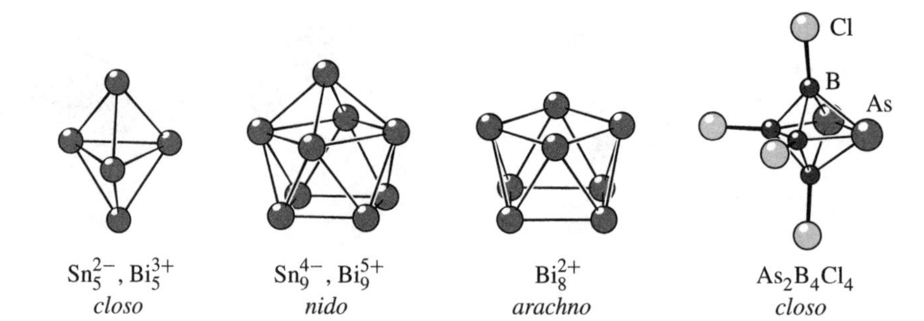

Fig. 13.14 Some WADE clusters: Sn_5^{2-}, Bi_5^{3+} *closo*; Sn_9^{4-}, Bi_9^{5+} *nido*; Bi_8^{2+} *arachno*; $As_2B_4Cl_4$ *closo*

trons; Fig. 13.14). The calculation also works if some of the atoms bear ligands (instead of lone pairs) and others have no ligands but lone electron pairs. Examples:

	n	g	$g-2n$		cluster type
Sn_5^{2-}, Bi_5^{3+}	5	22	12	$= 2n+2$	*closo*
Tl_6^{8-}	6	26	14	$= 2n+2$	*closo*
Sn_9^{4-}, Bi_9^{5+}	9	40	22	$= 2n+4$	*nido*
Bi_8^{2+}	8	38	22	$= 2n+6$	*arachno*
$As_2B_4Cl_4$	6	26	14	$= 2n+2$	*closo*

The examples should not give the misleading impression that bonding in clusters is a clear and simple matter. Next to many examples for which the WADE rules work well, they do not do so in many other cases, or additional assumptions have to be made.

KTl does not have the NaTl structure because the K^+ ions are too large to fit into the interstices of the diamond-like Tl^- framework. It is a cluster compound K_6Tl_6 with distorted octahedral Tl_6^{6-} ions. A Tl_6^{6-} ion could be formulated as an electron precise octahedral cluster, with 24 skeleton electrons and four $2c2e$ bonds per octahedron vertex. The thallium atoms then would have no lone electron pairs, the outside of the octahedron would have nearly no valence electron density, and there would be no reason for the distortion of the octahedron. Taken as a *closo* cluster with one lone electron pair per Tl atom, it should have two more electrons. If we assume bonding as in the $B_6H_6^{2-}$ ion (Fig. 13.11), but occupy the t_{2g} orbitals with only four instead of six electrons, we can understand the observed compression of the octahedra as a JAHN–TELLER distortion. Clusters of this kind, that have less electrons than expected according to the WADE rules, are known with gallium, indium and thallium. They are called hypoelectronic clusters; their skeleton electron numbers often are $2n$ or $2n-4$.

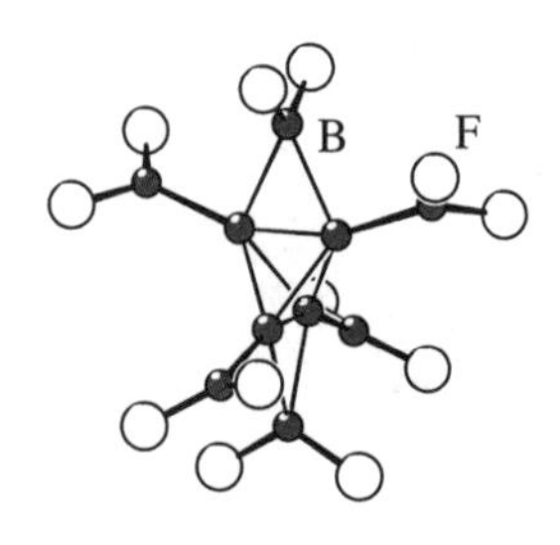

B_8Cl_8 has a dodecahedral B_8 *closo*-skeleton with $2n = 16$ electrons. In this case, the WADE rule neither can be applied, nor can it be interpreted as an electron precise cluster nor as a cluster with $3c2e$ bonds. $B_4(BF_2)_6$ has a tetrahedral B_4 skeleton with a radially bonded BF_2 ligand at each vertex, but it has two more BF_2 groups bonded to two tetrahedron edges. In such cases the simple electron counting rules fail.

WADE also extended the application of his rules to transition metal clusters; the further extension by D. M. P. MINGOS mainly concerns the bonding in metal carbonyl and metal phosphane clusters, *i.e.* organometallic compounds (WADE–MINGOS rules); these are beyond the scope of this book.

Clusters with Interstitial Atoms

Clusters derived from metals which have only a few valence electrons can relieve their electron deficit by incorporating atoms inside. This is an option especially for octahedral clusters which are able to enclose a binding electron pair anyway. The interstitial atom usually contributes all of its valence electrons to the electron balance. Nonmetal atoms such as H, B, C, N, and Si as well as metal atoms such as Be, Al, Mn, Fe, Co, and Ir have been found as interstitial atoms.

Transition metals of groups 3 and 4 form many octahedral clusters that are isostructural with those of the less electron-deficient elements of the following groups, but they contain additional atoms in their centers (Fig. 13.15). Starting from the above-mentioned Nb_6Cl_{14} (Fig. 13.10), we can substitute the niobium atoms by zirconium atoms; the number of available electrons is then reduced by six. This loss can partly be compensated by introducing a carbon atom in the Zr_6 octahedron. Despite the slightly inferior number of electrons the cluster in Zr_6CCl_{14} is stable due to some changes in the bonding. The more electronegative atom in the center of the cluster pulls electron density inwards, thus weakening the Zr–Zr bonds to some extent, but stronger bonding interactions with the C atom emerge.

On the other hand, the metal–metal bonds are strengthened when the interstitial atom is a metal atom. Nb_6F_{15}, for example, consists of Nb_6F_{12} clusters of the same kind as in the $Nb_6Cl_{12}^{2+}$ unit; they are linked by all six of their vertices via bridging fluorine atoms, forming a network. Th_6FeBr_{15} has the same kind of structure, but with an additional Fe atom in the octahedron center (Fig. 13.15). Nb_6F_{15} has one electron less than required for the eight $3c2e$ bonds; in Th_6Br_{15} a further six electrons are missing. The intercalated Fe atom (d^8) supplies these seven electrons; the eighth electron remains with the Fe atom.

Even the extremely electron-deficient alkali metals can form clusters when interstitial atoms contribute to their stabilization. Compounds of this kind are the alkali metal suboxides such as Rb_9O_2; it has two octahedra sharing a common face, and each is occupied by one O atom (Fig. 13.16). However, the electron deficiency is so severe that metallic bonding is needed between the clusters. In a way, these compounds are metals, but not with single metal ions as in the pure metal Rb^+e^-, but with a constitution $[Rb_9O_2]^{5+}(e^-)_5$, essentially with ionic bonding in the cluster.

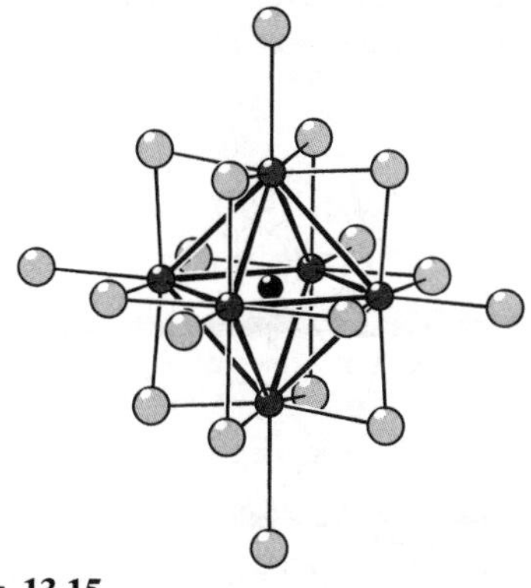

Fig. 13.15
Cluster unit with an interstitial atom in compounds such as Zr_6CCl_{14} and Th_6FeBr_{15}

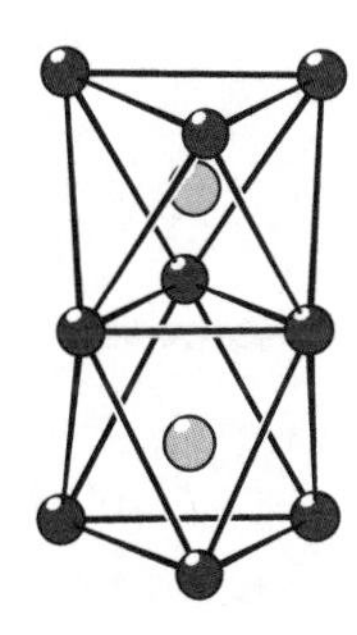

Fig. 13.16
Cluster in Rb_9O_2

I contrast to cages like $B_{12}X_{12}^{2-}$ or C_{60}, clusters with similar sizes consisting of metal atoms are not stable if they are hollow; the bonds at their surfaces are too weak. However, they can be stabilized by interstitial atoms, even if the interstitial atoms do not contribute with their electrons. Such clusters are called endohedral. Examples are the icosahedral

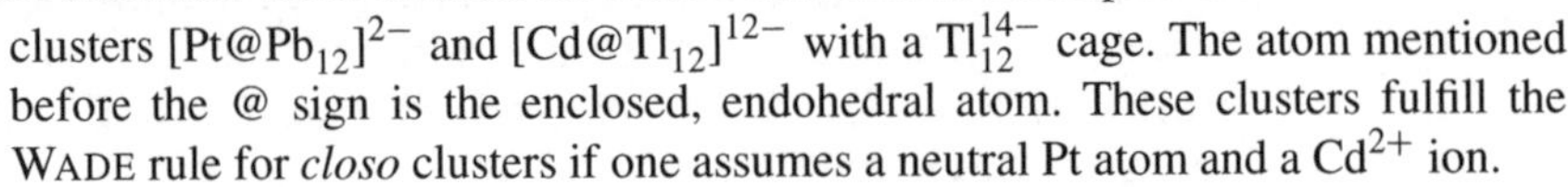

clusters $[Pt@Pb_{12}]^{2-}$ and $[Cd@Tl_{12}]^{12-}$ with a Tl_{12}^{14-} cage. The atom mentioned before the @ sign is the enclosed, endohedral atom. These clusters fulfill the WADE rule for *closo* clusters if one assumes a neutral Pt atom and a Cd^{2+} ion.

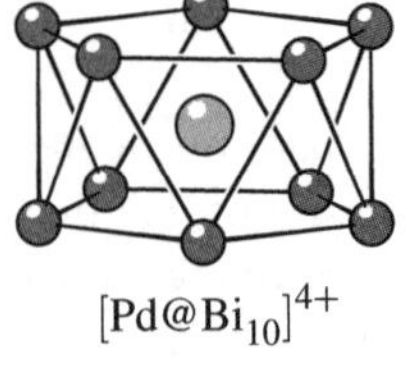

$[Pd@Bi_{10}]^{4+}$

$[Pd@Bi_{10}]^{4+}$ is an example of an *arachno* cluster in the compound $[Pd@Bi_{10}]^{4+}(BiBr_4^-)_4$. It has $2n+6$ skeleton electrons if one assumes one lone electron pair per Bi atom and a neutral Pd atom. The Bi atoms form a pentagonal antiprism which is the same as an icosahedron with two missing vertices.

Endohedral *closo* clusters can be regarded as intermediate stations on the way to the structures of metals. In a closest packing of spheres an atom is surrounded by 12 other atoms; that adds up to 13 atoms. With an additional covering of atoms, the total number of atoms is 55. A corresponding cluster is known in $Au_{55}(PPh_3)_{12}Cl_6$; the envelope of ligands prevents the condensation to the bulk metal. Metal clusters of different sizes can be stabilized by external ligands; as a rule, the metal atom arrangement corresponds to a section of the structure of the pure metal. Examples are: $[Al_{69}R_{18}]^{3-}$, $[Al_{77}R_{20}]^{2-}$, $[Ga_{19}R_6]^-$, $[Ga_{84}R_{20}]^{4-}$ with R = $N(SiMe_3)_2$ or $C(SiMe_3)_3$.

Condensed Clusters

Another possibility for relieving the electron deficiency consists of joining clusters to form larger building blocks. Among the known condensed clusters the majority consist of M_6 octahedra linked with each other. When joining M_6X_8 or M_6X_{12} units in such a way that metal atoms 'merge' with one another, some of the X atoms have to be 'merged' also.

Fig. 13.17 shows a possibility for the condensation of M_6X_8 clusters. Merging *trans* vertices of octahedra to a linear chain requires that opposite faces of the X_8 cubes also merge; every X atom is thus shared by two cubes. The resulting composition is M_5X_4. The relative arrangement of chains bundled in parallel allows the coordination of X atoms of one chain to the octahedra vertices of four adjacent chains, in a similar way as in the CHEVREL phases. Compounds with this structure are known with M = Ti, V, Nb, Ta, Mo and X = S, Se, Te, As, Sb, *e.g.* Ti_5Te_4. They have 12 (Ti_5Te_4) to 18 (Mo_5As_4) skeleton electrons per octahedron. Eight of the electrons form four 2*c*2*e* bonds at the four equatorial

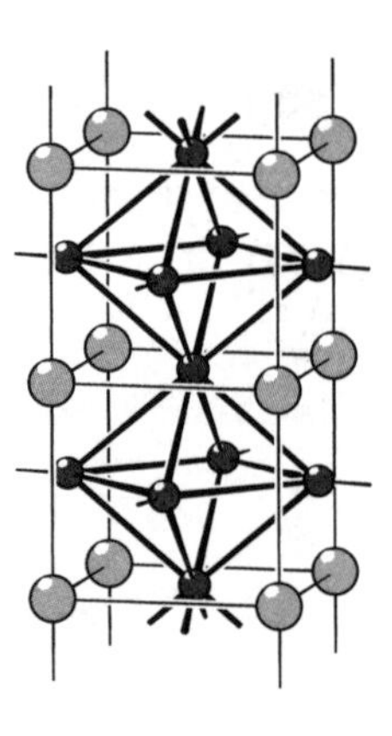

Fig. 13.17
Condensed M_6X_8 clusters in Ti_5Te_4

edges of the octahedron; the remaining electrons are oriented along the other octahedron edges, and their interaction in the chain direction results in metallic energy bands.

Chains with the composition $M_2M_{4/2}X_{8/2} = M_4X_4$ are the result of the condensation of M_6X_8 clusters by merging opposite octahedron edges. They are known for lanthanoid halides like Gd_2Cl_3; they have additional halogen atoms placed between the chains (Fig. 13.18). The clusters may contain interstitial atoms. For example, Sc_4BCl_6 has chains like Gd_2Cl_3 with a boron atom enclosed in each octahedron.

The cluster condensation can be carried on: the chains of octahedra sharing edges can be joined to double-strands and finally to layers of octahedra (Fig. 13.18). Every layer consists of metal atoms in two planes arranged in the same way as two adjacent layers of atoms in a closest-packing of spheres. This is simply a section from a metal structure. The X atoms occupy positions between the metal layers and act as 'insulating' layers. Substances like ZrCl that have this structure have metallic properties in two dimensions.

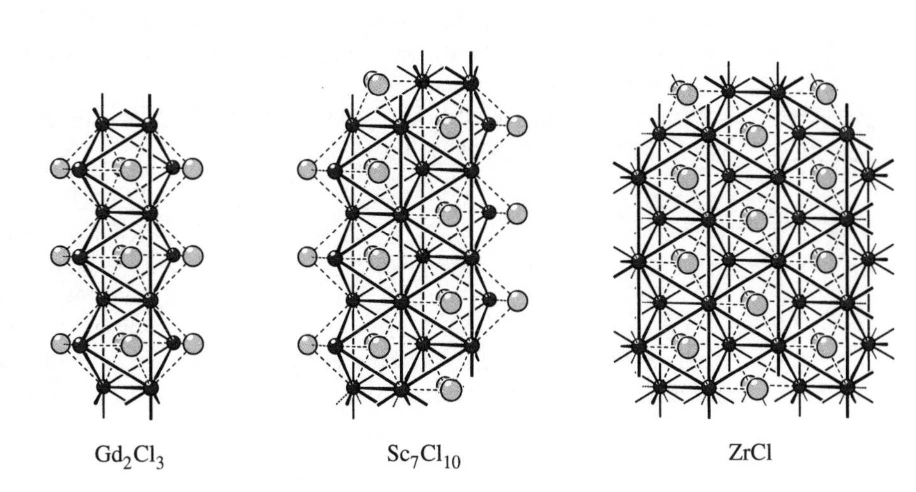

Fig. 13.18 Condensation of M_6X_8 clusters by sharing octahedra edges to yield chains in Gd_2Cl_3, double-strands in Sc_7Cl_{10} and layers in ZrCl. Every metal atom of Gd_2Cl_3 and Sc_7Cl_{10} is also coordinated to a chlorine atom of a neighboring chain

13.5 Problems

13.1 Use the extended $8-N$ rule to decide whether the following compounds are polyanionic, polycationic or simple ionic.
(a) Be_2C; (b) Mg_2C_3; (c) ThC_2; (d) Li_2Si; (e) In_4Se_3; (f) KSb; (g) Nb_3Cl_8; (h) TiS_2.

13.2 Which of the following compounds should be ZINTL phases?
(a) Y_5Si_3; (b) CaSi; (c) CaO; (d) K_3As_7; (e) NbF_4; (f) $LaNi_5$.

13.3 Draw valence bond formulas for the following ZINTL anions.
(a) $Al_2Te_6^{6-}$; (b) $[SnSb_3^{5-}]_\infty$; (c) $[SnSb^-]_\infty$; (d) $[Si^{2-}]_\infty$; (e) P_2^{4-}.

13.4 State which of the following clusters is electron precise, may have *3c2e* bonds or fulfills the WADE rule for *closo* clusters.
(a) $B_{10}C_2H_{12}$ (icosahedron); (b) $Re_6(\mu_3\text{-S})_4(\mu_3\text{-Cl})_4\mu_1\text{-Cl}_6$ (octahedron);
(c) $Pt_4(\mu_3\text{-H})_4(\mu_1\text{-H})_4(PR_3)_4$ (tetrahedron); (d) $Os_5(CO)_{16}$ (trigonal bipyramid);
(e) $Rh_6(CO)_{16}$ (octahedron).

14 Packings of Spheres. Metal Structures

Metals are materials in which atoms are held together by multicenter bonds. The entire set of atoms in a crystal contributes to the multicenter bonds; the valence electrons are delocalized throughout the crystal. More details are given in Chapter 10. The bonding forces act evenly on all atoms: usually there are no prevalent local forces that cause some specific atomic arrangement around an atom in such a way as in a molecule. In what way the atoms arrange themselves in a metallic crystal depends first on how a most dense packing can be achieved geometrically. However, second, the electronic configuration and the valence electron concentration do have some influence; they determine which of several possible packing variants will actually occur. In principle, band structure calculations can allow us to differentiate these variants.

If atoms are considered as hard spheres, the packing density can be expressed by the space filling *SF* of the spheres. It is:

$$SF = \frac{4\pi}{3V}\sum_i Z_i r_i^3 \qquad (14.1)$$

V = volume of the unit cell
r_i = radius of the i-th kind of sphere
Z_i = number of spheres of the i-th kind in the unit cell

If only one kind of sphere is present and all dimensions are referred to the diameter of one sphere, *i.e.* if we set the diameter to be 1 and the radius to be $r = \frac{1}{2}$, we obtain:

$$SF = \frac{\pi}{6} \cdot \frac{Z}{V} = 0.5236\frac{Z}{V}$$

14.1 Closest-packings of Spheres

In order to fill space in the most economical way with spheres of equal size, we arrange them in a closest-packing of spheres. The closest arrangement of spheres in a plane is a *hexagonal layer* of spheres (Fig. 14.1). In such a layer every sphere has six adjacent spheres; six voids remain between a sphere and its six adjacent spheres. The distance from one void to the *next but one* void is exactly the same as that between the centers of two adjacent spheres. Let us denote the position of the sphere centers by A as in Fig. 14.1, and the positions of the voids by B and C. The closest stacking of layers requires that a layer in the position A be followed by a layer having its spheres in hollows on top of either the voids B or the voids C. Altogether, there exist three possible layer positions in a closest stack of hexagonal layers; a layer can only be followed by another layer in a different position (A cannot be followed by A etc.).

The layer sequence *ABCABC*... is marked by arrows in Fig. 14.1. In this sequence all arrows point in the same direction. In a sequence *ABA* one arrow would point in one direction, and the next arrow would point in the opposite direction. If we designate the direction $A{\rightarrow}B = B{\rightarrow}C = C{\rightarrow}A$ by + and $A{\leftarrow}B = B{\leftarrow}C = C{\leftarrow}A$ by −, we can characterize the stacking sequence by a sequence of + and − signs (HÄGG, 1943). The

Inorganic Structural Chemistry, Second Edition Ulrich Müller

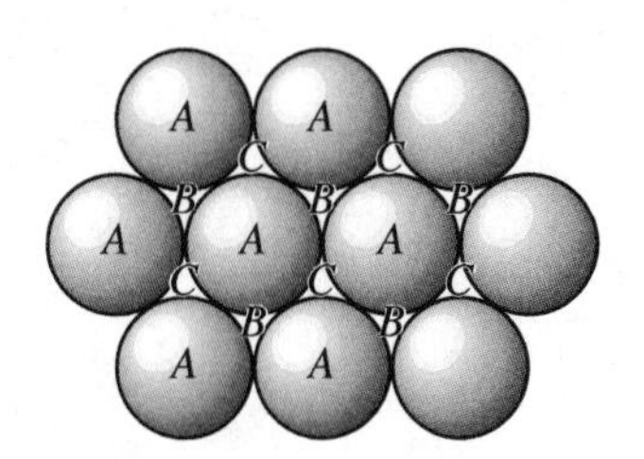

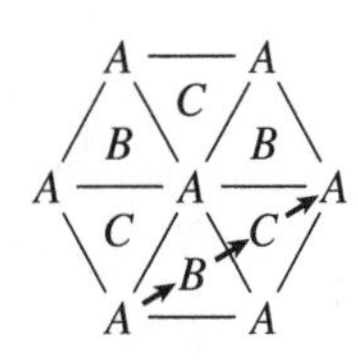

Fig. 14.1 Arrangement of spheres in a hexagonal layer and the relative position of the layer positions *A*, *B* and *C*

symbolism can be abbreviated according to ZHDANOV by a sequence of numbers, with every number specifying how many equal signs are side by side; only the numbers of one periodically repeating unit are given. Another frequently used symbolism is that by JAGODZINSKI: a layer having its two adjacent layers in different positions (*e.g.* the layer *B* in the sequence *ABC*), is designated by *c* (for cubic); if its two adjacent layers have the same position (*e.g.* *B* in the sequence *ABA*), the symbol is *h* (for hexagonal).

Although the number of possible stacking sequences is infinitely large, predominantly only the following two are observed:

	cubic closest-packing (c.c.p.)	hexagonal closest-packing (h.c.p.)
stacking sequence	...*ABCABC*...	...*ABABAB*...
HÄGG symbol	... ++++++...	... +−+−+−...
ZHDANOV symbol	∞	11
JAGODZINSKI symbol	*c*	*h*

Cubic closest-packing is also called copper type and hexagonal closest-packing is also called magnesium type. In the cubic closest-packing the spheres have a face-centered cubic (f.c.c.) arrangement (Fig. 14.2); the stacking direction of the hexagonal layers is perpendicular to either of the body diagonals across the cube. The coordination number of every sphere is 12 for both packings. The coordination polyhedron is a cuboctahedron for cubic closest-packing; a cuboctahedron can be regarded either as a truncated cube or as a truncated octahedron (*cf.* Fig. 2.2, p. 5). The coordination polyhedron for hexagonal closest-packing is an anticuboctahedron; it results when two opposite triangular faces of a cuboctahedron are mutually rotated by 30°.

More complicated stacking sequences occur less frequently. Some have been observed among the lanthanoids:

	stacking sequence	JAGODZINSKI	ZHDANOV
La, Pr, Nd, Pm	...*ABAC*...	*hc*	22
Sm	...*ABACACBCB*...	*hhc*	21

The *hc* packing is called *double-hexagonal closest-packing of spheres*. Gadolinium to thulium as well as lutetium form hexagonal closest-packings of spheres. The proportion of *h* layers hence increases with increasing number of *f* electrons. The electronic configuration controls the kind of packing adopted; the influence of the 4*f* shell decreases with the atomic number. Being in the interior of an atom, with increasing nuclear charge the *f* shell experiences a stronger contraction than the 5*d* and the 6*s* shells, *i.e.* the lanthanoid contraction shows up to a higher degree inside the atoms than can be seen in the atomic radii. The influence of the *f* electrons also expresses itself in the behavior of the lanthanoids

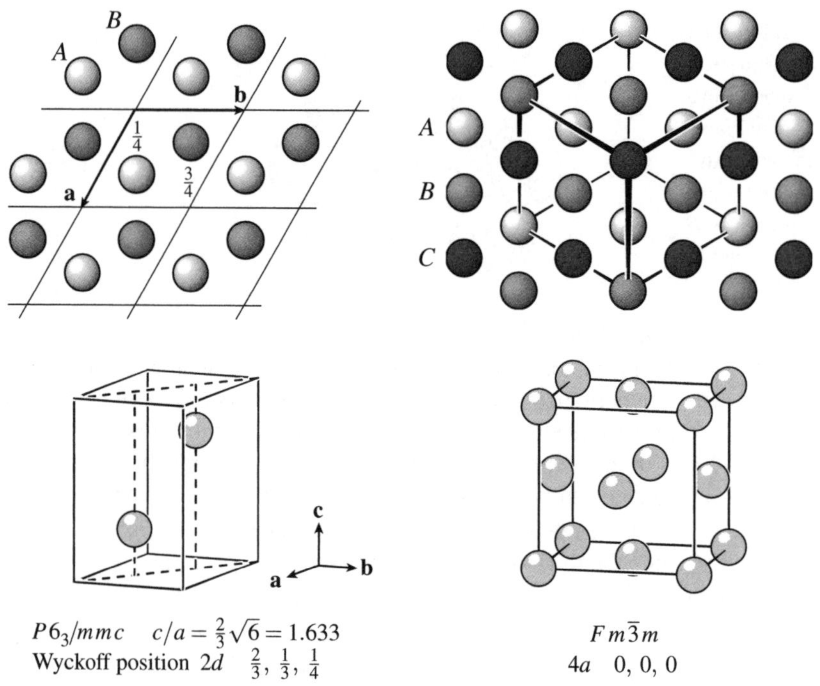

Fig. 14.2 Unit cells for hexagonal (left) and cubic closest-packing of spheres. Top row: projections in the stacking direction. Spheres are drawn smaller than their actual size. Spheres with the same coloring form hexagonal layers as in Fig. 14.1

$P6_3/mmc$ $c/a = \frac{2}{3}\sqrt{6} = 1.633$
Wyckoff position $2d$ $\frac{2}{3}, \frac{1}{3}, \frac{1}{4}$

$Fm\bar{3}m$
$4a$ 0, 0, 0

under pressure. When compressed, the outer shells are squeezed more than the inner ones, and the f electrons gain influence resulting in structures with more c layers:

	normal pressure	high pressure	higher pressure
La, Pr, Nd	*hc*	*c*	
Sm	*hhc*	*hc*	*c*
Gd, Tb, Dy, Ho, Tm	*h*	*hhc*	*hc*

Finally, the influence of the electronic configuration also shows up in the exceptions: europium and ytterbium, having $4f$ shells 'prematurely' half and completely filled, respectively, have structures which do not follow the sequence of the other metals (Table 14.2, p. 155; configuration for Eu $4f^7 6s^2$ instead of $4f^6 5d^1 6s^2$, for Yb $4f^{14} 6s^2$ instead of $4f^{13} 5d^1 6s^2$. These elements also have irregular atomic radii, *cf.* Table 6.2, p. 47). There is also an irregularity at the beginning of the series, since cerium adopts a cubic closest-packing.

The number of different possible stacking variants increases with increasing numbers of hexagonal layers in one periodically repeating slab of layers:

number of layers per slab:	2	3	4	5	6	7	8	9	10	11	12	20
number of stacking variants:	1	1	1	1	2	3	6	7	16	21	43	4625

The notable predominance of simple stacking forms is an expression of the *symmetry principle*:

Among several feasible structure types those having the highest symmetry are normally favored.

We discuss the reasons for and the importance of this principle in more detail in Section 18.2. The frequent observance of the *principle of most economic filling of space, i.e.* of purely geometric aspects, is also remarkable: of the 95 elements with known structures in the solid state, 46 adopt closest packings at ambient conditions. If we include low- and high-temperature and high-pressure modifications, closest-packings of spheres occur among 101 modifications of 75 elements.

Aside from the ordered stacking sequences we have considered so far, a more or less statistical sequence of hexagonal layers can also occur. Since there is some kind of an ordering principle on the one hand, but on the other hand the periodical order is missing in the stacking direction, this is called an *order–disorder* (OD) structure with stacking faults. In this particular case, it is a *one-dimensionally disordered* structure, since the order is missing only in one dimension. When cobalt is cooled from 500 °C it exhibits this kind of disorder.

The *space filling* is the same for all stacking variants of closest-packings of spheres. It amounts to $\pi/(3\sqrt{2}) = 0.7405$ or 74.05 %. That no packing of spheres can have a higher density was claimed in 1603 by J. KEPLER. However, conclusive proof was not furnished until 1998. Spheres surrounding a central sphere icosahedrally are not in contact with one another, *i.e.* there is slightly more space than for twelve neighboring spheres. However, icosahedra cannot be packed in a space-filling manner. Some nonperiodic packings of spheres have been described that have densities close to the density of closest-packings of spheres. Because no packing of spheres can have a higher density, one should not say 'close' packing, but *closest* packing.

14.2 Body-centered Cubic Packing of Spheres

The space filling in the body-centered cubic packing of spheres is less than in the closest packings, but the difference is moderate. The fraction of space filled amounts to $\frac{1}{8}\pi\sqrt{3} = 0.6802$ or 68.02 %. The reduction of the coordination number from 12 to 8 seems to be more serious; however, the difference is actually not so serious because in addition to the 8 directly adjacent spheres every sphere has 6 further neighbors that are only 15.5 % more distant (Fig. 14.3). The coordination number can be designated by $8 + 6$.

Corresponding to its inferior space filling, the body-centered cubic packing of spheres is less frequent among the element structures. None the less, 15 elements crystallize with this structure. As tungsten is one of them, the term tungsten type is sometimes used for this kind of packing.

The mentioned number of elements refers to ambient conditions. If we also include the modifications adopted by some elements at low and high temperatures and at high pres-

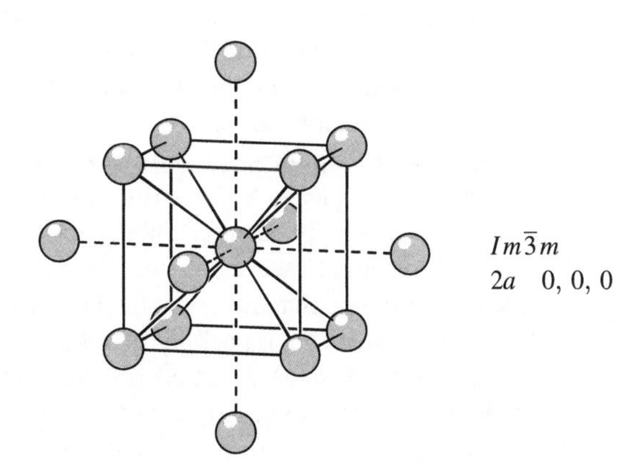

Fig. 14.3
Unit cell of the body-centered cubic packing of spheres and the coordination around one sphere

Table 14.1: Numbers of structures of the elements known until 2006 in the solid state at different conditions

	nonmetal structures	closest-packings of spheres†	body-centered cubic	other metal structures
nonmetals*				
ambient pressure	55	4‡	–	–
high pressure	4	8	7	31
metals				
up to 400 K	2	51	15	11
high temperature	–	7	23	6
high pressure	–	31	8	39
total	61	101	53	87

* including Si, Ge, As, Sb, Te. All temperature ranges
† including slightly distorted variants
‡ noble gases

sures, we obtain the statistics given in Table 14.1. As can be seen, the additional structures at high pressures are mainly closest-packings of spheres and 'exotic' metal structures. At high temperatures the body-centered packing becomes more important. This is in accordance with the GOLDSCHMIDT rule:
Increased temperatures favor structures with lower coordination numbers.

14.3 Other Metal Structures

The crystal structures of most metals correspond to the above-mentioned packings of spheres (with certain distortions in some cases; Table 14.2). Some metals, however, show structure types of their own: Ga, Sn, Bi, Po, Mn, Pa, U, Np, and Pu. For Sn, Bi, and Po refer to pp. 121, 109 and 107. Gallium has a rather unusual structure in which every Ga atom has coordination number 1 + 6; one of the seven adjacent atoms is significantly closer than the others (1× 244 pm, 6× 270 to 279 pm). This has been interpreted as a metal structure consisting not of single atoms, but of Ga–Ga pairs with a covalent bond. The notably low melting point of gallium (29.8 °C) shows this structure to be not especially stable; it seems to be only an 'expedient'. There also seems to be no optimal structure for Mn, U, Np, and Pu, as these elements form a remarkable number of polymorphic forms with rather peculiar structures. For example, the unit cell of α-Mn, the modification stable at room temperature, contains 58 atoms with four different kinds of coordination polyhedra having coordination numbers 12, 13, and 16.

A remarkable number of unusual structures are observed at high pressures, especially among the alkali and alkaline earth metals (Fig. 14.4). For example, caesium is first transformed from a body-centered cubic packing to a cubic closest-packing at 2.3 GPa, which is not surprising. However, at increasing pressures, three modifications follow with atoms having coordination numbers of 8 – 11, then 8 and then 10 – 11; finally, at 70 GPa, a closest (double-hexagonal) packing of spheres reappears. An electron transition from the 6s to the 5d band is the assumed reason for this behavior. Some of the caesium modifications are also adopted by rubidium; in addition, rubidium exhibits an incommensurate composite structure between 16 and 20 GPa which is similar to that of bismuth-III

(Fig. 11.11, p. 112). Incommensurate structures related to bismuth-III are also observed for strontium and barium. Magnesium, calcium and strontium are remarkable in that they transform from the normal closest-packing of spheres to a body-centered packing upon exertion of pressure. Even more remarkable is the following decrease of the coordination number to 6 for calcium and strontium (Ca-III, α-Po type; Sr-III, β-tin type).

Table 14.2: The element structures of the metals at ambient conditions
h = hexagonal closest-packing
c = cubic closest-packing
hc, *hhc* = other stacking variants of closest-packing
i = body-centered cubic packing
⋈ = structure type of its own
⋆ = slightly distorted
The solid noble gases also adopt closest-packings of spheres at low temperatures: Ne...Xe *c*; helium becomes solid only under pressure (depending on pressure, *c*, *h* or *i*)

Li *i*	Be *h*												
Na *i*	Mg *h*											Al *c*	
K *i*	Ca *c*	Sc *h*	Ti *h*	V *i*	Cr *i*	Mn ⋈	Fe *i*	Co *h*	Ni *c*	Cu *c*	Zn *h*⋆	Ga ⋈	
Rb *i*	Sr *c*	Y *h*	Zr *h*	Nb *i*	Mo *i*	Tc *h*	Ru *h*	Rh *c*	Pd *c*	Ag *c*	Cd *h*⋆	In *c*⋆	Sn ⋈
Cs *i*	Ba *i*	La *hc*	Hf *h*	Ta *i*	W *i*	Re *h*	Os *h*	Ir *c*	Pt *c*	Au *c*	Hg *c*⋆	Tl *h*	Pb *c*
Fr	Ra *i*	Ac *c*	Rf	Db	Sg	Bh	Hs	Mt	Ds	Rg			

Ce *c*	Pr *hc*	Nd *hc*	Pm *hc*	Sm *hhc*	Eu *i*	Gd *h*	Tb *h*	Dy *h*	Ho *h*	Er *h*	Tm *h*	Yb *c*	Lu *h*
Th *c*	Pa *i*⋆	U ⋈	Np ⋈	Pu ⋈	Am *hc*	Cm *hc*	Bk *c, hc*	Cf *h, hc*	Es	Fm	Md	No	Lr

14.4 Problems

14.1 State the JAGONDZINSKI and the ZHDANOV symbols for the closest-packings of spheres with the following stacking sequences:
(a) *ABABC*; (b) *ABABACAC*.

14.2 State the stacking sequence (by *A*, *B* and *C*) for the closest-packings of spheres with the following JAGONDZINSKI or ZHDANOV symbols:
(a) *hcc*; (b) *cchh*; (c) 221.

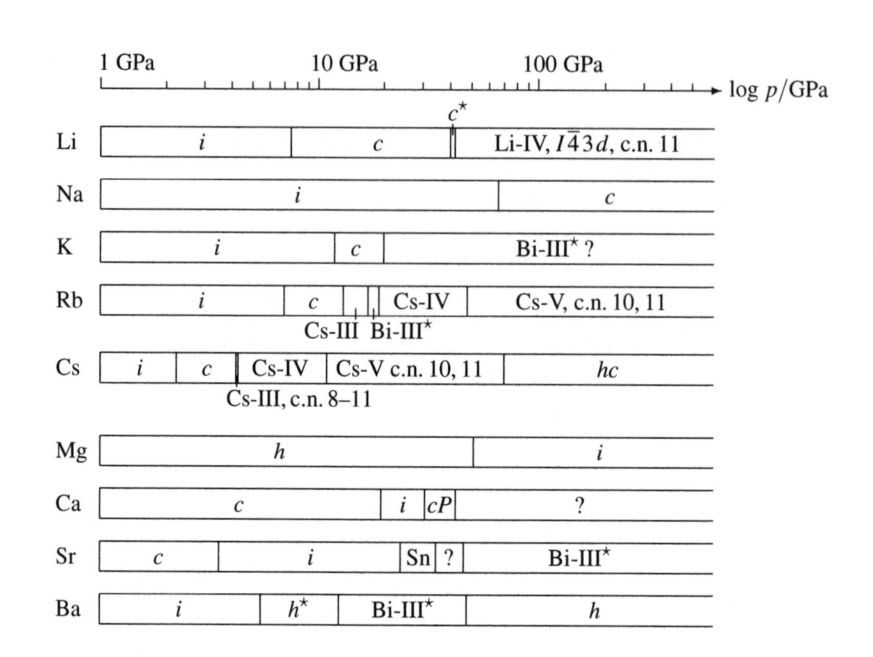

Fig. 14.4 Regions of stability for the structure types of the alkali and alkaline earth metals in dependence on pressure at room temperature. *h* = hexagonal closest-packing; *c* = cubic closest-packing; *hc* = double-hexagonal closest-packing; *i* = body-centered cubic; *cP* = cubic primitive (α-Po); * = slightly distorted; Cs-IV: c.n. = 8, arrangement as the Th atoms in $ThSi_2$ (Fig. 13.1, p. 131)

15 The Sphere-packing Principle for Compounds

The geometric principles for the packing of spheres do not only apply to pure elements. As might be expected, the sphere packings discussed in the preceding chapter are also frequently encountered when similar atoms are combined, especially among the numerous alloys and intermetallic compounds. Furthermore, the same principles also apply to many compounds consisting of elements which differ widely.

15.1 Ordered and Disordered Alloys

Different metals can very frequently be mixed with each other in the molten state, *i.e.* they form homogeneous solutions. A solid solution is obtained by quenching the liquid; in the *disordered alloy* obtained this way, the atoms are distributed randomly. When cooled slowly, in some cases solid solutions can also be obtained. However, it is more common that a segregation takes place, in one of the following ways:

1. The metals crystallize separately (complete segregation).
2. Two kinds of solid solutions crystallize, a solution of metal 1 in metal 2 and vice versa (limited miscibility).
3. An alloy with a specific composition crystallizes; its composition may differ from that of the liquid (formation of an intermetallic compound). The composition of the liquid can change during the crystallization process and further intermetallic compounds with other compositions may crystallize.

The phase diagram shows which of these possibilities applies and whether intermetallic compounds will eventually form (*cf.* Section 4.5, p. 34).

The tendency to form solid solutions depends mainly on two factors, namely the chemical relationship between the elements and the relative size of their atoms.

Two metals that are chemically related and that have atoms of nearly the same size form disordered alloys with each other. Silver and gold, both crystallizing with cubic closest-packing, have atoms of nearly equal size (radii 144.4 and 144.2 pm). They form solid solutions (mixed crystals) of arbitrary composition in which the silver and the gold atoms randomly occupy the positions of the sphere packing. Related metals, especially from the same group of the periodic table, generally form solid solutions which have any composition if their atomic radii do not differ by more than approximately 15%; for example Mo + W, K + Rb, K + Cs, but not Na + Cs. If the elements are less similar, there may be a limited miscibility as in the case of, for example, Zn in Cu (amount-of-substance fraction of Zn maximally 38.4%) and Cu in Zn (maximally 2.3% Cu); copper and zinc additionally form intermetallic compounds (*cf.* Section 15.4).

When the atoms differ in size or when the metals are chemically different, structures with ordered atomic distributions are considerably more likely. Since the transition from a disordered to an ordered state involves a decrease in entropy, and since the transition only takes place when $\Delta G = \Delta H - T\Delta S < 0$, the transformation enthalpy ΔH must be negative. The ordered structure therefore is favored energetically; the amount of its lattice energy is larger.

An ordered distribution of spheres of different sizes always allows a better filling of space; the atoms are closer together, and the attractive bonding forces become more effective. As for the structures of other types of compound, we observe the validity of the *principle of the most efficient filling of space*. A definite order of atoms requires a definite chemical composition. Therefore, metal atoms having different radii preferentially will combine in the solid state with a definite stoichiometric ratio: they will form an intermetallic compound.

Even when complete miscibility is possible in the solid state, ordered structures will be favored at suitable compositions if the atoms have different sizes. For example: copper atoms are smaller than gold atoms (radii 127.8 and 144.2 pm); copper and gold form mixed crystals of any composition, but ordered alloys are formed with the compositions AuCu and $AuCu_3$ (Fig. 15.1). The degree of order is temperature dependent; with increasing temperatures the order decreases continuously. Therefore, there is no phase transition with a well-defined transition temperature. This can be seen in the temperature dependence of the specific heat (Fig. 15.2). Because of the form of the curve, this kind of order–disorder transformation is also called a Λ type transformation; it is observed in many solid-state transformations.

15.2 Compounds with Close-packed Atoms

As in ionic compounds, the atoms in a binary intermetallic compound show a tendency, albeit less pronounced, to be surrounded by atoms of the other kind as far as possible. However, it is not possible to fulfill this condition simultaneously for both kinds of atoms if they form a closest-packed arrangement. For compositions MX_n with $n < 3$ it cannot be fulfilled for either the M or the X atoms: in every case every atom has to have some adjacent atoms of the same kind. Only with a higher content of X atoms, beginning with MX_3 ($n \geq 3$), are atomic arrangements possible in which every M atom is surrounded solely by X atoms; the X atoms, however, must continue to have other X atoms as neighbors.

Usually, the composition of the compound is fulfilled in all of the hexagonal layers. This facilitates a rational classification of the extensive data: it is only necessary to draw a sketch of the atomic arrangement in one layer and to specify the stacking sequence

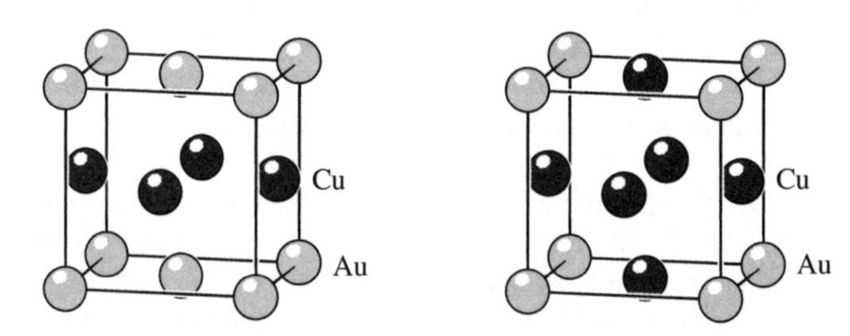

Fig. 15.1
The structures of the ordered alloys AuCu and $AuCu_3$. At higher temperatures they are transformed to alloys which have all atomic positions statistically occupied by the Cu and Au atoms

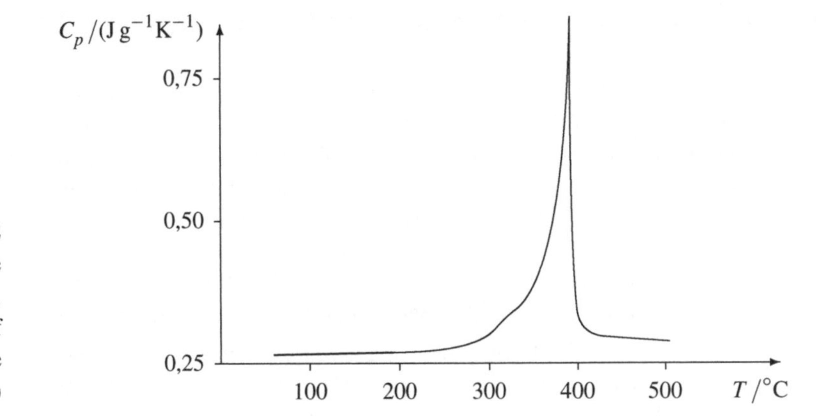

Fig. 15.2 Variation of the specific heat C_p with temperature of $AuCu_3$ (Λ-type transformation)

(ZHDANOV or JAGODZINSKI symbol). The most important structure types of this kind are the following (M atoms dark gray in the figures; positions of M atoms in the following layer marked by black circles):

1. MX_3 structures with hexagonal arrangement of M atoms in one layer

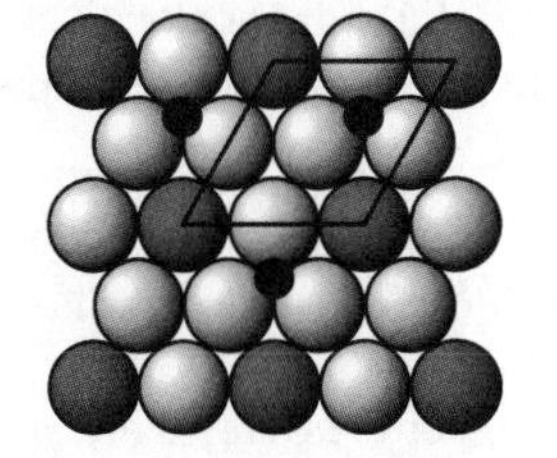

structure type	ZHDANOV symbol	JAGODZINSKI symbol
$AuCu_3$	∞	*c*
$SnNi_3$	11	*h*

2. MX_3 structures with rectangular arrangement of M atoms

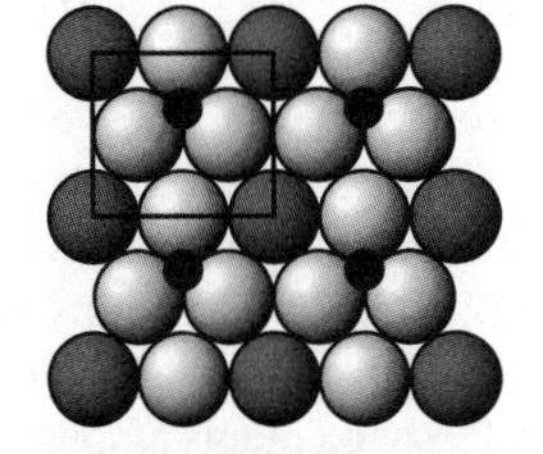

structure type	ZHDANOV symbol	JAGODZINSKI symbol
$TiAl_3$	∞	*c*
$TiCu_3$	11	*h*

3. MX structures with alternating strings of equal atoms

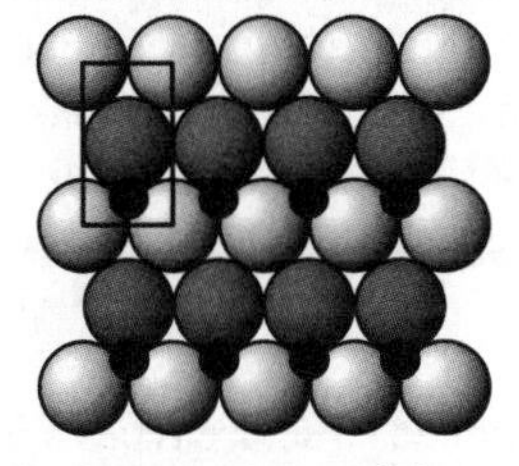

structure type	ZHDANOV symbol	JAGODZINSKI symbol
AuCu	∞	*c*
AuCd	11	*h*
TaRh	33	*hcc*

Because strings of the same atoms come to be adjacent when these layers are stacked, alternating layers of atoms of one kind each are formed. These layers are planar in AuCu; they are inclined relative to the plane of the paper; in the unit cell (Fig. 15.1) they are parallel to the base plane. The layers of equal atoms are undulated in the other two structure types.

15.3 Structures Derived of Body-centered Cubic Packing (CsCl Type)

Disordered alloys may form when two metals are mixed if both have body-centered cubic structures and if their atomic radii do not differ by much (*e.g.* K and Rb). The formation of ordered alloys, however, is usually favored; at higher temperatures the tendency towards disordered structures increases. Such an arrangement can even be adopted if metals are combined which do not crystallize with body-centered cubic packings themselves, on condition of the appropriate composition. β-Brass (CuZn) is an example; below 300 °C it has a CsCl structure, but between 300 °C and 500 °C a Λ type transformation takes place resulting in a disordered alloy with a body-centered cubic structure.

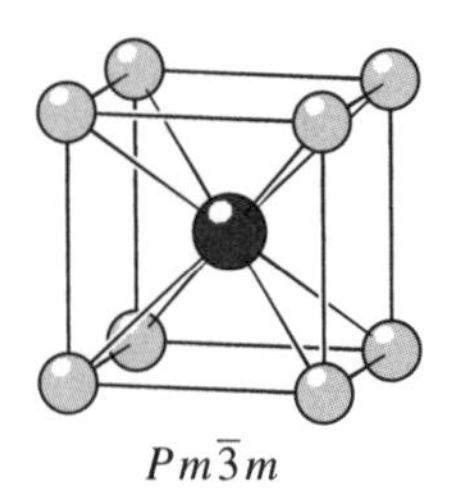

$Pm\bar{3}m$

The **CsCl type** offers the simplest way to combine atoms of two different elements in the same arrangement as in body-centered cubic packing: the atom in the center of the unit cell is surrounded by eight atoms of the other element in the vertices of the unit cell. In this way each atom only has adjacent atoms of the other element. This is a condition that cannot be fulfilled in a closest-packing of spheres (*cf.* preceding section).

Although the space filling of the body-centered cubic sphere packing is somewhat inferior to that of a closest-packing, the CsCl type thus turns out to be excellently suited for compounds with a 1:1 composition. Due to the occupation of the positions $0,0,0$ and $\frac{1}{2},\frac{1}{2},\frac{1}{2}$ with different kinds of atoms, the structure is not body-centered.

We presented the CsCl type in Chapter 7 as an important structure type for ionic compounds. Its importance, however, is by no means restricted to this class of compounds: only about 12 out of more than 200 compounds with this structure are salt-like (*e.g.* CsI, TlBr), although at higher temperatures or higher pressures there are some 15 more (*e.g.* NaCl, KCl at high pressure; TlCN at high temperature with rotating CN^- ions). More than 200 representatives are intermetallic compounds, *e.g.* MgAg, CaHg, AlFe and CuZn.

Superstructures of the CsCl type result when the unit cell of the CsCl structure is multiplied and the atomic positions are occupied by different kinds of atoms. If we double the cell edges in all three dimensions, we obtain a cell that consists of eight subcells, each of which contains one atom in its center (Fig. 15.3). The 16 atoms in the cell can be subdivided into four groups of four atoms each; each group has a face-centered arrangement. Depending on how we distribute atoms of different elements among these four groups, we obtain different structure types, as listed in Fig. 15.3. The list includes possibilities with certain vacant atomic positions (marked with the SCHOTTKY symbol □ in the table). This option reduces the space filling; however, as long as the positions *a* and *b* are occupied by different kinds of atoms than the positions *c* and *d*, each atom still has only atoms of a different kind as nearest neighbors. Consequently, the structure types are adequate for ionic compounds, including ZINTL phases with simple 'anions' such as As^{3-}, Sb^{3-} or Ge^{4-}.

The following series shows that the mentioned structure types are adopted by all kinds of compounds from purely ionic to purely metallic:

fluorite type and variants					Fe_3Al type and variants			W type
F_2Ca	Li_2O	Li_2Te	LiMgAs	Mg_2Sn	Cu_3Sb	Cu_2MnAl	Fe_3Al	Fe
ionic	⟶							metallic

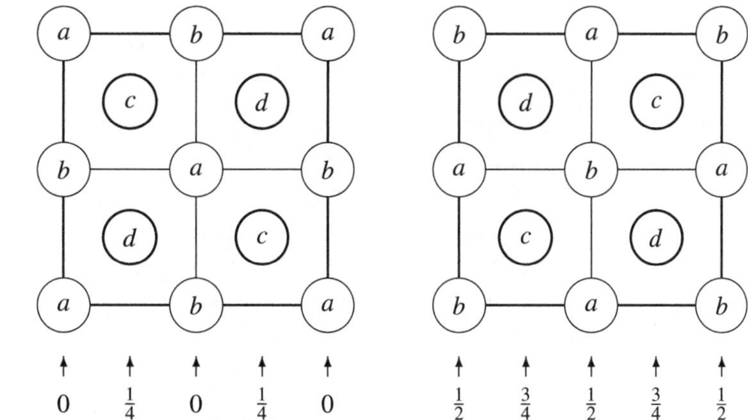

Fig. 15.3
Superstructure of the CsCl type with eightfold unit cell. Left, lower half and right, upper half of the cell in projection onto the plane of the paper. *a*, *b*, *c*, and *d* designate four different kinds of atomic sites that can be occupied in the following ways:

a	*b*	*c*	*d*	structure type	space group	examples
Al	Fe	Fe	Fe	Fe_3Al (Li_3Bi)	$Fm\bar{3}m$	Fe_3Si, Mg_3Ce, Li_3Au, Sr_3In
Al	Mn	Cu	Cu	$MnCu_2Al$ (Heusler alloy)	$Fm\bar{3}m$	$LiNi_2Sn$, $TiCo_2Si$
Tl	Na	Tl	Na	NaTl (Zintl phase)	$Fd\bar{3}m$	LiAl, LiZn
Ag	Li	Sb	Li	Li_2AgSb (Zintl phase)	$F\bar{4}3m$	Li_2AuBi, Na_2CdPb
Sn	Mg	Pt	Li	LiMgSnPt	$F\bar{4}3m$	
As	□	Mg	Ag	MgAgAs (Zintl phase)	$F\bar{4}3m$	LiMgAs, NiZnSb, BAlBe, SiCN
Ca	□	F	F	CaF_2 (fluorite)	$Fm\bar{3}m$	$BaCl_2$, ThO_2, TiH_2, Li_2O, Be_2C, Mg_2Sn
Zn	□	S	□	zinc blende	$F\bar{4}3m$	SiC, AlP, GaAs, CuCl
C	□	C	□	diamond	$Fd\bar{3}m$	Si, α-Sn
Na	Cl	□	□	NaCl	$Fm\bar{3}m$	LiH, AgF, MgO, TiC

When covalent bonds favor neighbors of the same element, the positions *c* and *d* can also be occupied by atoms of the same kind as in *a* or *b*. This applies to diamond and to the ZINTL phase NaTl; NaTl can be regarded as a network of Tl^- particles that form a diamond structure which encloses Na^+ ions (*cf.* Fig. 13.3, p. 134).

15.4 Hume–Rothery Phases

HUME–ROTHERY phases (brass phases, 'electron compounds') are certain alloys with the structures of the different types of brass (brass = Cu–Zn alloys). They are classical examples of the structure-determining influence of the valence electron concentration (VEC) in metals. VEC = (number of valence electrons)/(number of atoms). A survey is given in Table 15.1.

α-Brass is a solid solution of zinc in copper which has the structure of copper; the atoms statistically occupy the positions of the cubic closest-packing of spheres. In β-brass,

Table 15.1: Brass phases

	composition	VEC	structure type	examples
α	$Cu_{1-x}Zn_x$, $x=0$ to 0.38	1 to 1.38	Cu	
β	CuZn	1.50 = 21/14	W	AgZn, Cu_3Al, Cu_5Sn
γ	Cu_5Zn_8	1.62 = 21/13	Cu_5Zn_8	Ag_5Zn_8, Cu_9Al_4, $Na_{31}Pb_8$
ε	$CuZn_3$	1.75 = 21/12	Mg	$AgZn_3$, Cu_3Sn, Ag_5Al_3
η	Cu_xZn_{1-x}, $x=0$ to 0.02	1.98 to 2	Mg	

which is obtained by quenching the melt, the atoms also have a random distribution, and the packing is body-centered cubic. The composition is not exactly CuZn; this phase is stable only if the fraction of zinc atoms amounts to 45 to 48 %. The γ phase also has a certain range of compositions from $Cu_5Zn_{6.9}$ to $Cu_5Zn_{9.7}$. The γ-brass structure can be described as a superstructure of the body-centered cubic packing with tripled lattice constants, so that the unit cell has a volume enlarged by a factor $3^3 = 27$. However, the cell only contains 52 instead of $2 \times 27 = 54$ atoms; there are two vacancies. The distribution of the vacancies is ordered. There are four kinds of positions for the metal atoms in a ratio of 3 : 2 : 2 : 6, but a random distribution may occur to some extent. In Cu_5Zn_8 the distribution is 3Cu : 2Cu : 2Zn : 6Zn. A brass sample with a composition outside of the mentioned ranges consists of a mixture of the two neighboring phases.

Because of the permitted composition ranges, alloys with rather different compositions can adopt the same structure, as can be seen by the examples in Table 15.1. The determining factor is the valence electron concentration, which can be calculated as follows:

AgZn	$\frac{1+2}{2} = \frac{3}{2} = \frac{21}{14}$	Ag_5Zn_8	$\frac{5+16}{13} = \frac{21}{13}$	$AgZn_3$	$\frac{1+6}{4} = \frac{7}{4} = \frac{21}{12}$
Cu_3Al	$\frac{3+3}{4} = \frac{6}{4} = \frac{21}{14}$	Cu_9Al_4	$\frac{9+12}{13} = \frac{21}{13}$	Cu_3Sn	$\frac{3+4}{4} = \frac{7}{4} = \frac{21}{12}$
Cu_5Sn	$\frac{5+4}{6} = \frac{9}{6} = \frac{21}{14}$	$Na_{31}Pb_8$	$\frac{31+32}{39} = \frac{21}{13}$	Ag_5Al_3	$\frac{5+9}{8} = \frac{14}{8} = \frac{21}{12}$

A theoretical interpretation relating the valence electron concentration and the structure was put forward by H. JONES. If we start from copper and add more and more zinc, the valence electron concentration increases. The added electrons have to occupy higher energy levels, *i.e.* the energy of the FERMI limit is raised and comes closer to the limits of the first BRILLOUIN zone. This is approached at about VEC = 1.36. Higher values of the VEC require the occupation of antibonding states; now the body-centered cubic lattice becomes more favorable as it allows a higher VEC within the first BRILLOUIN zone, up to approximately VEC = 1.48.

15.5 Laves Phases

The term LAVES phases is used for certain alloys with the composition MM'_2, the M atoms being bigger than the M′ atoms. The classical representative is $MgCu_2$; its structure is shown in Fig. 15.4. It can be regarded as a superstructure of the CsCl type as in Fig. 15.3, with the following occupation of the positions *a*, *b*, *c*, and *d*:

a: Mg *b*: Cu_4 *c*: Mg *d*: Cu_4

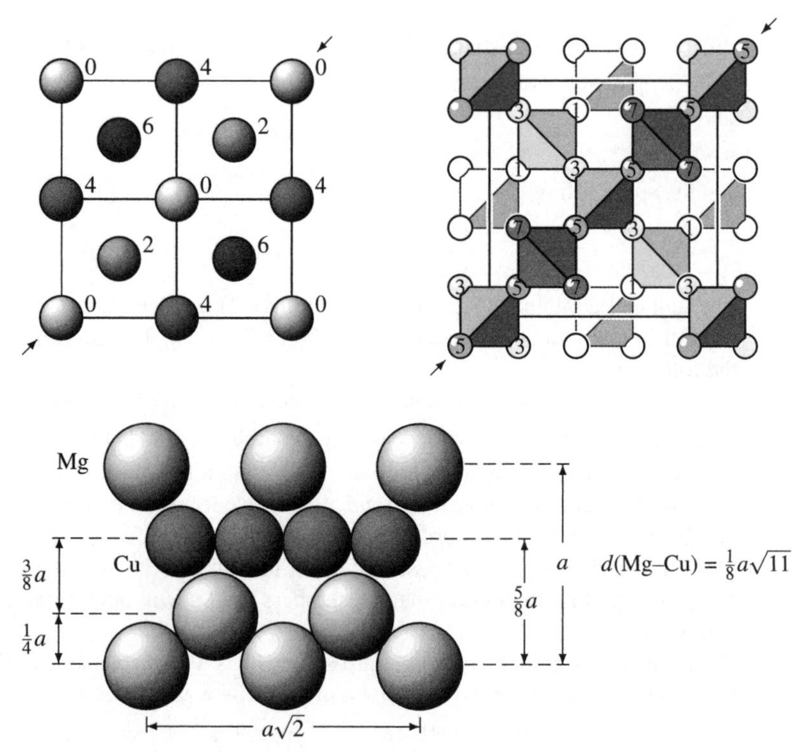

Fig. 15.4
Structure of the LAVES phase $MgCu_2$. Left: Mg partial structure. Right: Cu partial structure consisting of vertex-sharing tetrahedra. Numbers designate the heights in the unit cell as multiples of $\frac{1}{8}$. Bottom: section across the cell in the diagonal direction marked by arrows in the top row, plotted with atomic radii corresponding to atoms in contact with each other (on a smaller scale than the upper part)

We thus have placed a tetrahedron of four Cu atoms instead of a single atom in the position b; the same kind of Cu tetrahedron then also results at the position d. The magnesium atoms by themselves have the same arrangement as in diamond.

In addition to this cubic LAVES phase, a variant with magnesium atoms arranged as in hexagonal diamond exists in the $MgZn_2$ type, and further polytypes are known.

The copper atoms of the $MgCu_2$ type are linked to a network of vertex-sharing tetrahedra (Fig. 15.4), so that every Cu atom is linked with six other Cu atoms. If we assume an electron distribution according to the formula $Mg^{2+}(Cu^-)_2$, every copper atom attains a valence electron concentration of VEC(Cu) = $\frac{1}{2}(1 \cdot 2 + 2 \cdot 11)$ = 12. Taking equation (13.7) from p. 129, adapted to transition metals as b(X) = 18 – VEC(M), we calculate b(X) = 18 – 12 = 6 bonds per Cu atom. In other words, by linking Cu^- particles this way, copper attains the electron configuration of the next noble gas. $MgCu_2$, in a way, fulfills the rules for a ZINTL phase. Nevertheless, LAVES phases customarily are not considered to be ZINTL phases; some 170 intermetallic compounds having the $MgCu_2$ structure are known, and most of them do not fulfill ZINTL's valence rule (*e.g.* $CaAl_2$, YCo_2, $LiPt_2$).

The space filling in the $MgCu_2$ type can be calculated with the aid of equation (14.1) (p. 150); the geometric relations follow from the bottom image in Fig. 15.4: the four Cu spheres form a row along the diagonal of length $a\sqrt{2}$, therefore r(Cu) = $\frac{1}{8}\sqrt{2}a$; two Mg spheres along the space diagonal of the unit cell are at a distance of $\frac{1}{4}\sqrt{3}\,a$, therefore $r(\mathrm{Mg}) = \frac{1}{8}\sqrt{3}\,a$. The ideal radius ratio is therefore

$$\frac{r(\mathrm{Mg})}{r(\mathrm{Cu})} = \sqrt{\frac{3}{2}} = 1.225$$

and the space filling is

$$\frac{4}{3}\pi\frac{1}{a^3}[8(\frac{1}{8}\sqrt{3}a)^3 + 16(\frac{1}{8}\sqrt{2}a)^3] = 0.710$$

(the unit cell contains 8 Mg and 16 Cu atoms). The space filling of 71.0 % is somewhat inferior to that in a closest-packing of spheres (74.1 %). The coordination of the atoms is the following:

Mg: c.n. 16, 4 Mg at a distance of $\frac{1}{8}\sqrt{12}a$ and 12 Cu at a distance of $\frac{1}{8}\sqrt{11}a$;

Cu: c.n. 12, 6 Cu at a distance of $\frac{1}{8}\sqrt{8}a$ and 6 Mg at a distance of $\frac{1}{8}\sqrt{11}a$.

The coordination polyhedra are FRANK–KASPER polyhedra. These are polyhedra with equal or different triangular faces, and at least five triangles meeting at every vertex. Such polyhedra allow for the coordination numbers 12, 14, 15, and 16. Fig. 15.5 shows the two FRANK–KASPER polyhedra occurring in $MgCu_2$. FRANK–KASPER polyhedra and the corresponding high coordination numbers are known among numerous intermetallic compounds.

The sketched model assuming hard spheres has a flaw: the sum of the atomic radii of Mg and Cu is smaller than the shortest distance between these atoms:

$$\begin{aligned} r(\mathrm{Mg}) + r(\mathrm{Cu}) &= \tfrac{1}{8}(\sqrt{3}+\sqrt{2})a &= 0.393\,a \\ d(\mathrm{Mg–Cu}) &= \tfrac{1}{8}\sqrt{11}\,a &= 0.415\,a \end{aligned}$$

Whereas the Mg atoms are in contact with each other and the Cu atoms are in contact with each other, the Cu partial structure 'floats' inside the Mg partial structure. The hard sphere model proves to be insufficient to account for the real situation: atoms are not really hard. The principle of the most efficient filling space should rather be stated as the *principle of achieving the highest possible density.* Indeed, this shows up in the actual densities of the LAVES phases; they are greater than the densities of the components (in some cases up to 50 % more). For example, the density of $MgCu_2$ is 5.75 g cm^{-3}, which is 7% more than the mean density of 5.37 g cm^{-3} for 1 mole Mg + 2 moles Cu. Therefore,

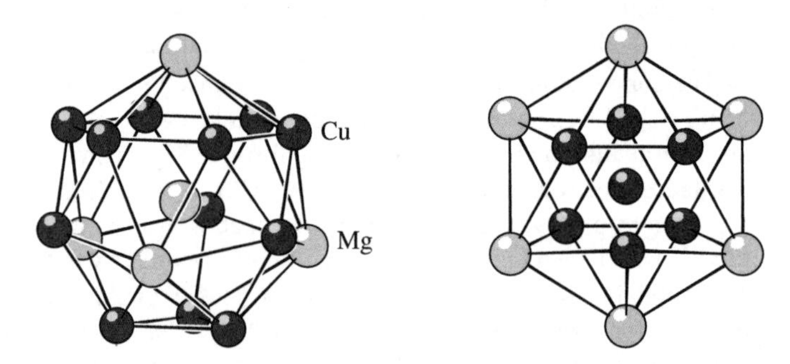

Fig. 15.5

FRANK–KASPER polyhedra in $MgCu_2$. The polyhedron around an Mg atom (c.n. 16) is composed of 12 Cu atoms and of four Mg atoms that form a tetrahedron by themselves; the Cu atoms form four triangles that are opposed to the Mg atoms. The polyhedron around a Cu atom (c.n. 12) is an icosahedron in which two opposite faces are occupied by Cu atoms

the atoms have a denser packing in $MgCu_2$ than in the pure elements, the atoms are effectively smaller. According to the hard sphere model, $MgCu_2$ should not be formed at all, as its space filling of 71 % is inferior to that of both magnesium and copper, both of which crystallize with closest-packings of spheres (74 % space filling).

It is mainly the Mg atoms that are affected by the compression of the atoms. The increase in density is the expression of the gain in lattice energy due to stronger bonding forces between the different kinds of atoms. These bonding forces have polar contributions since LAVES phases of this type experience a higher compression when the difference in the electronegativities of the atoms is higher. The polarity is an argument in favor of regarding LAVES phases in a similar way as ZINTL phases. More insight into the kind of bonding has been obtained by band structure calculations, which also allow the distinction of the electron counts at which the cubic $MgCu_2$ or the hexagonal $MgZn_2$ type is favored.

15.6 Problems

15.1 Use Table 14.2 to decide whether the following pairs of metals are likely to form disordered alloys of arbitrary composition with each other.
(a) Mg/Ca; (b) Ca/Sr; (c) Sr/Ba; (d) La/Ac; (e) Ti/Mn; (f) Ru/Os; (g) Pr/Nd; (h) Eu/Gd.

15.2 Draw a section of each of the structure types presented on p. 159 corresponding to a plane running in the vertical direction of the figure and perpendicular to the plane of the paper.

15.3 What structure types result when the atomic positions in Fig. 15.3 are occupied in the following manner (A, B, C and D refer to chemical elements)?
(a) *a* A, *b* A, *c* □, *d* B; (b) *a* A, *b* B, *c* C, *d* □; (c) *a* A, *b* A, *c* C, *d* D.

15.4 How is it possible that both Ag_5Zn_8 and Cu_9Al_4 have the γ-brass structure even though their compositions differ?

15.5 Can an icosahedron be considered to be a FRANK–KASPER polyhedron?

16 Linked Polyhedra

The immediate surroundings of single atoms can be rationalized quite well with the aid of coordination polyhedra, at least when the polyhedra show a certain degree of symmetry to a good approximation. The most important polyhedra are presented in Fig. 2.2 (p. 5). Larger structural entities can be regarded as a system of linked polyhedra. Two polyhedra can be linked by sharing a common vertex, a common edge or a common face, *i.e.* they share one, two or three (or more) common bridging atoms (Fig. 2.3, p. 6).

Depending on the kind of polyhedron and the kind of linking, the resulting bond angles at the bridging atoms have a definite value or values confined within certain limits. The bond angle is fixed by geometry in the case of face-sharing polyhedra. The bond angle can be varied within certain limiting values for vertex-sharing and in some cases for edge-sharing polyhedra by mutually rotating the polyhedra (Fig. 16.1; *cf.* also Fig. 16.18, p. 181). The values of the bond angles are listed in Table 16.1; they refer to undistorted tetrahedra and octahedra, and it is assumed that the closest contact of any two atoms corresponds to the distance of two adjacent atoms within a polyhedron. Distortions occur frequently and allow for an additional range of angles. Distortions may involve differing lengths of the polyhedron edges, but they may also come about by shifting the central atom out of the polyhedron center, thus changing the bond angles at the central atom and the bridging atoms even when the polyhedron edges remain constant.

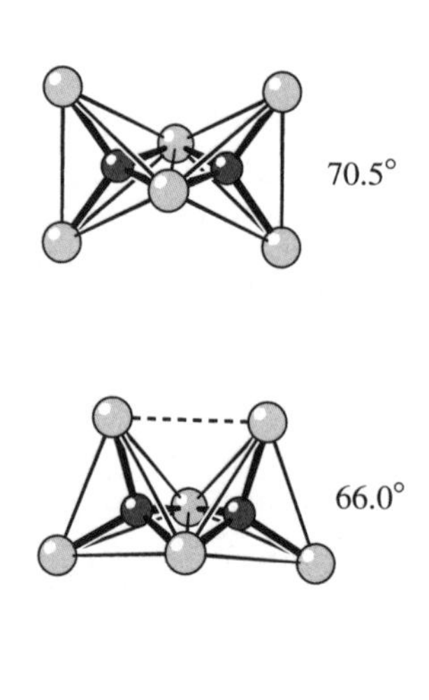

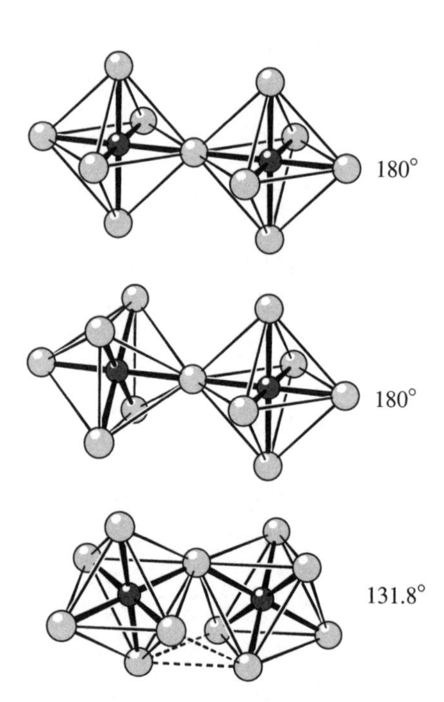

Fig. 16.1 Limits of the mutual rotation of vertex-sharing tetrahedra and of vertex-sharing octahedra and the resulting bond angles at the bridging atoms. The minimum distance between vertices of different polyhedra (dotted) was taken to be equal to the polyhedron edge

Inorganic Structural Chemistry, Second Edition Ulrich Müller

Table 16.1: Bond angles at the bridging atoms and distances between the central atoms M of linked tetrahedra and octahedra (disregarding possible distortions). The distances are given as multiples of the polyhedron edge length

		linking by		
		vertices	edges	faces
tetrahedron	bond angles	102.1 to 180°	66.0 to 70.5°	38.9°
	M–M distances	0.95 to 1.22	0.66 to 0.71	0.41
octahedron	bond angles	131.8 to 180°	90°	70.5°
	M–M distances	1.29 to 1.41	1.00	0.82

Distortions of coordination polyhedra can often be interpreted according to the GILLESPIE–NYHOLM rules and by taking into account the electrostatic forces. For example, a mutual repulsion of the Fe atoms can be perceived in the two edge-sharing tetrahedra of the $(FeCl_3)_2$ molecule; it can be ascribed to their positive partial charges. The Fe–Cl distances to the bridging atoms thus become longer than the remaining Fe–Cl bonds. The Cl atoms adjust their positions by a slight deformation of the tetrahedra (Fig. 16.2). If the bridging atoms have a more negative partial charge than the terminal atoms, they counterbalance this kind of distortion since they exert a stronger attraction towards the central atoms which now only experience a decreased shift from the polyhedron centers. $(FeSCl_2)_2^{2-}$, which is isoelectronic with $(FeCl_3)_2$, is an example (Fig. 16.2; in order to compare the electrostatic forces, in a simplified manner one can assume the existence of ions Fe^{3+}, Cl^- and S^{2-}).

The way in which polyhedra will join depends on several factors, which include:

1. Chemical composition. Only very definite patterns of linking polyhedra are consistent with a given composition.

2. The nature of the bridging atoms. They tend to attain certain bond angles and tolerate only bond angles within certain limits. Bridging sulfur, selenium, chlorine, bromine, and iodine atoms (having two lone electron pairs) favor angles close to 100°. This angle is compatible with vertex-sharing tetrahedra and with edge-sharing octahedra; however, examples with smaller angles among edge-sharing tetrahedra and face-sharing octahedra are known. Bridging oxygen and fluorine atoms allow for angles up to 180°; frequently observed values are in the range 130° to 150°.

3. Bond polarity. Very polar bonds do not harmonize with edge-sharing polyhedra and

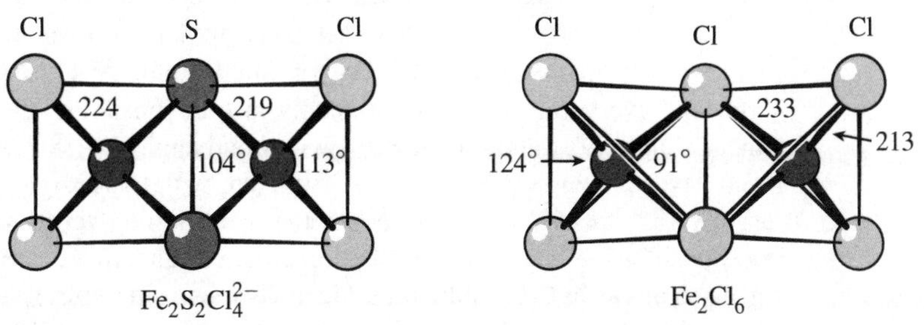

Fig. 16.2
Two edge-sharing tetrahedra showing only minor distortions in the $Fe_2S_2Cl_4^{2-}$ ion and two more distorted tetrahedra in the Fe_2Cl_6 molecule. The distortions can be ascribed mainly to the electrostatic repulsion between the Fe atoms. Bond lengths in pm

especially with face-sharing polyhedra because of the increased electrostatic repulsion between the central atoms (PAULING's third rule, p. 59). Central atoms in high oxidation states therefore favor vertex-sharing. If there are two kinds of central atom, those with the higher oxidation state will avoid having their polyhedra linked with one another (PAULING's fourth rule).

4. Interactions between the central atoms of the linked polyhedra. When a direct bond between the central atoms is advantageous, they tend to come close together. This favors edge-sharing or face-sharing arrangements. For example, the face-sharing of two octahedra in the $[W_2Cl_9]^{3-}$ ion renders the formation of a W≡W triple bond possible; in this way every tungsten atom gains electrons in addition to its electronic configuration d^3 and the electrons supplied by the ligands, thus attaining noble gas configuration (18 valence electrons).

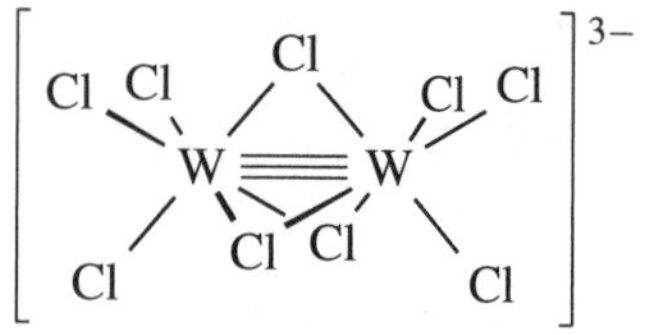

With our present knowledge, we often cannot understand, let alone predict, the more profound details concerning the kind of linking. Why does BiF_5 form linear, polymeric chains, SbF_5 tetrameric molecules and AsF_5 monomeric molecules? Why are there chloro and not sulfur bridges in $(WSCl_4)_2$? Why does no modification of TiO_2 exist which has the quartz structure?

The composition of a compound is intimately related to the way of linking the polyhedra. An atom X with coordination number c.n.(X) that acts as a common vertex to this number of polyhedra makes a contribution of 1/c.n.(X) to every polyhedron. If a polyhedron has n such atoms, this amounts to n/c.n.(X) for this polyhedron. This can be expressed with NIGGLI formulae, as shown in the following sections. To specify the coordination polyhedra, the formalism presented at the end of Section 2.1 and in Fig. 2.2 (p. 5) is useful.

16.1 Vertex-sharing Octahedra

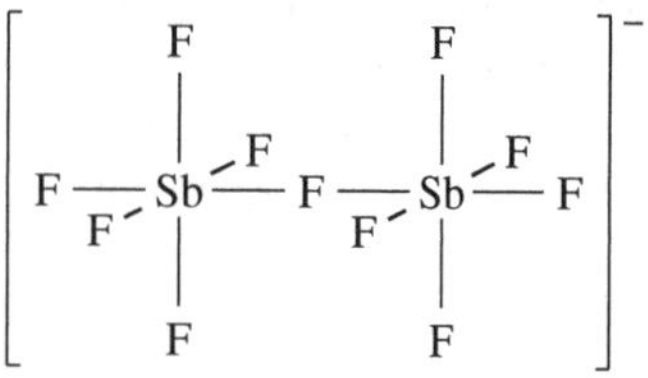

A single octahedral molecule has the composition MX_6. Two octahedra with a common vertex can be regarded as a unit MX_6 to which a unit MX_5 has been added, so that the composition is M_2X_{11}. If the addition of MX_5 units is continued, one obtains chain-like or ring-like molecules of composition $(MX_5)_n$ (Fig. 16.3). In these, every octahedron has four terminal atoms and two atoms that act as common vertices to other octahedra, corresponding to the NIGGLI formula $MX_{4/1}X_{2/2}$. If the two bridging vertices of every octahedron are mutually in *trans* positions, the result is a chain; it can either be entirely straight as in BiF_5, with bond angles of 180° at the bridging atoms, or it can have a zigzag shape as in the CrF_5^{2-} ion, with bond angles between 132° and 180° (usually 132° to 150° for fluorides). If the two bridging vertices of every octahedron are in *cis* positions, a large number of geometrical arrangements are possible. Among these, zigzag chains as in CrF_5 and tetrameric molecules as in $(NbF_5)_4$ are of importance. Again, the bond angles at the bridging atoms can have values from 132° to 180°. In pentafluorides, the frequent occurrence of angles of either 132° or 180° has to do with the packing of the molecules in the crystal: these two values result geometrically when the fluorine atoms for themselves form a hexagonal or a cubic closest-packing of spheres, respectively

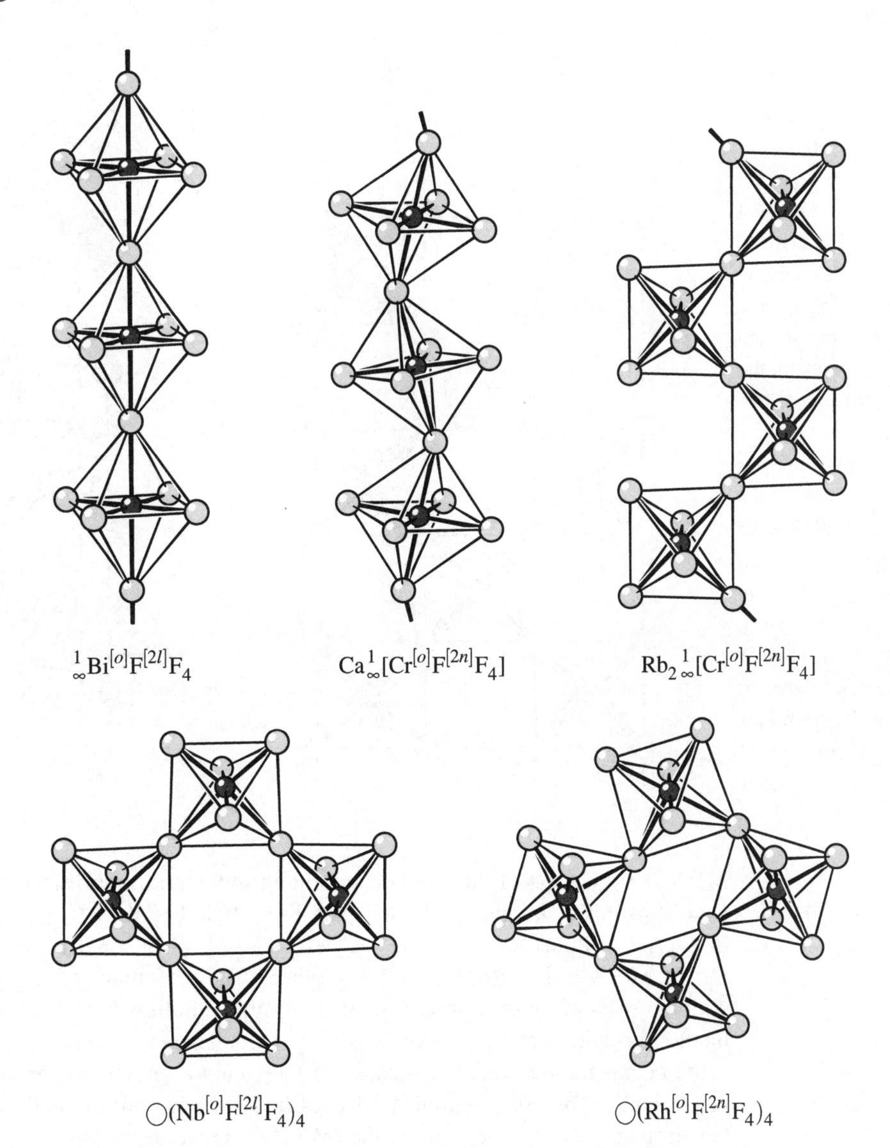

Fig. 16.3 Some possibilities for joining octahedra via common vertices to form MX_5 chains and rings

(this is discussed in more detail in Chapter 17). The most important linking patterns for pentafluorides, pentafluoro anions and oxotetrahalides are:

	octahedron configuration	bond angle at bridging atom approx.	examples
rings $(MF_5)_4$	*cis*	180°	$(NbF_5)_4$, $(MoF_5)_4$
rings $(MF_5)_4$	*cis*	132°	$(RuF_5)_4$, $(RhF_5)_4$
linear chains	*trans*	180°	BiF_5, UF_5, $WOCl_4$
zigzag chains	*trans*	150°	$Ca[CrF_5]$, $Ca[MnF_5]$
zigzag chains	*cis*	180°	$Rb_2[CrF_5]$
zigzag chains	*cis*	152°	VF_5, CrF_5, $MoOF_4$

The layer shown in the left part of Fig. 16.4 represents the most important way to join octahedra by sharing four vertices each; the composition is $MX_4 = MX_{2/1}X_{4/2}$ or

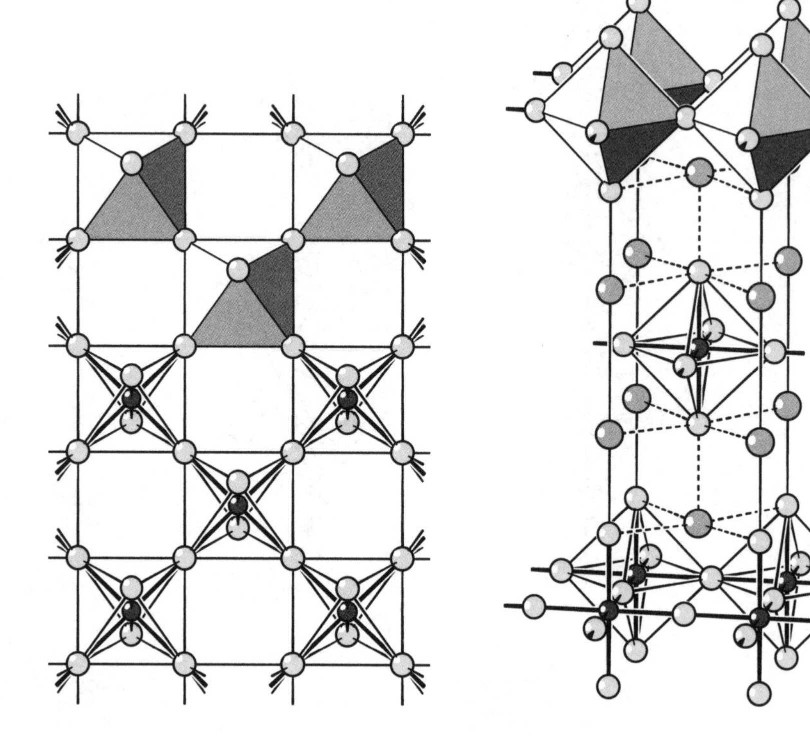

Fig. 16.4 MX_4 layer of vertex-sharing octahedra, and the packing of such layers in the K_2NiF_4 type. The packing in SnF_4 is obtained by leaving out the K^+ ions and shifting the layers towards each other in such a way that every octahedron apex of one layer comes to be between four apexes of the next layer

$M^{[o]}X_2^{[2l]}X_2$. Layers of this kind occur among some tetrafluorides such as SnF_4 and PbF_4 as well as in the anions of $Tl[AlF_4]$ and $K_2[NiF_4]$. The K_2NiF_4 type has been observed in a series of fluorides and oxides: K_2MF_4 with M = Mg, Zn, Co, Ni; Sr_2MO_4 with M = Sn, Ti, Mo, Mn, Ru, Rh, Ir, and some others. The preference of K^+ and Sr^{2+} has to do with the sizes of these cations: they just fit into the hollow between four F or O atoms of the non-bridging octahedron vertices (Fig. 16.4). Larger cations such as Cs^+ or Ba^{2+} fit if the octahedra are widened because of large central atoms, as for example in Cs_2UO_4 or Ba_2PbO_4. The composition A_2MX_4 is fulfilled when all the hollows between the octahedron apexes on either side of the $[MX_4]^{2n-}$ layer are occupied with A^{n+} ions. In the stacking of this kind of layer, the A^{n+} ions of one layer are placed exactly above the X atoms of the preceding layer. Every A^{n+} ion then has coordination number 9 (four of the bridging atoms in the layer, the four X atoms of the surrounding octahedron apexes and the one X atom of the next layer); the coordination polyhedron is a capped square antiprism.

If we stack MX_4 layers with octahedron apex on top of octahedron apex and amalgamate the apexes with each other, the result is the network of the ReO_3 structure with connections in three dimensions (Fig. 16.5). In this structure every octahedron shares all of its vertices with other octahedra; the bond angles at the bridging atoms are 180°. The centers of eight octahedra form a cube which corresponds to the unit cell. There is a rather large cavity in the center of the unit cell. This cavity can be occupied by a cation, which gives the perovskite type (perovskite = $CaTiO_3$); this structure type is rather frequent among compounds of the compositions AMF_3 and AMO_3, and because of its importance we discuss it separately (Section 17.4, p. 202).

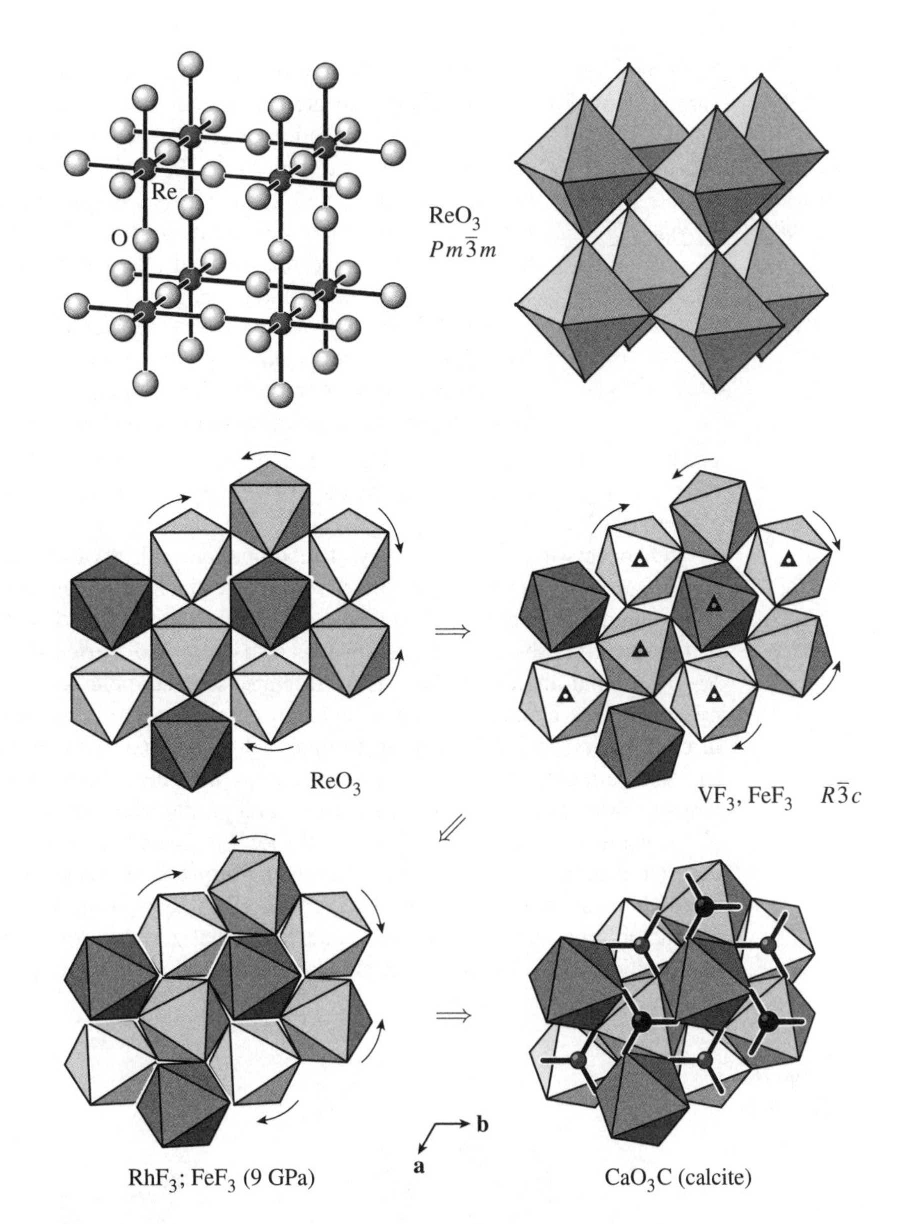

Fig. 16.5 Top: framework of octahedra sharing all vertices = ReO_3 type. Center and bottom: by rotating the octahedra, the ReO_3 type is converted to the VF_3, RhF_3 and calcite type. The unit cell of the VF_3 type is spanned by the threefold rotation axes running through the octahedra drawn in light gray

The degree of space filling of the ReO_3 type can be increased by rotating the octahedra about the direction of one of the space diagonals of the cubic unit cell (Fig. 16.5). Thereby the large cavity in the ReO_3 cell becomes smaller, the octahedra come closer to each other, and the bond angles at the bridging atoms decrease from 180° to 132°. Once this value is reached, we have the RhF_3 type, in which the F atoms are arranged as in a hexagonal closest-packing. A number of trifluorides crystallize with structures between these two extreme cases, with bond angles of about 150° at the F atoms: GaF_3, TiF_3, VF_3, CrF_3, FeF_3, CoF_3, and others. Some of them, such as ScF_3, are near to the ReO_3 type; others, such as MoF_3, are nearer to the hexagonal closest-packed structure. The mutual rotation of the octahedra can be continued to a 'superdense' sphere packing which contains groups which have three squeezed atoms each. This corresponds to the structure of calcite ($CaCO_3$); the carbon atom of the carbonate ion is located in the center between three squeezed O atoms.

Observed octahedron rotation angles for FeF_3

p/GPa	a/pm	c/pm	angle/°
10^{-4}	521	1332	17.0
1.5	504	1341	21.7
4.0	485	1348	26.4
6.4	476	1348	28.2
9.0	470	1349	29.8

The described rotation of the octahedra can actually be performed. At ambient pressure ($p = 10^{-4}$ GPa), the octahedra of FeF_3 are turned by 17.0°, as compared to ReO_3. As listed in the table in the margin, they rotate under high pressure up to nearly 30°, which corresponds to the ideal RhF_3 type. Simultaneously, the lattice parameter a is reduced, whereas c shows only minor changes.

The $LiSbF_6$ type results when the sites of the metal atoms in the VF_3 type are occupied alternately by atoms of two different elements; this structure type is frequent among compounds AMF_6, *e.g.* $ZnSnF_6$. Similarly, two kinds of metal atoms can alternate in the metal positions of RhF_3 packing; this applies to PdF_3 (and PtF_3), which has to be regarded as $Pd^{II}Pd^{IV}F_6$ or $Pd^{2+}[PdF_6]^{2-}$, as can be seen by the different Pd–F bond lengths of 217 pm (Pd^{II}) and 190 pm (Pd^{IV}).

WO_3 occurs in a greater variety of modifications, all of which are distorted forms of the ReO_3 type (with W atoms shifted from the octahedron centers and with varying W–O bond lengths). In addition, a form exists that can be obtained by dehydrating $WO_3 \cdot \frac{1}{3}H_2O$; its framework is shown in Fig. 16.6. This also consists of vertex-sharing octahedra, with W–O–W bond angles of 150°. This structure is remarkable because of the channels it contains. These channels can be occupied by potassium ions in varying amounts, resulting in compositions K_xWO_3 ranging from $x = 0$ to $x = 0.33$ (rubidium and caesium ions can also be included). These compounds are termed hexagonal *tungsten bronzes*. Cubic tungsten bronzes have the ReO_3 structure with partial occupation of the voids by Li^+ or Na^+, *i.e.* they are intermediate between the ReO_3 type and the perovskite type. Tetragonal tungsten bronzes are similar to the hexagonal bronzes, but have narrower four and five-sided channels that can take up Na^+ or K^+ (Fig. 16.6). Tungsten bronzes are metallic conductors, and have metallic luster and colors that go from gold to black, depending on composition. They are very resistant chemically and serve as industrial catalysts and as pigments in 'bronze colors'.

Fig. 16.6 Linking of the octahedra in hexagonal and tetragonal tungsten bronzes M_xWO_3. In the direction of view, the octahedra are arranged one on top of the other with common vertices. The channels in this direction contain varying amounts of alkali metal ions

16.2 Edge-sharing Octahedra

Two octahedra sharing one edge correspond to the composition $(MX_5)_2$ or $(MX_{4/1}X_{2/2})_2$. This is the kind of structure common among pentahalides and ions $[MX_5]_2^{n-}$ when X = Cl, Br or I:

$(SbCl_5)_2$	$(NbCl_5)_2$	$(TaCl_5)_2$	$(MoCl_5)_2$	$(WCl_5)_2$	$(ReCl_5)_2$	$(OsCl_5)_2$	$(UCl_5)_2$
$< -54\,°C$	$(NbBr_5)_2$	$(TaBr_5)_2$		$(WBr_5)_2$			$(UBr_5)_2$
	$(NbI_5)_2$	$(TaI_5)_2$					$(PaBr_5)_2$
	$[TiCl_5]_2^{2-}$	$[ZrCl_5]_2^{2-}$	$[MoCl_5]_2^{2-}$	$[WCl_5]_2^{2-}$		$[OsBr_5]_2^{2-}$	$(PaI_5)_2$

There are some exceptions in which the metal atoms are not coordinated octahedrally: $SbCl_5$ (monomeric above $-54\,°C$), PCl_5 (ionic $PCl_4^+PCl_6^-$), and PBr_5 (ionic $PBr_4^+Br^-$). $(MX_5)_2$ molecules can be packed very efficiently in such a way that the X atoms for themselves form a closest-packing of spheres.

If the linking is continued to form a string of edge-sharing octahedra, the resulting composition is $MX_{2/1}X_{4/2}$, *i.e.* MX_4. Every octahedron then has two common edges with other octahedra in addition to two terminal X atoms. If the two terminal X atoms have a mutual *trans* arrangement, the chain is linear (Fig. 16.7). This kind of chain occurs among tetrachlorides and tetraiodides if metal–metal bonds form in pairs between the M atoms of adjacent octahedra; the metal atoms are then shifted from the octahedron centers in the direction of the corresponding octahedron edge, and the octahedra experience some distortion. Examples are $NbCl_4$, NbI_4, WCl_4. The same kind of chain, but with metal atoms in the octahedron centers, has been observed for $OsCl_4$.

If the two terminal X atoms of an octahedron in an MX_4 chain have a *cis* arrangement, the chain can have a large variety of configurations. The most frequent one is a zigzag chain (Fig. 16.7); known examples include $ZrCl_4$, $TcCl_4$, $PtCl_4$, PtI_4, and UI_4. Chains with other configurations are rare; ZrI_4 is such an exception. Six edge-sharing octahedra can also join to form a ring (Fig. 16.7), but this kind of a structure is known for only one modification of $MoCl_4$.

By linking edge-sharing octahedra to form a layer as in Fig. 16.8, all X atoms act as bridging agents, and every one of them simultaneously belongs to two octahedra. This kind of layer encloses voids that have an octahedral shape (see also the figure on the cover). The composition of the layer is MX_3 $(MX_{6/2})$. Numerous trichlorides, tribromides and triiodides and also some trihydroxides are composed of layers of this kind. The layers

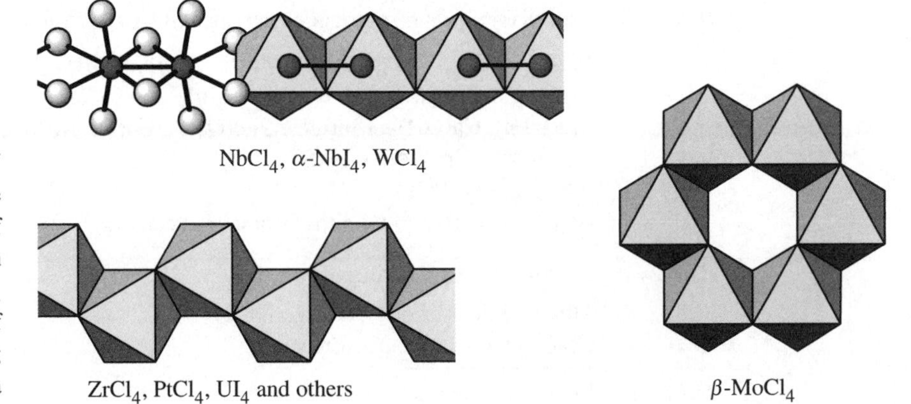

Fig. 16.7 Some configurations of chains with composition MX_4 consisting of edge-sharing octahedra

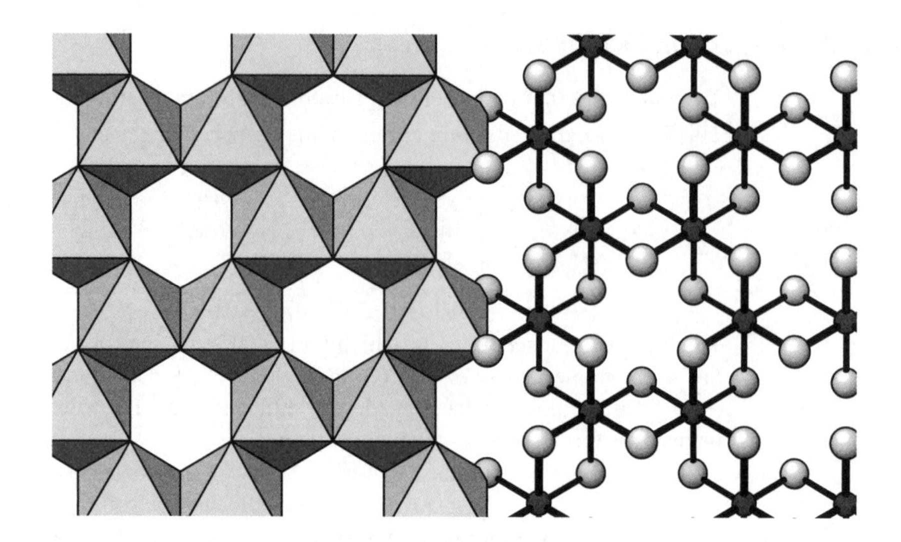

Fig. 16.8 Layer of edge-sharing octahedra in the BiI_3 and $AlCl_3$ type

are stacked in such a way that the X atoms themselves form a closest-packing of spheres, namely:

BiI_3 type: hexagonal closest-packing of X atoms
$FeCl_3$, $CrBr_3$, $Al(OH)_3$ (bayerite) and others

$AlCl_3$ type: cubic closest-packing of X atoms
YCl_3, $CrCl_3$ (high temperature) and others

The same kind of layers also occur in a second modification of $Al(OH)_3$, hydrargillite (gibbsite), but with a stacking in which adjacent O atoms of two layers are exactly one on top of the other; they are joined via hydrogen bridges.

The $CdCl_2$ and the CdI_2 type are also layer structures (Fig. 16.9; because there exist numerous stacking variants of CdI_2, some authors prefer the term $Cd(OH)_2$ type instead of CdI_2 type). The octahedra in the layer share six edges each. The structure of the layer is the same as in an MX_3 layer if the voids in the MX_3 layer were occupied by M atoms. Every halogen atom is shared by three octahedra ($MX_{6/3}$). The stacking variants of the layers are:

CdI_2 type ($Cd(OH)_2$ type): hexagonal closest-packing of X atoms

$MgBr_2$, $TiBr_2$, VBr_2, $CrBr_2{}^{\dagger}$, $MnBr_2$, $FeBr_2$, $CoBr_2$, $NiBr_2$, $CuBr_2{}^{\dagger}$
MgI_2, CaI_2, PbI_2, TiI_2, VI_2, $CrI_2{}^{\dagger}$, MnI_2, FeI_2, CoI_2
$Mg(OH)_2$, $Ca(OH)_2$, $Mn(OH)_2$, $Fe(OH)_2$, $Co(OH)_2$, $Ni(OH)_2$, $Cd(OH)_2$
SnS_2, TiS_2, ZrS_2, NbS_2, PtS_2
$TiSe_2$, $ZrSe_2$, $PtSe_2$
Ag_2F, Ag_2O (F and O in the octahedron centers)

$CdCl_2$ type: cubic closest-packing of X atoms

$MgCl_2$, $MnCl_2$, $FeCl_2$, $CoCl_2$, $NiCl_2$
Cs_2O (O in the octahedron centers)

† distorted by Jahn–Teller effect

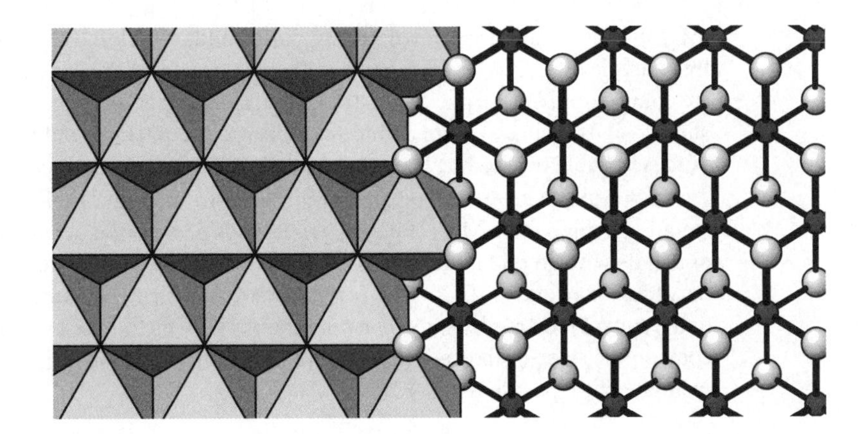

Fig. 16.9 Edge-sharing octahedra in a layer of the CdI_2 and $CdCl_2$ type

Among hydroxides such as $Mg(OH)_2$ (brucite) and $Ca(OH)_2$ the packing of the O atoms deviates from an ideal hexagonal closest-packing in that the layers are somewhat flattened; the bond angles M–O–M in the layer are larger than the ideal 90° for undistorted octahedra (*e.g.* 98.5° in $Ca(OH)_2$).

16.3 Face-sharing Octahedra

Two octahedra sharing a common face correspond to a composition M_2X_9 (Fig. 16.10). This structure is known for some molecules, for example $Fe_2(CO)_9$, and especially for some ions with trivalent metals. In some cases, the reason for the face-sharing is the presence of metal–metal bonds, for example in the $[W_2Cl_9]^{3-}$ ion; its small magnetic moment suggests a W≡W bond (*cf.* the structure shown on p. 168). The $[Cr_2Cl_9]^{3-}$ ion has the same structure, but nevertheless, it exhibits the paramagnetism that is to be expected for the electron configuration d^3; $[Mo_2Cl_9]^{3-}$ is intermediate in its behavior. The bond angles at the bridging atoms also reflect these differences: 58° in $[W_2Cl_9]^{3-}$, 77° in $[Cr_2Cl_9]^{3-}$.

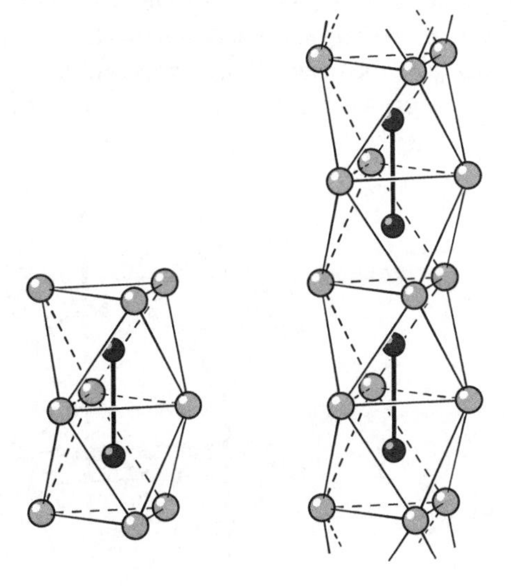

Fig. 16.10 Two face-sharing octahedra in ions $[M_2X_9]^{3-}$ and a string of face-sharing octahedra in ZrI_3

$[M_2X_9]^{3-}$ ions are also known among compounds which have no metal–metal bonding, for example $[Tl_2Cl_9]^{3-}$ or $[Bi_2Br_9]^{3-}$. The occurrence of this kind of ion often depends on the counter-ion, *i.e.* the packing in the crystal is an important factor. For example, Cs^+ ions and Cl^- ions, being of comparable size, allow close packing and facilitate the occurrence of these double octahedra. They also occur with large cations such as $P(C_6H_5)_4^+$.

If opposite faces of the octahedra are used to continue their linking, the result is a strand of composition MX_3 (Fig. 16.10). Strands of this kind occur among some trihalides of metals with an odd number of *d* electrons. Between pairs of adjacent octahedra metal–metal bonds occur: β-$TiCl_3$, ZrI_3 (d^1), $MoBr_3$ (d^3), $RuCl_3$, $RuBr_3$ (d^5). Anionic strands of the same kind are also known in compounds such as $Cs[NiCl_3]$ or $Ba[NiO_3]$; again, the comparable sizes of the cations Cs^+ and Ba^{2+} and the anions Cl^- and O^{2-}, respectively, facilitate close packing.

16.4 Octahedra Sharing Vertices and Edges

Units $(MX_5)_2$ can be joined via common vertices to form double strands as shown in Fig. 16.11. Since every octahedron still has two terminal atoms, the composition $MX_{2/1}X_{2/2}Z_{2/2}$ or MX_3Z results, the atoms in the common vertices being designated by Z. The same structure also results when two parallel strands of the BiF_5 type are joined via common edges. Compounds such as $NbOCl_3$, $MoOBr_3$ or WOI_3 have this kind of structure, with oxygen atoms taking the positions of the common vertices.

In rutile every O atom is common to three octahedra, as expressed by the formula $TiO_{6/3}$. As can be seen in Fig. 16.12, linear strands of edge-sharing octahedra are present, as in compounds MX_4. Parallel strands are joined by common octahedron vertices. Compared with the layers of the CdI_2 type, the number of common edges is reduced, namely two instead of six per octahedron. According to PAULING's third rule (p. 59), this favors the rutile type for electrostatic reasons. Compounds MX_2 with octahedrally coordinated M atoms therefore prefer the rutile type to the $CdCl_2$ or CdI_2 type if they are very polar: among dioxides and difluorides the rutile type is very common, examples being GeO_2, SnO_2, CrO_2, MnO_2, and RuO_2 as well as MgF_2, FeF_2, CoF_2, NiF_2, and ZnF_2.

In rutile the metal atoms in a strand of edge-sharing octahedra are equidistant. In some dioxides, however, alternating short and long M–M distances occur, *i.e.* the metal atoms are shifted in pairs from the octahedron centers towards each other. This phenomenon occurs (though not always) when the metal atoms still have *d* electrons and thus can engage

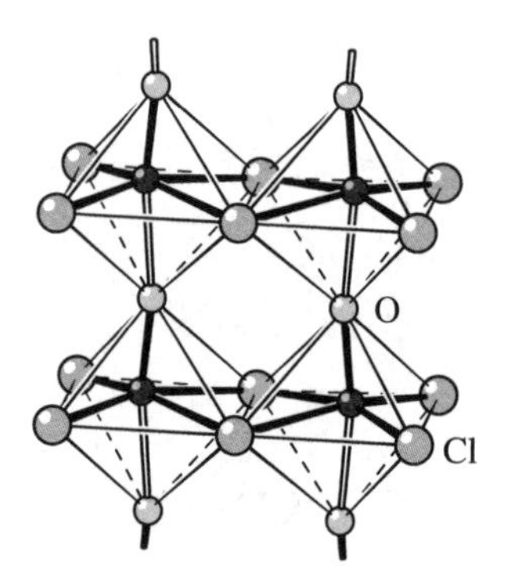

Fig. 16.11
Linking of octahedra in $NbOCl_3$ with alternating short and long Nb–O bonds

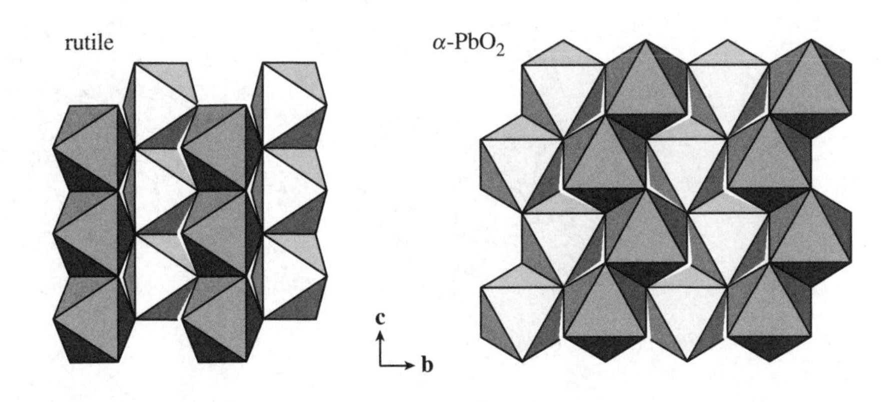

Fig. 16.12 Strands of edge-sharing octahedra running parallel to **c** are joined with each other in rutile and α-PbO_2 via common vertices

in metal–metal bonds, for example in the low-temperature modifications of VO_2, NbO_2, MoO_2, and WO_2 (the high-temperature forms have the normal rutile structure).

The zigzag chains of edge-sharing octahedra that occur among compounds MX_4 can also be joined by common vertices, resulting in the α-PbO_2 type (Fig. 16.12). This structure type is less frequent.

Octahedra linked in different ways often occur when different kinds of metal atom are present. Li_2ZrF_6 offers an example. The Li and the F atoms are arranged in layers of the same kind as in BiI_3. The layers are joined by single ZrF_6 octahedra which are placed below and on top of the voids of the layer (Fig. 16.13). The octahedra of the Li_2F_6 layer share common edges with one another and they share vertices with the ZrF_6 octahedra.

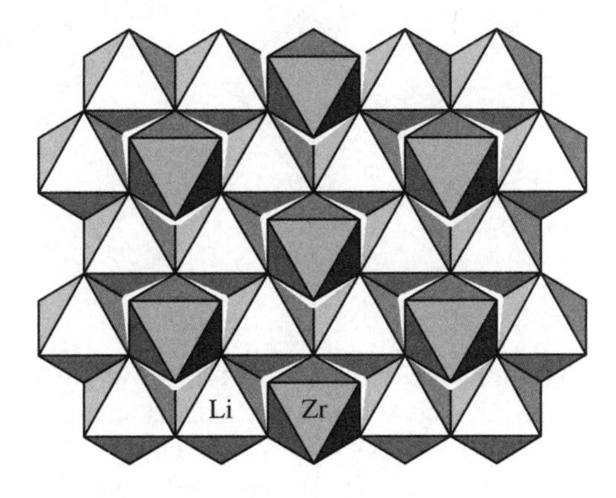

Fig. 16.13 Linking of octahedra in Li_2ZrF_6 and Sn_2PbO_6

Isopoly and Heteropoly Acids

Numerous linking patterns, some of which are very complex, consisting mainly of octahedra sharing vertices and edges are known among polyvanadates, niobates, tantalates, molybdates, and tungstates. If only one of these elements occurs in the polyhedra, they are also called isopoly acids or isopoly anions. If additional elements also form part of the structures, they are called heteropoly acids; the additional atoms can be coordinated tetrahedrally, octahedrally, square-antiprismatically or icosahedrally. The dodecamolybdatophosphate $[PO_4Mo_{12}O_{36}]^{3-}$ is the classical example of this compound class; the precipitation of its ammonium salt serves as an analytical proof for phosphate ions. It has the KEGGIN structure: four groups consisting of three edge-sharing octahedra are

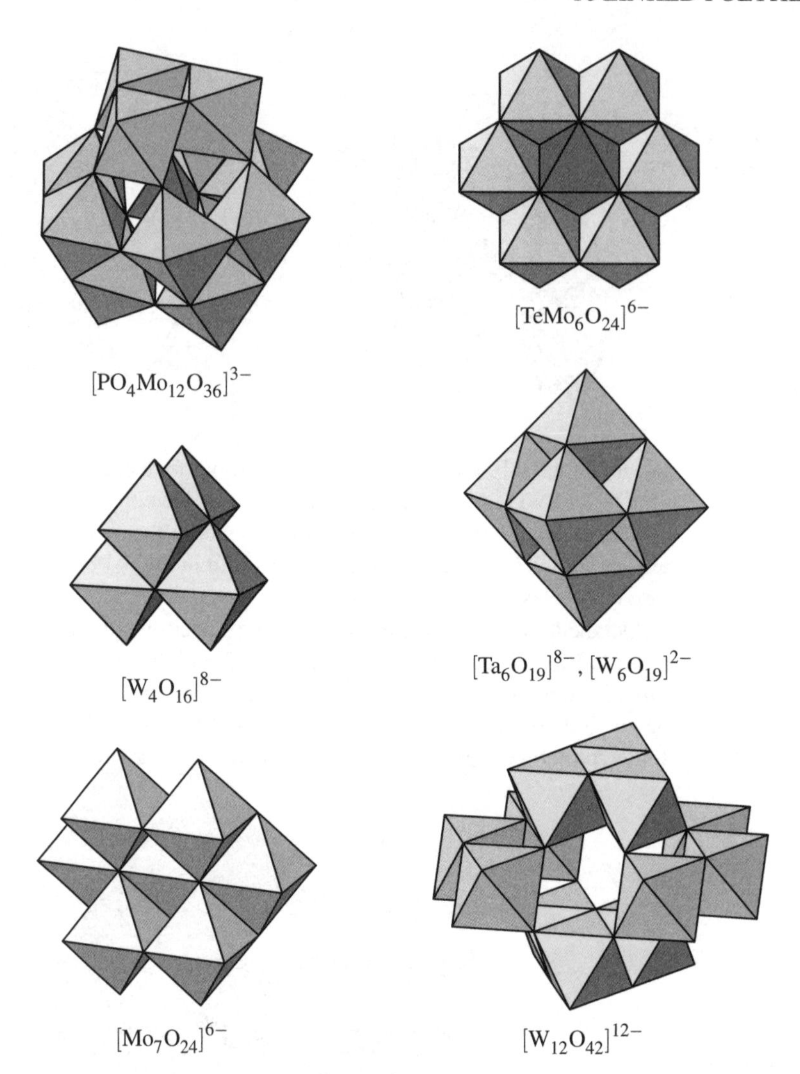

Fig. 16.14 Structures of some heteropoly and isopoly anions

linked by common vertices, so that the twelve MoO_6 octahedra form a cage (Fig. 16.14). The tetrahedrally coordinated phosphorus atom is in the interior of the cage; it can be replaced by Al(III), Si(IV), As(V) and others. Octahedrally coordinated heteroatoms are found in the ions $[EMo_6O_{24}]^{n-}$, *e.g.* $[TeMo_6O_{24}]^{5-}$ [E = Te(VI), I(VII), Mn(IV)].

Some isopoly anions consist of a compact system of edge-sharing octahedra; a few examples are shown in Fig. 16.14. Oxygen atoms with high coordination numbers are situated in their interior; for example, the O atom at the center of the $[W_6O_{19}]^{2-}$ ion has coordination number 6. Other representatives with part of their octahedra sharing only vertices can have more or less large cavities in their interior, such as, for example, the $[W_{12}O_{42}]^{2-}$ ion. Isopoly anions are formed in aqueous solutions, depending on the pH value. Molybdate solutions, for example, contain MoO_4^{2-} ions at high pH values, $[Mo_7O_{24}]^{6-}$ ions at pH ≈ 5 and even larger aggregates in more acidic solutions. 'Giant wheels' can be obtained by reduction of part of the molybdenum atoms to Mo(IV), as in $H_{48}Mo_{176}O_{536}$; the molecule consists of groups of vertex-sharing MoO_6 octahedra that share edges with pentagonal bipyramids, altogether forming a hoop with an inner diameter of 2.3 nm.

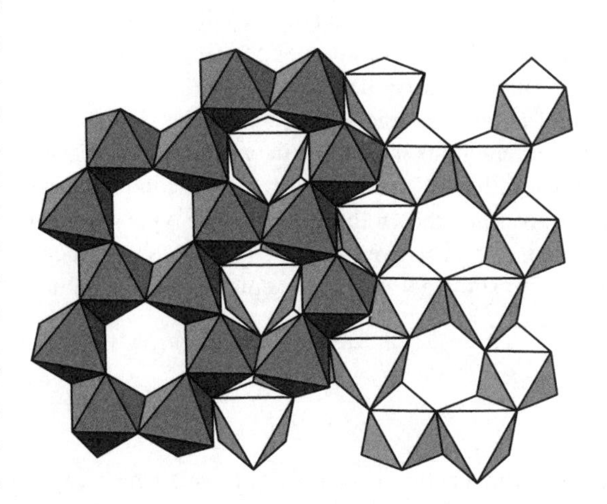

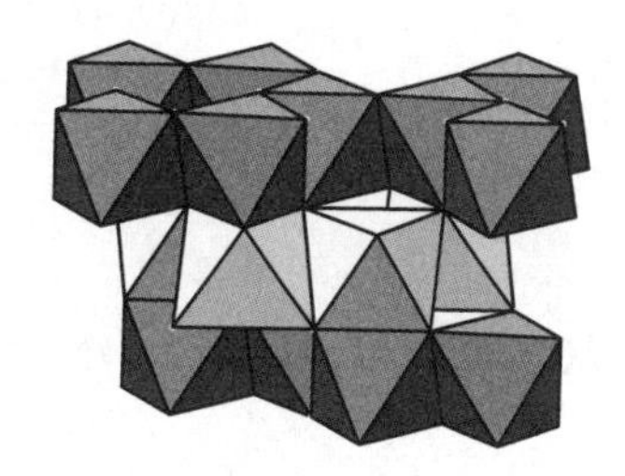

Fig. 16.15 Linked octahedra in corundum (α-Al_2O_3) and in ilmenite ($FeTiO_3$); Fe light, Ti dark octahedra. Left: Plan view of two layers (both layers are shown only in the central part). Right: Side view of sections of three layers with face-sharing octahedra

16.5 Octahedra Sharing Edges and Faces

The corundum structure (α-Al_2O_3) is the result of linking layers of the BiI_3 kind one on top of the other; the layers are mutually shifted as shown in Fig. 16.15. There are pairs of face-sharing octahedra and, in addition, every octahedron shares three edges within a layer and three vertices with octahedra from the adjacent layer to which it has no face-sharing connection. This structure type is adopted by some oxides M_2O_3 (*e.g.* Ti_2O_3, Cr_2O_3, α-Fe_2O_3).

Alternate layers can be occupied by two different kinds of metal atom, then every pair of the face-sharing octahedra contains two different metal atoms; this is the ilmenite type ($FeTiO_3$). Ilmenite is, along with perovskite, another structure type for the composition $A^{II}M^{IV}O_3$. The space for the A^{2+} ion is larger in perovskite. Which structure type is preferred can be estimated with the aid of the ionic radius ratio:

$r(A^{2+})/r(O^{2-}) < 0.7$ ilmenite
$r(A^{2+})/r(O^{2-}) > 0.7$ perovskite

Another criterion for the same purpose is discussed on p. 203.

The **nickel arsenide type** (NiAs) is the result of linking layers of the kind as in cadmium iodide. Continuous strands of face-sharing octahedra perpendicular to the layers

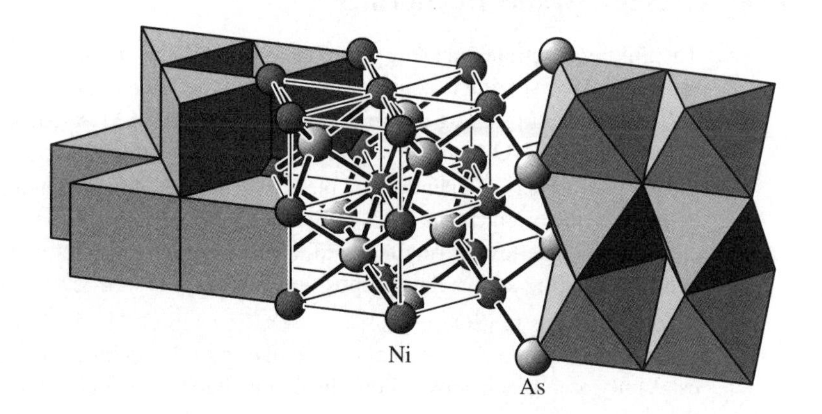

Fig. 16.16 Octahedra and trigonal prisms in the NiAs structure

arise from this connection. The same structure results when strands of the kind found in ZrI_3 are linked from all sides by common edges (Fig. 16.16). The nickel atoms in the octahedron centers form a primitive hexagonal lattice; every arsenic atom is surrounded by a trigonal prism of Ni atoms. Since the atoms in the face-sharing octahedra are quite close to one another, interactions must exist between them. This shows up in the electric properties: compounds with the NiAs structure are semiconductors or metallic conductors. Numerous representatives are known for this structure type: the metallic component is an element of the titanium to nickel groups, and Ga, Si, P, S, and their heavier homologous elements are adequate to substitute for the arsenic. Examples include TiS, TiP, CoS, and CrSb.

16.6 Linked Trigonal Prisms

In NiAs the Ni atoms form a network of trigonal prisms which contain the As atoms, and the Ni atoms have octahedral coordination. Metal atoms with trigonal-prismatic coordination are present in MoS_2. The S atoms form hexagonal planes with the stacking sequence *AABBAABB*... or *AABBCC*... (or some additional stacking variants). In every pair of congruent planes, *e.g.* *AA*, there are edge-sharing trigonal prisms which contain the Mo atoms and form a layer (Fig. 16.17). Between the layers, *i.e.* between sulfur atom planes having different positions, *e.g.* *AB*, the only attractive forces are weak VAN DER WAALS interactions. The MoS_2 layers can easily slip as in graphite; for this reason MoS_2 is used as a lubricant. Further similarities to graphite include the anisotropic electrical conductivity and the ability to form intercalation compounds, *e.g.* $K_{0.5}MoS_2$.

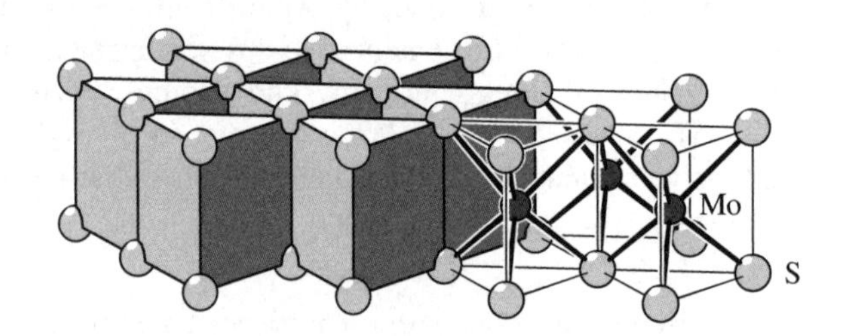

Fig. 16.17 Layer of edge-sharing trigonal prisms in MoS_2

16.7 Vertex-sharing Tetrahedra. Silicates

The linking of tetrahedra takes place predominantly by sharing vertices. Edge-sharing and especially face-sharing is considerably less frequent than among octahedra.

Two tetrahedra sharing a common vertex form a unit M_2X_7. This unit is known among oxides such as Cl_2O_7 and Mn_2O_7 and among several anions, *e.g.* $S_2O_7^{2-}$, $Cr_2O_7^{2-}$, $P_2O_7^{4-}$, $Si_2O_7^{6-}$ and $Al_2Cl_7^-$. Depending on the conformation of the two tetrahedra, the bond angle at the bridging atom can have values between 102.1° and 180° (Fig. 16.18).

A chain of vertex-sharing tetrahedra results when every tetrahedron has two terminal and two bridging atoms; the composition is $MX_{2/1}X_{2/2}$ or MX_3. The chain can be closed to form a ring as in $[SO_3]_3$, $[PO_3^-]_3$, $[SiO_3^{2-}]_3$ or $[SiO_3^{2-}]_6$. Endless chains have different shapes depending on the mutual conformation of the tetrahedra (Fig. 16.19). They occur especially among silicates, where the chain shape is also determined by the interactions

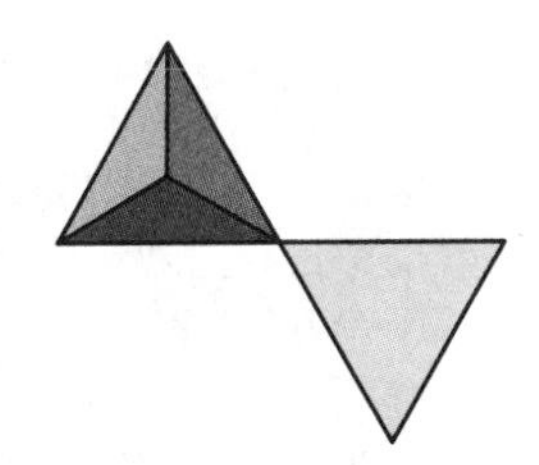
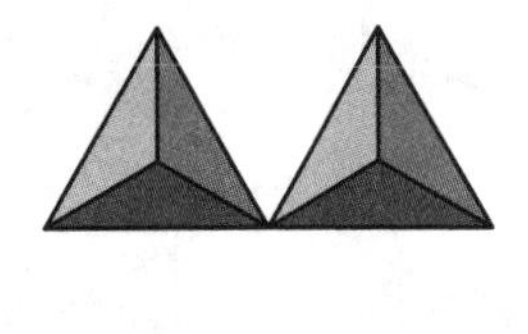
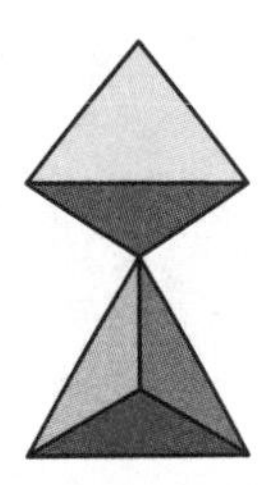

Fig. 16.18 Different conformations of two vertex-sharing tetrahedra

with the cations. In silicates of the composition $MSiO_3$ with octahedrally coordinated M^{2+} ions (M^{2+} = Mg^{2+}, Ca^{2+}, Fe^{2+}, and others, ionic radii 50 to 100 pm), the coordination octahedra of the metal ions are arranged to form layers as in $Mg(OH)_2$. Thus the octahedra share edges, and their vertices are also shared with vertices of the SiO_3^{2-} chains corresponding to terminal O atoms of the chain; these O atoms thus link tetrahedra with octahedra. Different chain conformations occur, depending on the kind of cation, *i.e.* octahedron size (Fig. 16.19). Compounds of this kind such as enstatite, $MgSiO_3$, are termed pyroxenes if the silicate chain is a *zweierkette*, *i.e.* if the chain pattern repeats after two tetrahedra; pyroxenoids have more complicated chain forms, for example the *dreierkette* in wollastonite, $CaSiO_3$.*

Linked tetrahedra in P_4O_{10}

Tetrahedra linked via three vertices correspond to a composition $MX_{1/1}X_{3/2}$ or $MX_{2.5}$ = M_2X_5. Small units consisting of four tetrahedra are known in P_4O_{10}, but most important are the layer structures in the numerous sheet silicates and aluminosilicates with anions of the compositions $[Si_2O_5^{2-}]_\infty$ and $[AlSiO_5^{3-}]_\infty$. Because the terminal vertices of the single

*From the German *zwei* = two, *drei* = three, *zweier* = two-membered, *dreier* = three-membered, *Kette* = chain.

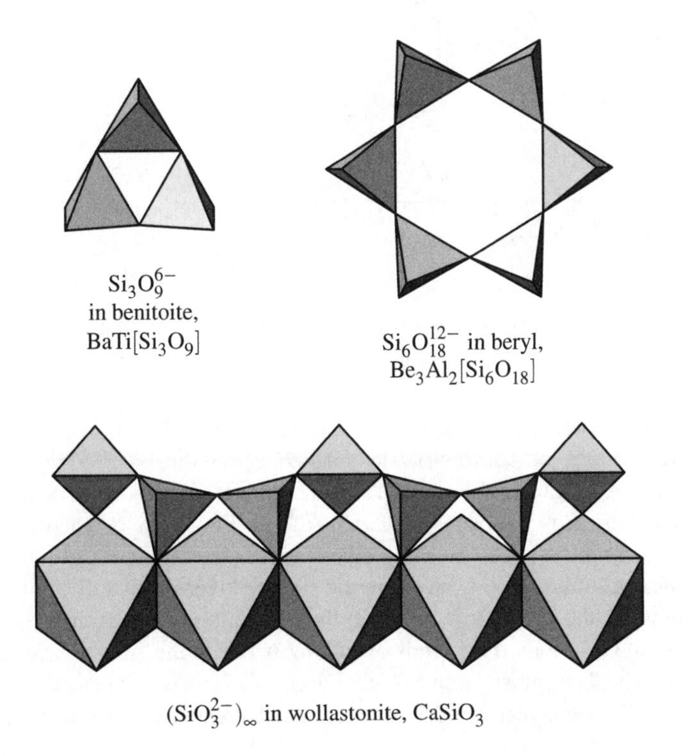

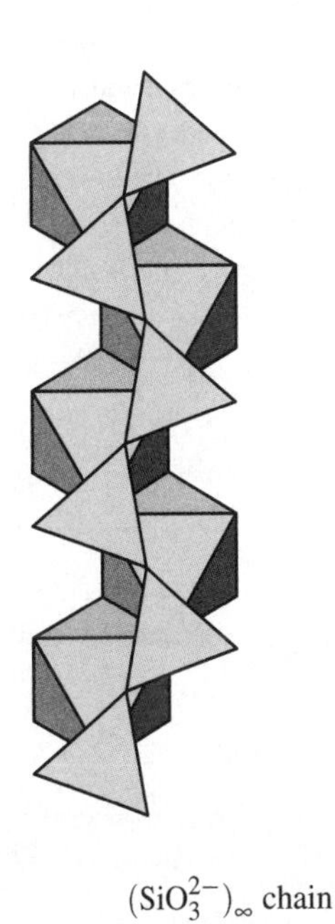

Fig. 16.19 Some forms of rings and chains of vertex-sharing tetrahedra in silicates. How the chain conformations adapt to the size of the cation octahedra is shown for two chains (the octahedron chain is a section of a layer)

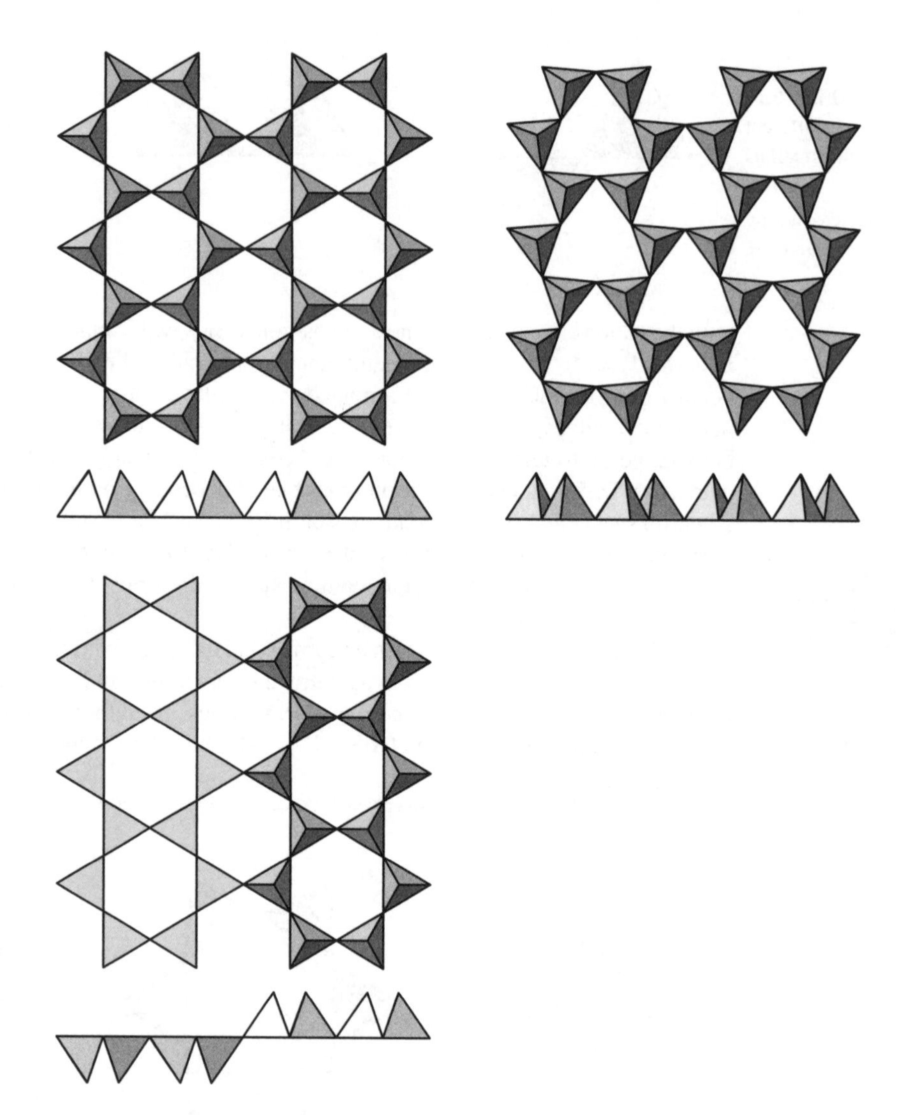

Fig. 16.20 Some arrangements of tetrahedra in sheet silicates. The lower image in each case represents a side view

tetrahedra can be oriented in different sequences to one or the other side of the layer, a large number of structural varieties are possible; moreover, the layers can be corrugated (Fig. 16.20).

Sheet silicates are of frequent natural occurrence, the most important ones being clay minerals (prototype: kaolinite), talc (soapstone) and micas (prototype: muscovite). In these minerals the terminal O atoms of a silicate layer are bonded with octahedrally coordinated cations; these are mainly Mg^{2+}, Ca^{2+}, Al^{3+} or Fe^{2+}. The octahedra are linked with each other by common edges, forming layers as in $Mg(OH)_2$ ($\widehat{=}$ CdI_2) or $Al(OH)_3$ ($\widehat{=}$ BiI_3). The number of terminal O atoms in the silicate layer is not sufficient to provide all of the O atoms for the octahedron layer, so the remaining octahedron vertices are occupied by additional OH^- ions. Two kinds of linking between the silicate layer and the octahedron layer can be distinguished: in *cation-rich sheet silicates* an octahedron layer is linked with only one silicate layer on one of its sides, and the result is an octahedron–tetrahedron sheet;

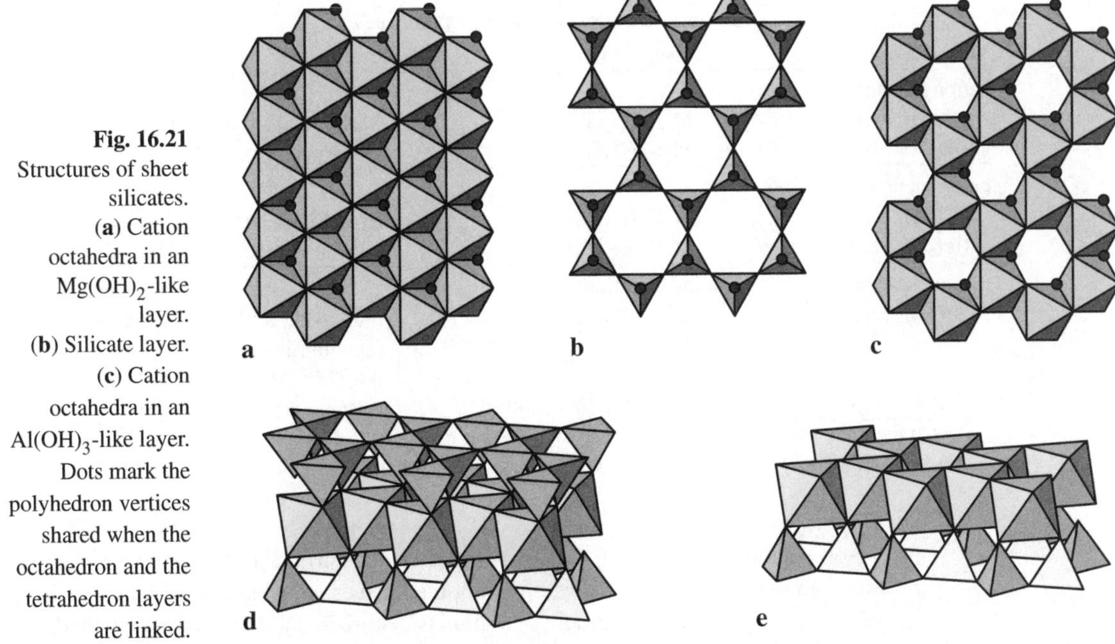

Fig. 16.21 Structures of sheet silicates. (**a**) Cation octahedra in an $Mg(OH)_2$-like layer. (**b**) Silicate layer. (**c**) Cation octahedra in an $Al(OH)_3$-like layer. Dots mark the polyhedron vertices shared when the octahedron and the tetrahedron layers are linked. Connection of the layers in: (**d**) cation-poor sheet silicates, (**e**) cation-rich sheet silicates. Octahedron vertices that do not act as common vertices with tetrahedra are occupied by OH^- ions

in *cation-poor sheet silicates* there are tetrahedron-octahedron-tetrahedron sheets, an octahedron layer being linked to two silicate layers on either side (Fig. 16.21). Depending on whether the cation layer is of the $Mg(OH)_2$ or the $Al(OH)_3$ type, certain numerical relations result between the number of cations in the octahedron layer and the number of silicate tetrahedra; the number of OH^- ions needed to complete the octahedra is also fixed by geometry. Additional cations may be intercalated between the sheets (Table 16.2).

The sheets consisting of tetrahedron-octahedron-tetrahedron layers in cation-poor sheet silicates are completely planar due to the symmetrical environment of the cation layer. If the sheets are electrically neutral as in talc, the attractive forces between them are weak; as a consequence, the crystals are soft and easy to cleave. The use of talc as powder, lubricating agent, polishing material and filling material for paper is due to these properties.

Micas are cation-poor sheet silicates consisting of electrically charged sheets that are held together by intercalated, unhydrated cations. For this reason the sheets cannot slip as in talc, but the crystals can be cleaved parallel to the sheets. The crystals usually form thin, stiff plates. Larger plates (in sizes from centimeters to meters) are used industrially because of their ruggedness, transparency, electrical insulating properties, and chemical and thermal resistance (muscovite up to approximately 500 °C, phlogopite, $KMg_3(OH)_2[AlSi_3O_{10}]$, up to approximately 1000 °C).

Clay materials show a different behavior. They are either cation-poor or cation-rich sheet silicates. They can swell by taking up varying amounts of water between the sheets. If the intercalated cations are hydrated as in montmorillonite, they act as cation exchangers. Montmorillonite, especially when it has intercalated Ca^{2+} ions, has thixotropic properties and is used to seal up drill holes. The effect is due to the charge distribution on

Table 16.2: Different kinds of sheet silicates

cation layer	composition	examples
cation-rich sheet silicates		
$Al(OH)_3$ type	$M_2(OH)_4[T_2O_5]$	kaolinite, $Al_2(OH)_4[Si_2O_5]$
$Mg(OH)_2$ type	$M_3(OH)_4[T_2O_5]$	chrysotile, $Mg_3(OH)_4[Si_2O_5]$
cation-poor sheet silicates		
$Al(OH)_3$ type	$M_2(OH)_2[T_4O_{10}]$	pyrophyllite, $Al_2(OH)_2[Si_4O_{10}]$
$Mg(OH)_2$ type	$M_3(OH)_2[T_4O_{10}]$	talc (soapstone), $Mg_3(OH)_2[Si_4O_{10}]$
cation-poor sheet silicates with intercalations		
$Al(OH)_3$ type	$A\{M_2(OH)_2[T_4O_{10}]\}$	muscovite, $K\{Al_2(OH)_2[AlSi_3O_{10}]\}$
	$A_x\{M_2(OH)_2[T_4O_{10}]\} \cdot nH_2O$	montmorillonite, $Na_x\{Mg_xAl_{2-x}(OH)_2[Si_4O_{10}]\} \cdot nH_2O$
$Mg(OH)_2$ type	$M_3(OH)_2[T_4O_{10}] \cdot nH_2O$	vermiculite, $Mg_3(OH)_2[Si_4O_{10}] \cdot nH_2O$

T = tetrahedrally coordinated Al or Si
$M = Mg^{2+}, Ca^{2+}, Al^{3+}, Fe^{2+}$ etc. $A = Na^+, K^+, Ca^{2+}$ etc.

the crystal platelets: they bear a negative charge on the surface and a positive charge on the edges. In suspension they therefore orient themselves edge against surface, resulting in a jelly. Upon agitation the mutual orientation is disturbed and the mass is liquefied.

Swollen clay materials are soft and easy to mould. They serve to produce ceramic materials. High quality fire-clay has a high kaolinite content. Upon firing, the intercalated water is removed first at approximately 100 °C. Then, beginning at 450 °C, the OH groups are converted to oxidic O atoms by liberation of water, and after some more intermediate steps, mullite is formed at approximately 950 °C. Mullite is an aluminum aluminosilicate, $Al_{(4-x)/3}[Al_{2-x}Si_xO_5]$ with $x \approx 0.6$ to 0.8.

Because the dimensions of an octahedron and a tetrahedron layer usually do not coincide exactly, the unilateral linking of the layers in cation-rich sheet silicates leads to tensions. If the dimensions do not deviate too much, the tension is relieved by slight rotations of the tetrahedra and the sheets remain planar. This applies to kaolinite, which has only Al^{3+} ions in the cation layer. With the larger Mg^{2+} ions the metric fit is inferior; the tension then causes a bending of the sheets. This can be compensated for by tetrahedron apexes periodically pointing to one side and then to the other side of the tetrahedron layer, as in antigorite (Fig. 16.22). If the bending is not compensated for, the sheets curl up to form tubes in the way shown in Fig. 16.22, corresponding to the structure of chrysotile, $Mg_3(OH)_4[Si_2O_5]$. Because the sheet only tolerates curvatures within certain limits, the tubes remain hollow and they cannot exceed some maximum diameter. The inner diameter in chrysotile is about 5 nm, the outer one 20 nm. The tubular building blocks explain the fibrous properties of chrysotile which used to be the most important asbestos mineral.

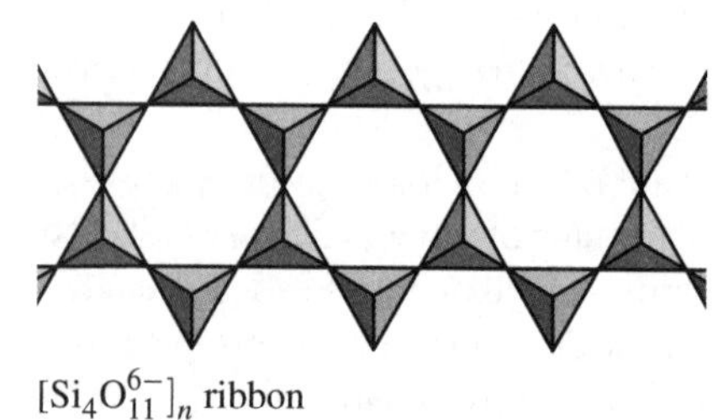
$[Si_4O_{11}^{6-}]_n$ ribbon

A layer of tetrahedra can be considered as being built up by linking parallel chains. That this is not a mere formalism is shown by the existence of intermediate stages. Two linked silicate chains result in a ribbon of the composition $[Si_4O_{11}^{6-}]_n$; it has two kinds of tetrahedra, one kind being joined via three and the other kind via two vertices, $[SiO_{1/1}O_{3/2}SiO_{2/1}O_{2/2}]^{3-}$. Silicates of this type are termed amphiboles. They are fibrous and also used to be used as asbestos.

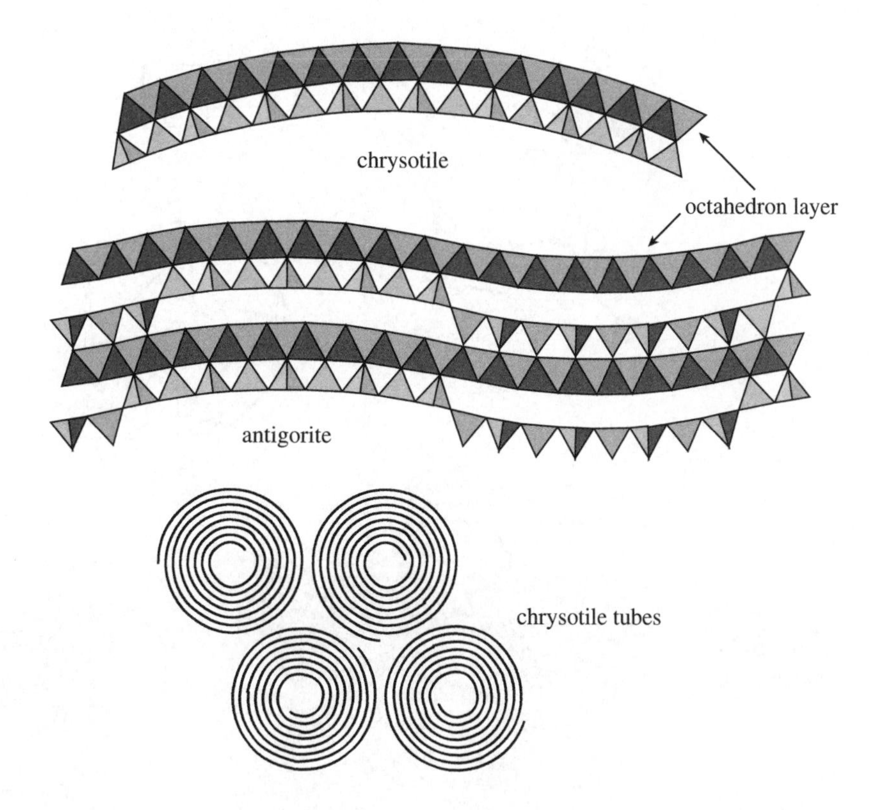

Fig. 16.22 Bent sheets of linked octahedron and tetrahedron layers in chrysotile and antigorite. Bottom: rolled sheets forming tubes in chrysotile

Linking Tetrahedra by All Four Vertices. Zeolites

Mercury iodide offers an example of a layer structure consisting of tetrahedra sharing all of their vertices (Fig. 16.23). Much more frequent are framework structures; they include the different modifications of SiO_2 and the aluminosilicates that are discussed in Section 12.5. Another important class of aluminosilicates are the zeolites. They occur as minerals, but are also produced industrially. They have structures consisting of certain polyhedra that are linked in such a way that hollows and channels of different sizes and shapes are present.

Fig. 16.24 shows the structure of the methyloctasiloxane, $Me_8Si_8O_{12}$, which can be made by hydrolysis of $MeSiCl_3$. Its framework is a cube of silicon atoms linked via oxygen

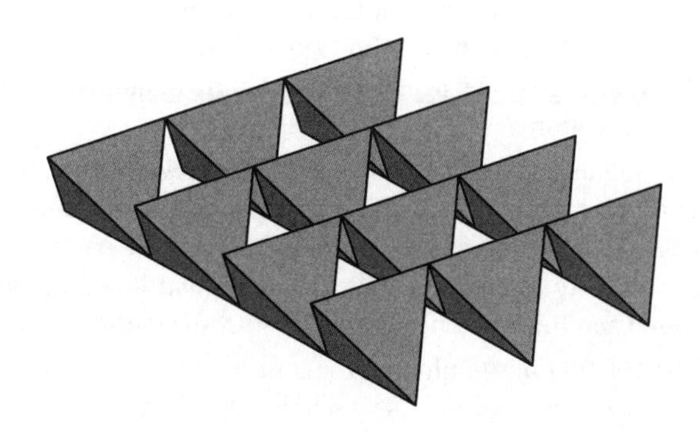

Fig. 16.23 Section of a layer in HgI_2

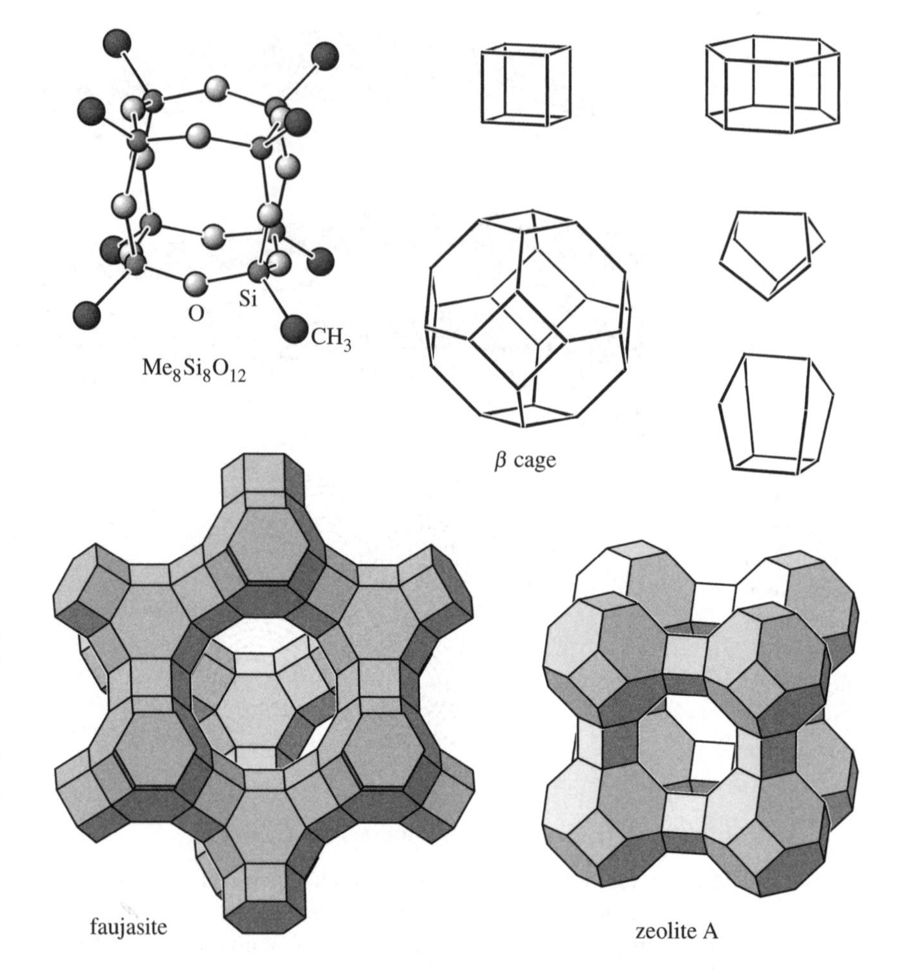

Fig. 16.24 Structure of $Me_8Si_8O_{12}$ and schematic representations of some Si–O polyhedra; linking of these polyhedra to the frameworks of two zeolites

atoms placed on each cube edge. The O atoms are situated slightly to one side of the edges, thus allowing for a framework without tensions and bond angles of 109.5° at the Si atoms and of 148.4° at the O atoms. This framework has been drawn schematically in the rest of Fig. 16.24 as a simple cube. It is one of several possible building units occurring in zeolites; the place of the methyl groups is taken by O atoms which mediate the connection to other Si atoms. In addition to the cube, other polyhedra occur, some of which are shown in Fig. 16.24. Every vertex of these polyhedra is occupied by an Si or Al atom, and in the middle of each edge there is an O atom which joins two of the atoms in the vertices. In a zeolite four edges meet at every vertex, corresponding to the four bonds of the tetrahedrally coordinated atoms.

The linking pattern of two zeolites is shown in Fig. 16.24. They have the 'β-cage' as one of their building blocks, that is, a truncated octahedron, a polyhedron with 24 vertices and 14 faces. In the synthetic zeolite A (Linde A) the β-cages form a cubic primitive lattice, and are joined by cubes. β-Cages distributed in the same manner as the atoms in diamond and linked by hexagonal prisms make up the structure of faujasite (zeolite X).

The fraction of aluminum atoms in the framework is variable. For each of them there is one negative charge. As a whole, the framework is thus a polyanion; the cations occupy places in the hollows. This in principle applies also to other aluminosilicates, but the framework of the zeolites is much more open. This is the basis of the characteristic properties of the zeolites: they act as cation exchangers and absorb and release water easily. A zeolite that has been dehydrated by heating it *in vacuo* is highly hygroscopic and can be used to remove water from solvents or gases. In addition to water, it can also absorb other molecules; the size and shape of the molecules relative to the size and shape of the hollow spaces in the zeolite determine how easily this occurs and how tightly the guest molecules are retained by the host framework. Different types of zeolite differ widely with regard to their hollows and channels, and they can be made to measure in order to take up certain molecules. This effect is applied for the selective separation of compounds, and therefore zeolites are also termed molecular sieves. For example, they can separate unbranched and branched alkanes, which is important for petroleum refineries. Even the separation of O_2 and N_2 is possible. The channels can also accommodate different molecules simultaneously, the shape of the channels forcing the molecules to adopt some definite mutual orientation. As a consequence, zeolites can act as selective catalysts. Synthetic zeolite ZSM-5, for example, serves to catalyze the hydrogenation of methanol to alkanes.

Zeolites are structurally related to colorless sodalite, $Na_4Cl[Al_3Si_3O_{12}]$, and to deeply colored ultramarines. These have aluminosilicate frameworks that enclose cations but no water molecules (Fig. 16.25). Their special feature is the additional presence of anions in the hollows, *e.g.* Cl^-, SO_4^{2-}, S_2^-, or S_3^-. The two last-mentioned species are colored radical ions (green and blue, respectively) that are responsible for the brilliant colors. The best-known representative is the blue mineral lapis lazuli, $Na_4S_x[Al_3Si_3O_{12}]$, which is also produced industrially and serves as color pigment.

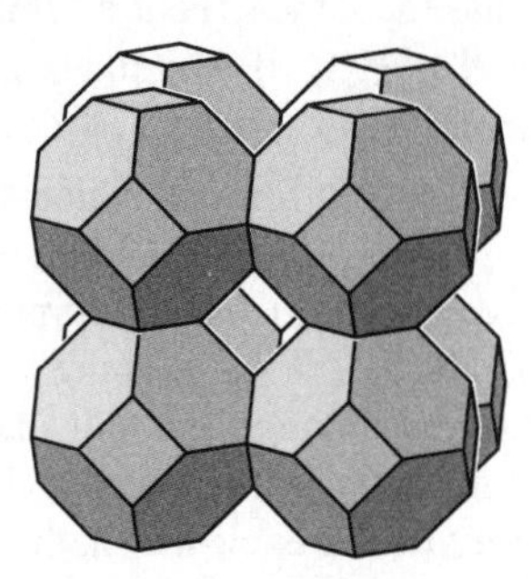

Fig. 16.25
Sodalite and ultramarine framework

Framework silicates are also termed tectosilicates. Their common feature is the three-dimensional connection of tetrahedra sharing all four vertices. They are subdivided into:

1. Pyknolites, which have a framework with relatively small cavities that are filled with cations; for example: feldspars $M^+[AlSi_3O_8^-]$ and $M^{2+}[Al_2Si_2O_8^{2-}]$ such as $K[AlSi_3O_8]$ (orthoclase, sanidine) or plagioclase, $Ca_{1-x}Na_x[Al_{2-x}Si_{2+x}O_8]$ which includes $Na[AlSi_3O_8]$ (albite; $x = 1$) and $Ca[Al_2Si_2O_8]$ (anorthite; $x = 0$). Feldspars, especially plagioclase, are by far the most abundant minerals in Earth's crust.
2. Clathrasils, which have polyhedral cavities, but with windows that are too small to allow the passage of other molecules, so that enclosed ions or foreign molecules cannot escape. Examples are ultramarines and melanophlogite $(SiO_2)_{46}{\cdot}8\,(N_2, CO_2, CH_4)$.
3. Zeolites with polyhedral cavities which are connected by wide windows or channels that permit the diffusion of foreign ions or molecules.

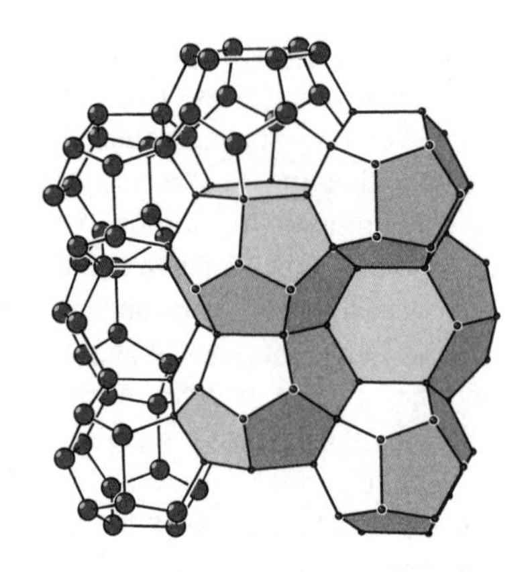

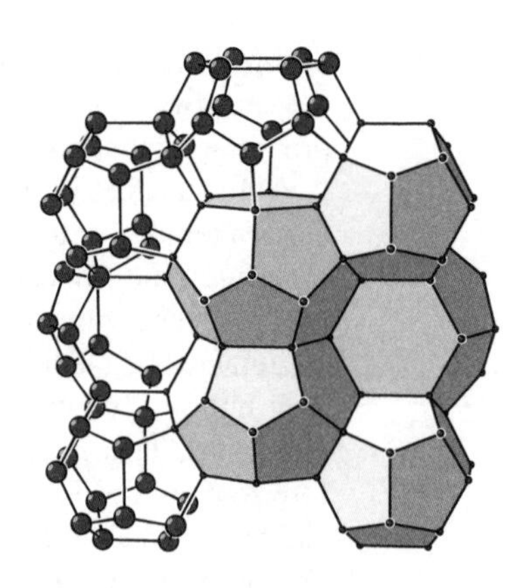

Fig. 16.26 Section of the framework in gas hydrates of type I. Each vertex represents an O atom, and along each edge there is an H atom (stereo image)

The structural relationship between SiO_2 and H_2O (*cf.* Section 12.5) also shows up in the clathrates (inclusion compounds); they include clathrasils that enclose foreign molecules. Water forms analogous clathrate hydrates which consist of foreign molecules enclosed by a framework of H_2O molecules. As in ice, every O atom is surrounded by four H atoms. The structures are only stable in the presence of the enclosed molecules, otherwise the hollow, wide-meshed framework would collapse. The gas hydrates are among the best-known species of this kind. They have particles such as Ar, CH_4, H_2S or Cl_2 enclosed by a framework which has two dodecahedral and six larger tetracaidecahedral (polyhedron with 14 faces) cavities per 46 molecules of H_2O (Fig. 16.26; the above-mentioned melanophlogite has the same structure). If all of the cavities are occupied, the composition is $(H_2O)_{46}X_8$ or $X{\cdot}5\frac{3}{4}H_2O$; if only the larger cavities are occupied, as in the Cl_2 hydrate, the composition is $(H_2O)_{46}(Cl_2)_2$ or $Cl_2 \cdot 7\frac{2}{3}H_2O$. Different frameworks with larger cavities form with larger foreign molecules. Examples include $(CH_3)_3CNH_2{\cdot}9\frac{3}{4}H_2O$, $HPF_6{\cdot}6H_2O$, and $CHCl_3{\cdot}17H_2O$. Clathrates such as $C_3H_8{\cdot}17H_2O$, which has a melting point of 8.5 °C, can crystallize from humid natural gas during cold weather and obstruct pipelines. $(H_2O)_{46}(CH_4)_8$ is stable at the pressures that exist in the oceans at depths below 600 m. There it occurs in huge amounts that surpass the deposits of natural gas; however, the extraction is not worthwhile because the energy needed for this purpose exceeds the combustion energy of the enclosed methane. The clathrate structure also occurs among the compounds Na_8Si_{46}, K_8Si_{46}, K_8Ge_{46}, and K_8Sn_{46}, the Si atoms taking the positions of the water molecules and thus having four bonds each. The alkali metal ions occupy the cavities, and their electrons contribute to a metallic electron gas.

16.8 Edge-sharing Tetrahedra

Two tetrahedra sharing one edge lead to the composition M_2X_6, as in Al_2Cl_6 (in the gaseous state or in solution) (Fig. 16.2, p. 167). Continuation of the linking using opposite edges results in a linear chain, with all X atoms having bridging functions. Chains of this kind are known in $BeCl_2$ and SiS_2 as well as in the anion of $K[FeS_2]$ (Fig. 16.27).

If tetrahedra are joined via four of their edges, the resulting composition is $MX_{4/4}$ or MX. This kind of linking corresponds to the structure of the red modification of PbO, in which O atoms occupy the tetrahedron centers and the Pb atoms the vertices (Fig. 16.28). This rather peculiar structure may be regarded as a consequence of the steric influence of

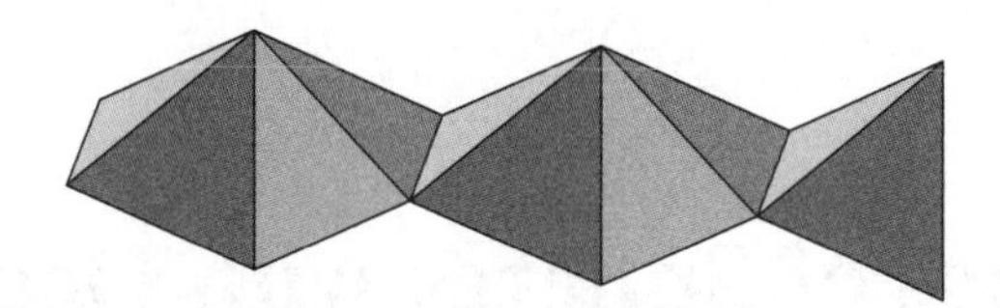

Fig. 16.27
Linked tetrahedra in SiS_2

the lone electron pair at the Pb(II) atom; if we include the electron pair, the coordination polyhedron of a lead atom is a square pyramid. The layer can be described as a checker board having O atoms at its cross-points; the Pb atoms are placed above the black and under the white fields of the board.

The CaF_2 structure can be regarded as a network of three-dimensionally linked, edge-sharing FCa_4 tetrahedra (*cf.* Fig. 17.3 b).

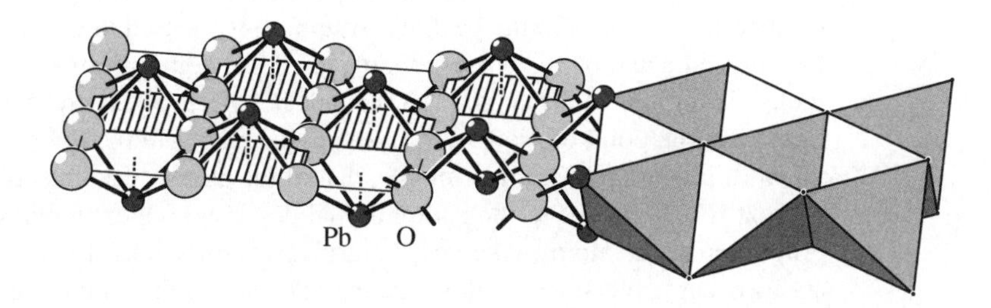

Fig. 16.28
Section of a layer in red PbO

16.9 Problems

16.1 Take $W_4O_{16}^{8-}$ ions (Fig. 16.14) and pile them to form a column consisting of pairs of edge-sharing octahedra that alternate crosswise. What is the composition of the resulting column?

16.2 Take pairs of face-sharing coordination octahedra and join them by common vertices to form a chain, with every octahedron taking part in one common vertex not belonging to the shared face. What is the composition of the resulting chain?

16.3 What is the composition of a column of square antiprisms joined by common square faces?

16.4 Which of the following compounds could possibly form columns of face-sharing octahedra as in ZrI_3?
InF_3, $InCl_3$, MoF_3, MoI_3, TaS_3^{2-}

16.5 Take the network of vertex-sharing tetrahedra of the Cu atoms in $MgCu_2$ (Fig. 15.4) and assume that there is an additional atom inside of every tetrahedron. What structure type would this be?

17 Packings of Spheres with Occupied Interstices

The packing of polyhedral building blocks is mentioned repeatedly in Chapter 16, for example with respect to the difference between the $CdCl_2$ and CdI_2 types. These both consist of the same kind of layers of edge-sharing octahedra. The layers are stacked in such a way that the halogen atoms, taken by themselves, form a cubic closest-packing in the $CdCl_2$ type and a hexagonal closest-packing in the CdI_2 type. The metal atoms occupy octahedral interstices of the packing. Attention is focused in Chapter 16 on the linking of the polyhedra and on the corresponding chemical compositions, while the packing of the molecules or ions in the crystal is a secondary aspect. In this chapter we develop the same facts from the point of view of the general packing. We restrict the discussion mainly to the most important packing principle, that of the closest-packings of spheres.

That which applies to the cadmium halides also applies to many other compounds: a proportion of the atoms, taken by themselves, form a closest-packing, and the remaining atoms occupy interstices in this packing. The atoms forming the packing do not have to be the same, but they must have similar sizes, in the sense outlined in the Sections 15.1 and 15.2 concerning compounds with closest-packed atoms. In perovskite, $CaTiO_3$, for example, the calcium and the oxygen atoms together form a cubic closest-packing, and the titanium atoms occupy certain octahedral interstices. Due to the space requirements of the atoms in the interstices and to their bonding interactions with the surrounding atoms, the sphere packing frequently experiences certain distortions, but these are often surprisingly small. Moreover, it is possible to include atoms that are too large for the interstices if the packing is expanded; strictly speaking, the packing is then no longer a closest-packing (the spheres have no contact with each other), but their relative arrangement in principle remains unchanged.

17.1 The Interstices in Closest-packings of Spheres

Octahedral Interstices in the Hexagonal Closest-packing

Fig. 17.1(a) shows a section of two superimposed hexagonal layers in a closest-packing of spheres. This representation has the disadvantage that the spheres of layer *A* are largely concealed by the layer *B*. In all the following figures we will therefore use the representation shown in Fig. 17.1(b); it shows exactly the same section of the packing, but the spheres are drawn smaller. Of course, since the real size of the spheres is larger, the points of contact between the spheres can no longer be perceived, but we now gain an excellent impression of the sites of the octahedral interstices in the packing: they appear as the large holes surrounded by six spheres. The edges of two octahedra are plotted in Fig. 17.1(b); these two octahedra share a common edge. Fig. 17.1(c) represents a side view of a hexagonal closest-packing (looking towards the edges of the hexagonal layers); the two plotted octahedra share a common face. The two octahedra shown in Fig. 17.1(d) are next to each other at different heights, they share a common vertex.

Inorganic Structural Chemistry, Second Edition Ulrich Müller

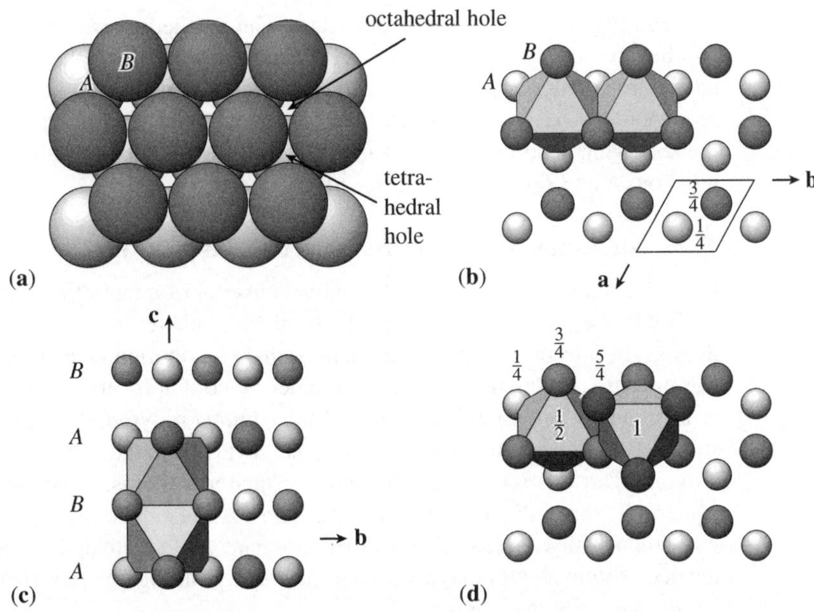

Fig. 17.1
(**a**) Relative positions of two hexagonal layers in a closest-packing of spheres. (**b**) The same layers with spheres drawn to a smaller scale; two edge-sharing octahedra and the unit cell of the hexagonal closest-packing are shown. (**c**) Side view of the hexagonal closest-packing; two face-sharing octahedra are shown. (**d**) Two vertex-sharing octahedra in the hexagonal closest-packing. Numbers: z coordinates of the spheres and octahedron centers

From Fig. 17.1 we can see how adjacent octahedra are linked in a hexagonal closest-packing:

- face-sharing when the octahedra are located one on top of the other in the direction c;
- edge-sharing when they are adjacent in the a–b plane;
- vertex-sharing when they are adjacent at different heights.

The bond angles at the bridging X atoms in the common octahedron vertices are fixed by geometry (angles M–X–M, M in the octahedron centers):

70.5° for face-sharing;
90.0° for edge-sharing;
131.8° for vertex-sharing.

The number of octahedral holes in the unit cell can be deduced from Fig. 17.1(c): two differently oriented octahedra alternate in direction c, *i.e.* it takes two octahedra until the pattern is repeated. Hence there are two octahedral interstices per unit cell. Fig. 17.1(b) shows the presence of two spheres in the unit cell, one each in the layers A and B. The number of spheres and of octahedral interstices are thus the same, *i.e. there is exactly one octahedral interstice per sphere.*

The size of the octahedral interstices follows from the construction of Fig. 7.2 (p. 53). There, it is assumed that the spheres are in contact with one another just as in a packing of spheres. A sphere with radius 0.414 can be accommodated in the hole between six octahedrally arranged spheres with radius 1.

From Fig. 17.1 we realize another fact. The octahedron centers are arranged in planes parallel to the *a–b* plane, half-way between the layers of spheres. The position of the octahedron centers corresponds to the position *C* which does not occur in the stacking sequence *ABAB*... of the spheres. We designate octahedral interstices in this position in the following sections by γ. By analogy, we will designate octahedral interstices in the positions *A* and *B* by α and β, respectively.

Tetrahedral Interstices in the Hexagonal Closest-packing

Fig. 17.2 shows sections of the hexagonal closest-packing in the same manner as in Fig. 17.1, but displaying tetrahedra made up of four spheres each. The tetrahedra share vertices in the *a–b* plane. Pairs of tetrahedra share a common face in the stacking direction, and the pairs are connected with each other by common vertices. A pair can be regarded as a trigonal bipyramid. The center of the trigonal bipyramid is identical with the interstice between three atoms in the hexagonal layer; from the center, the axial atoms of the bipyramid are 41 % more distant than the equatorial atoms. If we only consider the three equatorial atoms, the interstice is triangular; if we also take into account the axial atoms, it is trigonal-bipyramidal. The tetrahedral interstices are situated above and below this interstice. Within a pair of layers *AB* a tetrahedron pointing upwards shares edges with three tetrahedra pointing downwards.

The bond angles M–X–M at the bridging atoms between two occupied tetrahedra are:

56.7° for face-sharing;
70.5° for edge-sharing;
109.5° for vertex-sharing.

As can be seen from Fig. 17.2(b), there is one tetrahedral interstice above and one below every sphere, *i.e. there are two tetrahedral interstices per sphere.*

According to the calculation of Fig. 7.2 (p. 53), a sphere with radius 0.225 fits into the tetrahedral hole enclosed by four spheres of radius 1.

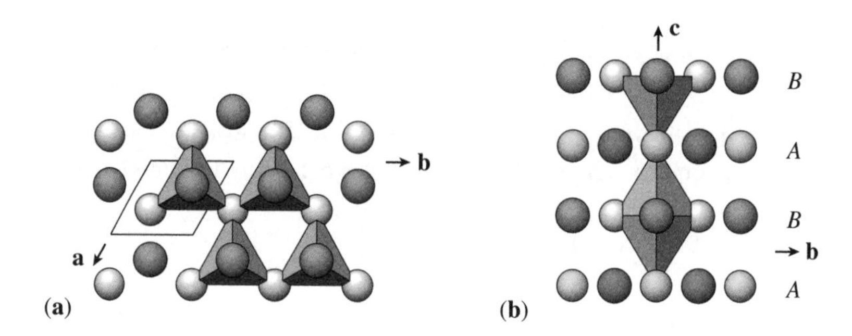

Fig. 17.2 Tetrahedra in hexagonal closest-packing: (**a**) view of the hexagonal layers; (**b**) view parallel to the hexagonal layers (stacking direction upwards)

Octahedral and Tetrahedral Interstices in the Cubic Closest-packing

In cubic closest-packing, consideration of the face-centered unit cell is a convenient way to get an impression of the arrangement of the interstices. The octahedral interstices are situated in the center of the unit cell and in the middle of each of its edges [Fig. 17.3(a)]. The octahedra share vertices in the three directions parallel to the unit cell edges. They share edges in the directions diagonal to the unit cell faces. There are no face-sharing octahedra.

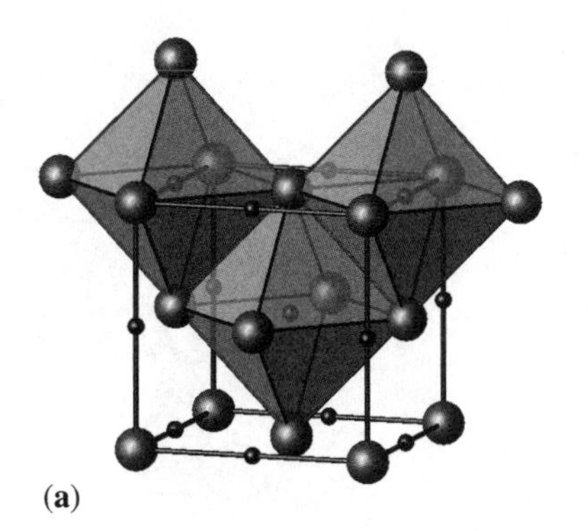

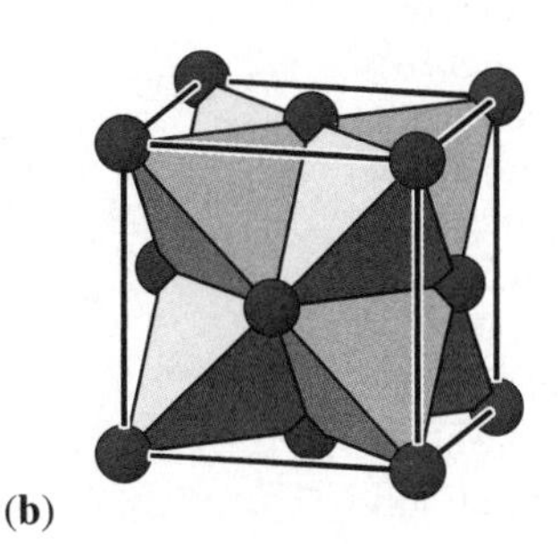

Fig. 17.3 Face-centered unit cell of cubic closest-packing; (**a**) with octahedral interstices (small spheres); (**b**) with tetrahedral interstices

If we consider the unit cell to be subdivided into eight octants, we can perceive one tetrahedral interstice in the center of every octant [Fig. 17.3(b)]. Two tetrahedra share an edge when their octants have a common face. They share a vertex if their octants only have a common edge or a common vertex. There are no face-sharing tetrahedra.

There are four spheres, four octahedral interstices and eight tetrahedral interstices per unit cell. Therefore, their numerical relations are the same as for hexagonal closest-packing, as well as for any other stacking variant of closest-packings: one octahedral and two tetrahedral interstices per sphere. Moreover, the sizes of these interstices are the same in all closest-packings of spheres.

The bond angles M–X–M at the bridging atoms between two polyhedra occupied by M atoms are:

edge-sharing octahedra	90.0°	vertex-sharing tetrahedra in octants with a common edge	109.5°
vertex-sharing octahedra	180.0°	vertex-sharing tetrahedra in octants with a common vertex	180.0°
edge-sharing tetrahedra	70.5°		

The hexagonal layers with the stacking sequence *ABCABC*... are perpendicular to the space diagonals of the unit cell. The layers in one pair of layers, say *AB*, have the same mutual arrangement as in Fig. 17.1(b). The position of the following layer *C* is situated exactly over the octahedral interstices between *A* and *B*. The pattern of the edge-sharing octahedra within *one pair of layers* is independent of the stacking sequence. The sequence of the positions of the octahedron centers in the stacking direction is $\gamma\gamma\ldots$ in the hexagonal closest-packing and it is $\gamma\alpha\beta\gamma\alpha\ldots$ in the cubic closest-packing (Fig. 17.4).

The crystallographic data for the two packings of spheres are summarized in Table 17.1.

Table 17.1: Crystallographic data of the hexagonal and cubic closest-packings of spheres. $+F$ means $+(\frac{1}{2},\frac{1}{2},0)$, $+(\frac{1}{2},0,\frac{1}{2})$, $+(0,\frac{1}{2},\frac{1}{2})$ (face centering). Values given as 0 or fractional numbers are fixed by space-group symmetry (special positions)

	space group	position of the spheres	centers of the octahedral voids	centers of the tetrahedral voids	c/a
hexagonal closest-packing	$P6_3/mmc$	2*d* $\frac{2}{3},\frac{1}{3},\frac{1}{4}$; $\frac{1}{3},\frac{2}{3},\frac{3}{4}$	2*a* 0, 0, 0; 0, 0, $\frac{1}{2}$	4*f* $\pm(\frac{2}{3},\frac{1}{3},0.625)$; $\pm(\frac{1}{3},\frac{2}{3},0.125)$	$\frac{2}{3}\sqrt{6}=$ 1.633
cubic closest-packing	$Fm\bar{3}m$	4*a* 0, 0, 0 $+F$	4*b* 0, 0, $\frac{1}{2}$ $+F$	8*c* $\frac{1}{4},\frac{1}{4},\frac{1}{4}$; $\frac{1}{4},\frac{1}{4},\frac{3}{4}$ $+F$	

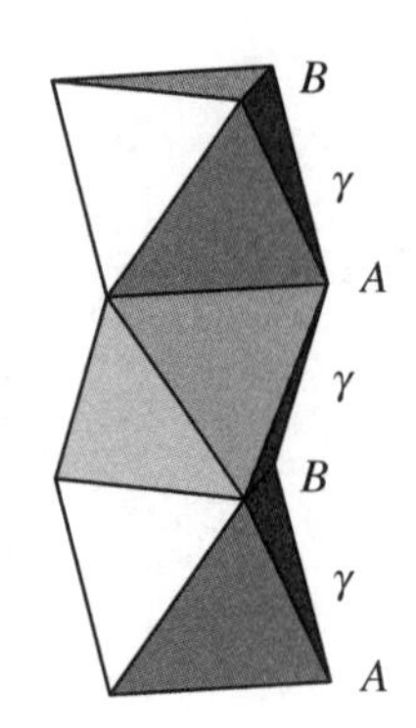

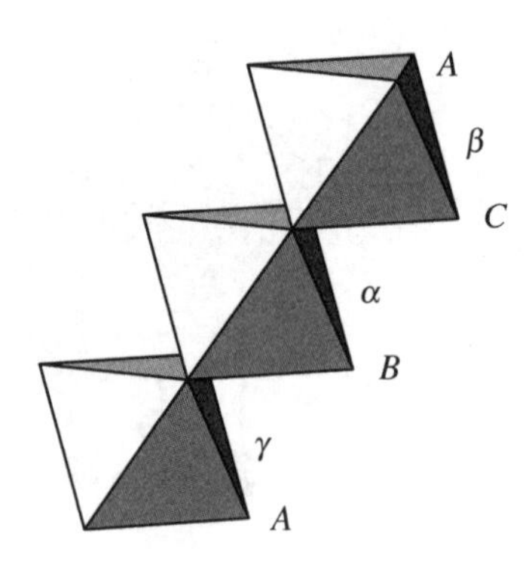

Fig. 17.4 Relative arrangement of the octahedra in hexagonal and in cubic closest-packing in the direction of stacking of the hexagonal layers

17.2 Interstitial Compounds

The concept of intercalating atoms in the interstices between the spheres is not just an idea; with some elements it can actually be performed. The uptake of hydrogen by certain metals yielding metal hydrides is the most familiar example. During the absorption of the hydrogen the properties of the metal experience significant changes and usually phase transitions take place, *i.e.* the packing of the metal atoms in the final metal hydride is usually not the same as that of the pure metal. However, as a rule, it still is one of the packings typical for metals. For this reason the term interstitial hydrides has been coined. The hydrogen content is variable, it depends on pressure and temperature; we have to deal with nonstoichiometric compounds.

Interstitial hydrides are known for the transition metals (including lanthanoids and actinoids). Magnesium hydride can also be included, since magnesium can take up hydrogen under pressure up to the composition MgH_2; upon heating the hydrogen is released. The limiting composition is MH_3 for most of the lanthanoids and actinoids, otherwise it is MH_2 or less. In some cases the compounds are unstable for certain composition ranges (*e.g.* only cubic $HoH_{1.95}$ to $HoH_{2.24}$ and hexagonal $HoH_{2.64}$ to $HoH_{3.00}$ are stable).

The typical structure for the composition MH_2 is a cubic closest-packing of metal atoms in which all tetrahedral interstices are occupied by H atoms; this is the CaF_2 type. The surplus hydrogen in the lanthanoid hydrides MH_2 to MH_3 is placed in the octahedral interstices (Li_3Bi type for LaH_3 to NdH_3, *cf.* Fig. 15.3, p. 161).

The interstitial hydrides of transition metals differ from the salt-like hydrides of the alkali and alkaline-earth metals MH and MH_2, as can be seen from their densities. While the latter have higher densities than the metals, the transition metal hydrides have expanded metal lattices. Furthermore, the transition metal hydrides exhibit metallic luster and are semiconducting. Alkali metal hydrides have NaCl structure; MgH_2 has rutile structure.

The packing density of the H atoms is very high in all hydrogen-rich metal hydrides. For example, in MgH_2 it is 55 % higher than in liquid hydrogen. Years of efforts to develop a storage medium for hydrogen using magnesium have not met with success. The alloy $LaNi_5$ can also absorb and release hydrogen easily; it is used as electrode material in metal hydride batteries.

The carbides and nitrides of the elements Ti, Zr, Hf, V, Nb, Ta, Cr, Mo, W, Th, and U are considered to be typical interstitial compounds. Their compositions frequently correspond to one of the approximate formulas M_2X or MX. As a rule, they are nonstoichiometric compounds with compositions ranging within certain limits. This fact, the limitation to a few similar structure types, and very similar properties show that the geometric packing conditions of the atoms have fundamental importance in this class of compounds.

The nitrides can be prepared by heating a metal powder in an N_2 or NH_3 atmosphere to temperatures above 1100 °C. The carbides form upon heating mixtures of the metal powders with carbon to temperatures of about 2200 °C. Both the nitrides and carbides can also be made by chemical transport reactions by the VAN ARKEL–DE BOER method if the metal deposition takes place in an atmosphere of N_2 or a hydrocarbon. Their remarkable properties are:

- Very high hardness with values of 8 to 10 on the MOHS scale; in some cases they approach the hardness of diamond (*e.g.* W_2C).
- Extremely high melting points, for example (values in °C):

(Ti	1660)	TiC	3140	TiN	2950	VC	2650
(Zr	1850)	ZrC	3530	ZrN	2980	NbC	2600
(Hf	2230)	HfC	3890	HfN	3300	TaC	3880

  Values for comparison: melting point of W 3420 °C (highest melting metal), sublimation point of graphite approximately 3350 °C.
- Metallic electrical conductivity, in some cases also superconductivity at low temperatures (*e.g.* NbC, transition temperature 10.1 K).
- High chemical resistance, except to oxidizing agents at high temperatures (such as atmospheric oxygen above 1000 °C or hot concentrated nitric acid).

The intercalation of C or N into the metal thus involves an increased refractoriness with preservation of metallic properties.

The structures can be considered as packings of metal atoms which have incorporated the nonmetal atoms in their interstices. Usually, the metal atom packings are not the same as those of the corresponding pure metals. The following structure types have been observed:

M_2C and M_2N	hexagonal closest-packing of M atoms, C or N atoms in half of the octahedral interstices
MC and MN	cubic closest-packing of M atoms, C or N atoms in all octahedral interstices = NaCl type (not for Mo, W)
MoC, MoN, WC, WN	WC type

In the WC type the metal atoms do not have a closest-packing, but a hexagonal-primitive packing; the metal atoms form trigonal prisms that are occupied by the C atoms.

For the structures of M_2C and M_2N the question arises: is there an ordered distribution of occupied and unoccupied octahedral holes? There are several possibilities for an ordered distribution, some of which actually occur. For example, in W_2C occupied and unoccupied octahedral holes alternate in layers; this is the CdI_2 type. In β-V_2N there are alternating layers in which the octahedral holes are one-third and two-thirds occupied. The question of ordered distributions of occupied interstices is the subject of the following sections.

17.3 Structure Types with Occupied Octahedral Interstices in Closest-packings of Spheres

We focus attention here on the binary compounds MX_n, the X atoms being arranged in a closest-packed manner and the M atoms occupying the octahedral interstices. Since the number of octahedral interstices coincides with the number of X atoms, exactly the fraction $1/n$ of them has to be occupied to ensure the correct composition. As outlined above, in the following we denote the positions of the layers of X atoms by A, B and C, and the intermediate planes of octahedral interstices by α (between B and C), β (between C and A) and γ (between A and B). Fractional numbers indicate the fraction of octahedral holes that are occupied in the corresponding intermediate plane; a completely unoccupied intermediate plane is marked by the SCHOTTKY symbol □.

Compounds MX

structure type	stacking sequence	examples
NaCl	$A\gamma B\alpha C\beta$	LiH, KF, AgCl, MgO, PbS, TiC, CrN
NiAs	$A\gamma B\gamma$	CrH, TiS, CoS, CoSb, AuSn

In both the NaCl and the NiAs structure types all octahedral interstices are occupied in a cubic closest-packing or hexagonal closest-packing, respectively. The coordination number is 6 for all atoms. In the NaCl type all atoms have octahedral coordination, and it does not matter whether the structure is regarded as a packing of Na^+ spheres with intercalated Cl^- ions or vice versa. The situation is different for the NiAs type; only the arrangement of the As atoms is that of a closest-packing, while the nickel atoms in the octahedral interstices (γ positions) are stacked one on top the other (Fig. 17.5). Only the nickel atoms have octahedral coordination; the coordination polyhedron of the arsenic atoms is a trigonal prism. The structure can also be considered as a primitive hexagonal lattice of Ni atoms; in this lattice the only occurring polyhedra are trigonal prisms, their number being twice the number of the Ni atoms. One half of these prisms are occupied by As atoms (*cf.* also Fig. 16.16, p. 179).

The above-mentioned examples show that the NaCl type occurs preferentially in salt-like (ionic) compounds, some oxides and sulfides, and the interstitial compounds discussed in the preceding section. For electrostatic reasons the NaCl type is well-suited for very polar compounds, since every atom only has atoms of the other element as closest neighbors. Sulfides, selenides, and tellurides, as well as phosphides, arsenides and antimonides, with NaCl structure have been observed with alkaline earth metals and with elements of the third transition metal group (MgS, CaS, ..., MgSe, ..., BaTe; ScS, YS, LnS, LnSe, LnTe; LnP, LnAs, LnSb with Ln = lanthanoid). However, with other transition metals, the NiAs type and its distorted variants are preferred. The Ni atoms in the face-sharing octahedra are rather close to each other (Ni–Ni distance 252 pm, just slightly longer than the Ni–As distance of 243 pm). This suggests the presence of bonding metal–metal interactions, particularly since this structure type only occurs if the metal atoms still have d electrons available. The existence of metal–metal interactions is also supported by the following observations: metallic luster and conductivity, variable composition, and the dependence of the lattice parameters on the electronic configuration, for example:

ratio c/a of the hexagonal unit cell

TiSe	VSe	CrSe	$Fe_{1-x}Se$	CoSe	NiSe
1.68	1.67	1.64	1.64	1.46	1.46

Even smaller c/a ratios are observed for the more electron-rich arsenides and antimonides (e.g. 1.39 for NiAs). Since the ideal c/a ratio of hexagonal closest-packing is 1.633, there is a considerable compression in the c direction, *i.e.* in the direction of the closest contacts among the metal atoms.

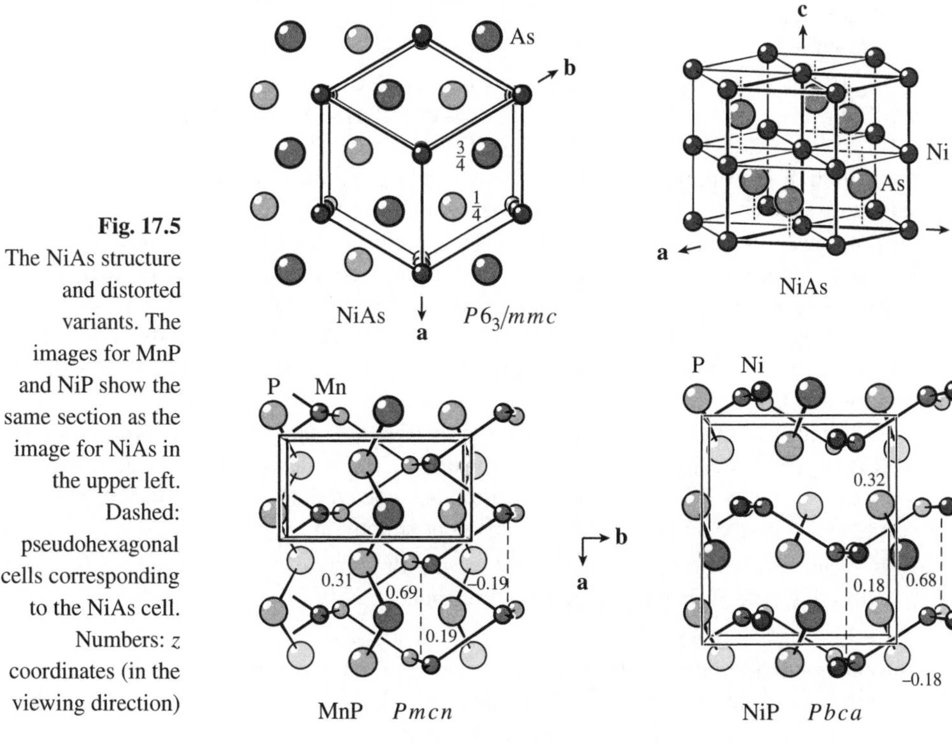

Fig. 17.5 The NiAs structure and distorted variants. The images for MnP and NiP show the same section as the image for NiAs in the upper left. Dashed: pseudohexagonal cells corresponding to the NiAs cell. Numbers: *z* coordinates (in the viewing direction)

The structure of MnP is a distorted variant of the NiAs type: the metal atoms also have close contacts with each other in zigzag lines parallel to the *a*–*b* plane, which amounts to a total of four close metal atoms (Fig. 17.5). Simultaneously, the P atoms have moved up to a zigzag line; this can be interpreted as a $(P^-)_\infty$ chain in the same manner as in ZINTL phases. In NiP the distortion is different, allowing for the presence of P_2 pairs (P_2^{4-}). These distortions are to be taken as PEIERLS distortions. Calculations of the electronic band structures can be summarized in short: 9–10 valence electrons per metal atom favor the NiAs structure, 11–14 the MnP structure, and more than 14 the NiP structure (phosphorus contributes 5 valence electrons per metal atom); this is valid for phosphides. Arsenides and especially antimonides prefer the NiAs structure also for the larger electron counts.

Compounds which have the NiAs structure often exhibit a certain phase width in that metal atom positions can be vacant. The composition then is $M_{1-x}X$. The vacancies can have a random or an ordered distribution. In the latter case we have to deal with superstructures of the NiAs type; they are known, for example, among iron sulfides such as Fe_9S_{10} and $Fe_{10}S_{11}$. If metal atoms are removed from every other layer, we have a continuous series from $M_{1.0}X$ with the NiAs structure down to $M_{0.5}X$ (= MX_2) with the CdI_2 structure; phases of this kind are known for $Co_{1-x}Te$ (CoTe: NiAs type; $CoTe_2$: CdI_2 type).

Compounds MX_2

In compounds MX_2, half of the octahedral interstices are occupied. There are several possibilities for the distribution of occupied and vacant interstices in the intermediate planes:

1. Fully occupied and vacant intermediate planes alternate. In the occupied planes the octahedra share common edges (Fig. 16.9, p. 175).

structure type	stacking sequence	examples
$CdCl_2$	$A\gamma B\square C\beta A\square B\alpha C\square$	$MgCl_2$, $FeCl_2$, Cs_2O
CdI_2	$A\gamma B\square$	$MgBr_2$, PbI_2, SnS_2, $Mg(OH)_2$, $Cd(OH)_2$, Ag_2F

In addition, further polytypes exist, *i.e.* structures having other stacking sequences of the halogen atoms. Especially for CdI_2 itself a large number of such polytypes are now known; for this reason the term CdI_2 type is nowadays considered unfortunate, and the terms $Mg(OH)_2$ (brucite) or $Cd(OH)_2$ type are preferred by some authors. The H atoms of the hydroxides are oriented into the tetrahedral interstices between the layers, and do not act as H bridges. Botallackite, $Cu_2(OH)_3Cl$, has a structure like CdI_2, every other layer of the packing of spheres consisting of Cl atoms and OH groups (another modification with this composition is atacamite, which is mentioned below).

Distorted variants, similar to the distorted variants of the NiAs type, are known for the CdI_2 type. For example, ZrI_2 has a distorted CdI_2 structure in which the Zr atoms form zigzag chains. Therefore, every Zr atom is involved in two Zr–Zr bonds which is in accordance with the d^2 configuration of divalent zirconium.

2. The intermediate planes are alternately two-thirds and one-third occupied.

structure type	stacking sequence	examples
ε-Fe_2N	$A\gamma_{2/3}B\gamma'_{1/3}$	β-Nb_2N, Li_2ZrF_6

The intermediate planes with two-thirds occupation have octahedra sharing edges with the honeycomb pattern as in BiI_3; the octahedra in the intermediate planes with one-third occupation are not directly connected with one another, but they have common vertices with octahedra of the adjacent layers. In Li_2ZrF_6 the Zr atoms are those in the intermediate plane with one-third occupation (*cf.* Fig. 16.13, p. 177).

3. The intermediate planes are alternately one-quarter and three-quarters occupied. This is the arrangement in atacamite, a modification of $Cu_2(OH)_3Cl$ with the stacking sequence: $A\gamma_{1/4}B\alpha_{3/4}C\beta_{1/4}A\gamma'_{3/4}B\alpha'_{1/4}C\beta'_{3/4}$.

4. Every intermediate plane is half occupied.

structure type	stacking sequence	examples
$CaCl_2$	$A\gamma_{1/2}B\gamma'_{1/2}$	$CaBr_2$, ε-$FeO(OH)$, Co_2C
α-PbO_2	$A\gamma_{1/2}B\gamma''_{1/2}$	TiO_2 (high pressure)
α-$AlO(OH)$	$A\gamma_{1/2}B\gamma'''_{1/2}$	α-$FeO(OH)$ (goethite)

In $CaCl_2$ linear strands of edge-sharing octahedra are present, and the strands are joined by common octahedron vertices (Fig. 17.6). Marcasite is a modification of FeS_2 related to the $CaCl_2$ type, but distorted by the presence of S_2 dumbbells. The joining of adjacent S atoms to form dumbbells is facilitated by a mutual rotation of the octahedron strands (Fig. 17.6). Several compounds adopt this structure type, for example $NiAs_2$ and $CoTe_2$.

If the octahedron strands are rotated in the opposite direction, the rutile type results (Fig. 17.6). Due to this relation with the $CaCl_2$ type, a hexagonal closest-packing of O atoms has frequently been ascribed to the rutile type. However, the deviations from this kind of packing are quite significant. For one thing, every O atom is no longer in contact with twelve other O atoms, but only with eleven; moreover, the 'hexagonal' layers are considerably corrugated. By the formalism of group theory it is also not permissible to regard the tetragonal rutile as a derivative of the hexagonal closest-packing (*cf.* Section

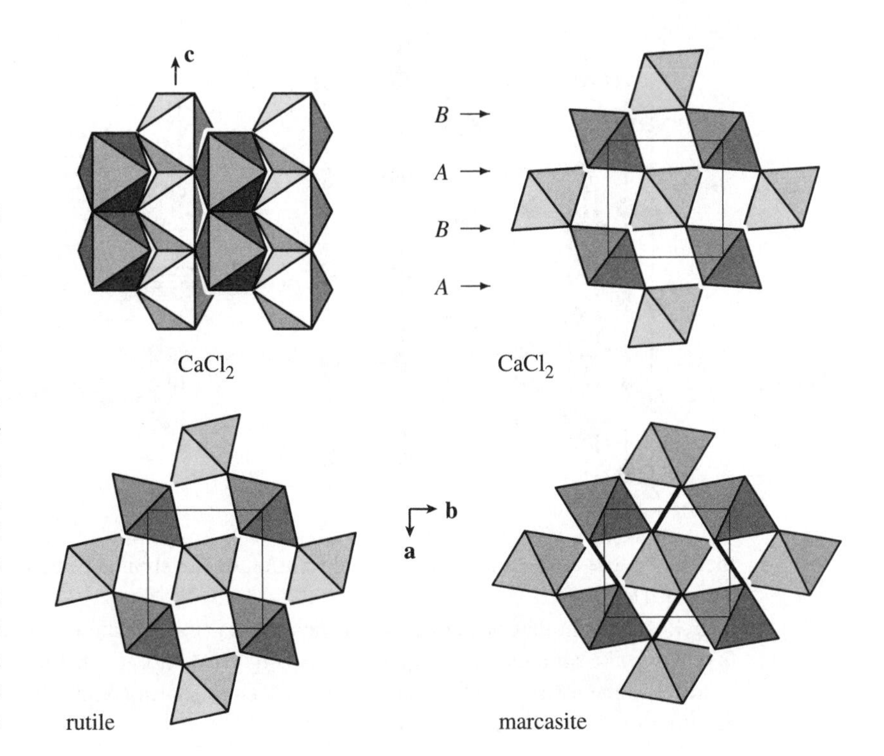

Fig. 17.6 Top left: $CaCl_2$ structure, view perpendicular to the hexagonal layers. Top right and bottom: view along the strands of edge-sharing octahedra in $CaCl_2$, rutile and marcasite; the hexagonal layers in $CaCl_2$ are marked by *A* and *B*. Thick lines: S–S bonds in marcasite

18.3). In fact, the arrangement of the O atoms in rutile is that of a tetragonal close-packing of spheres. This is a packing filling space to 71.9 %, which is only slightly less than in a closest-packing. Consider a ladder which has spheres at the joints of the rungs (Fig. 17.7). Set up such ladders vertical, but mutually rotated around the vertical axis in such a way that the spheres of one ladder come to be next to the gaps between the rungs of the adjacent ladders. In the packing obtained in this way every sphere has coordination number 11: $2+4+2$ spheres from three neighboring ladders and 3 spheres within the ladder. The gaps between the rungs correspond to the octahedral interstices that are occupied by Ti atoms in rutile, and the ladders correspond to the strands of edge-sharing octahedra. Compared to a closest-packing of spheres (c.n. 12), the coordination number is reduced by 8 %, but the space filling is reduced by only 3 %; this gives an idea of why the rutile type is the preferred packing for highly polar compounds (dioxides, difluorides).

The similarity of the structures of rutile, $CaCl_2$ and marcasite also comes to light by comparison of their crystal structure data (Table 17.2). The space groups of $CaCl_2$ and marcasite (both *Pnnm*) are subgroups of the space group of rutile. The tetragonal sym-

Table 17.2: Crystal data of rutile, $CaCl_2$ and marcasite

	space group	a / pm	b / pm	c / pm	M atom x	y	z	anion x	y	z
rutile	$P4_2/mnm$	459.3	459.3	295.9	0	0	0	0.305	0.305	0
$CaCl_2$	$Pnnm$	625.9	644.4	417.0	0	0	0	0.275	0.325	0
marcasite	$Pnnm$	444.3	542.4	338.6	0	0	0	0.200	0.378	0

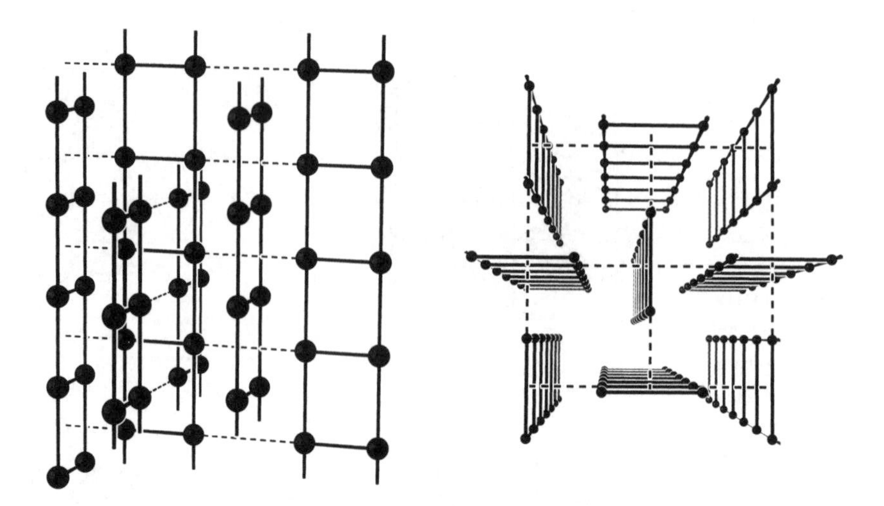

Fig. 17.7
Tetragonal close-packing of spheres

metry of rutile is broken by the mutual rotation of the strands of octahedra (*cf.* p. 33 and Section 18.4).

α-PbO_2 is another structure which has mutually connected strands of edge-sharing octahedra; the strands have zigzag shape (Fig. 16.12, p. 177). Linear strands of edge-sharing octahedra as in $CaCl_2$, but which form edge-sharing double-strands, are present in diaspore, α-AlO(OH) (Fig. 17.8).

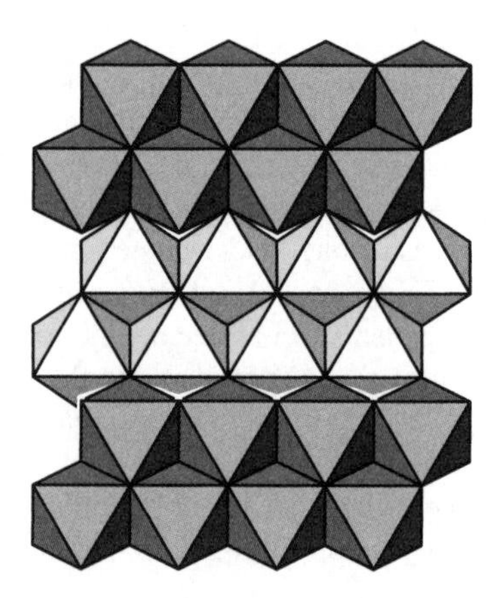

Fig. 17.8
Double strands of edge-sharing octahedra in diaspore, α-AlO(OH)

Compounds MX_3

In compounds MX_3, one-third of the octahedral interstices are occupied. Again, there are several possible distributions of vacant and occupied interstices in the intermediate planes:

1. Every third intermediate plane is fully occupied, and the others are unoccupied. The octahedra within the occupied planes share edges as in CdI_2. This is the structure of Cr_2AlC, in which the sequence of layers is:

$$A_{Cr}\gamma B_{Cr}\square A_{Al}\square B_{Cr}\gamma A_{Cr}\square B_{Cr}\square$$

The carbon atoms take positions inside octahedra made up from only one kind of atom, those of the transition metal.

2. Every other intermediate plane is two-thirds occupied.

structure type	stacking sequence	examples
$AlCl_3$	$A\gamma_{2/3}B\Box C\beta_{2/3}A\Box B\alpha_{2/3}C\Box$	YCl_3, $CrCl_3$ (high temperature)
BiI_3	$A\gamma_{2/3}B\Box$	$FeCl_3$, $CrCl_3$ (low temperature)

Both structure types have the same kind of layer of edge-sharing octahedra (Fig. 16.8). Among the trihalides with this kind of layer structure, stacking disorder is quite common, *i.e.* the stacking sequence of the hexagonal halogen-atom layers is not strictly *AB* or *ABC*, but stacking faults occur frequently. This also applies to $AlCl_3$ and BiI_3 themselves. The frequency of the stacking faults depends on the growth conditions of the specific single crystal. For example, one crystal of BiI_3 that had been obtained by sublimation essentially had the hexagonal stacking sequence *hhh*..., but in a random sequence one out of 16 layers was a *c* layer.

3. Every intermediate layer is one-third occupied.

structure type	stacking sequence	examples
$RuBr_3$	$A\gamma_{1/3}B\gamma_{1/3}$	β-$TiCl_3$, ZrI_3, $MoBr_3$
RhF_3	$A\gamma_{1/3}B\gamma'_{1/3}$	IrF_3, PdF_3, $TmCl_3$

In the $RuBr_3$ type a succession of face-sharing octahedra form a strand in the *c* direction. The metal atoms in adjacent octahedra are shifted in pairs from the octahedron centers, forming metal–metal bonds (Fig. 16.10, p. 175). This seems to be the condition for the existence of this structure type, *i.e.* it only occurs with transition metals that have an odd-numbered *d* electron configuration.

In the RhF_3 type all octahedra share vertices, and corresponding to the hexagonal closest-packing of the F atoms the Rh–F–Rh angles are approximately 132°. By mutual rotation of the octahedra the angle can be widened up to 180°, but then the packing is less dense. This has been observed for the VF_3 type (V–F–V angle approximately 150°), which occurs with some trifluorides (GaF_3, TiF_3, FeF_3 etc.; Fig. 16.5). In PdF_3 the Pd–F lengths in the octahedra alternate (217 and 190 pm) in accordance with the formula $Pd^{II}Pd^{IV}F_6$.

Compounds M_2X_3

Two-thirds of the octahedral interstices are occupied. In a way the possible structure types are the 'inverse' of the MX_3 structures, since in these two-thirds of the octahedral interstices are vacant. If we take an MX_3 type, clear the occupied interstices and occupy the vacant ones, the result is an M_2X_3 structure. The kind of linking between the occupied octahedra, however, is different. The arrangement of the vacant octahedral interstices of the RhF_3 type corresponds to the occupied interstices in corundum, Al_2O_3; its occupied octahedra share edges and faces (Fig. 16.15, p. 179). The layer sequence is:

$$A\gamma_{2/3}B\gamma'_{2/3}A\gamma''_{2/3}B\gamma_{2/3}A\gamma'_{2/3}B\gamma''_{2/3}$$

Compounds MX_4, MX_5, and MX_6

$\frac{1}{4}$, $\frac{1}{5}$ and $\frac{1}{6}$ of the octahedral interstices are occupied, respectively. There are various possibilities for the distribution of the occupied sites, and the specification of a layer sequence alone is not very informative. Fig. 17.9 shows some examples which also allow us to recognize an important principle concerning the packing of molecules: all octahedral interstices that immediately surround a molecule must be vacant, and then occupied interstices have to follow; otherwise either the molecules would be joined to polymeric assemblies or not all of the atoms of the sphere packing would be part of a molecule. These

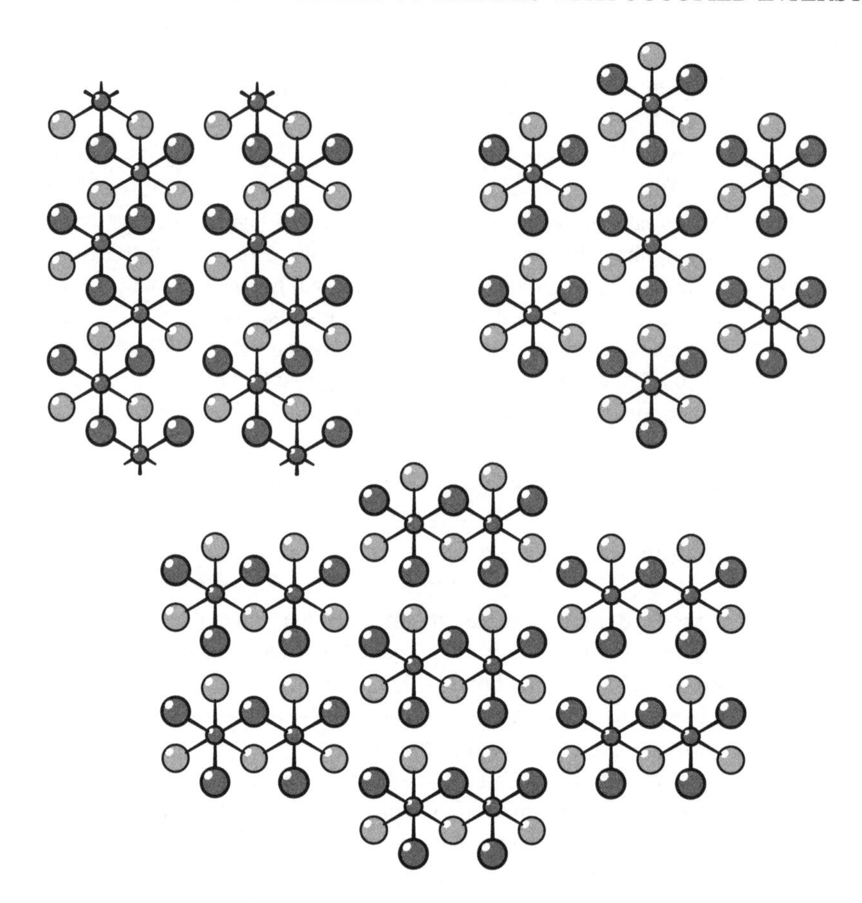

Fig. 17.9 Some examples for the packings of compounds MX_4, MX_5 and MX_6

statements may seem self-evident; however, they imply a severe restriction on the number of possible packing variants for a given kind of molecule. For tetrahalides consisting of chains of edge-sharing octahedra, we noted on page 173 that numerous chain configurations are conceivable. However, because of the packing necessities, some of them are not compatible with a closest-packing; no examples are known for them and it is unlikely that any will ever be observed.

17.4 Perovskites

Among the MX_4 compounds γ'-Fe_4N is a special case: its Fe atoms form a cubic closest-packing, and one-quarter of the octahedral interstices are occupied with N atoms. The occupied octahedra share vertices and form a framework; however, only three out of four Fe atoms take part in this framework; the fourth Fe atom is not bonded to an N atom.

This structure is nothing else than the perovskite type ($CaTiO_3$; Fig. 17.10). The Ca and O atoms jointly form the cubic closest-packing in an arrangement corresponding to the ordered alloy $AuCu_3$ (Fig. 15.1, p. 158). The atomic order in the hexagonal layers of spheres is that shown for $AuCu_3$ on page 159. Being a part of the packing of spheres, a Ca^{2+} ion has coordination number 12. The titanium atoms occupy one-quarter of the octahedral voids, namely those which are surrounded solely by oxygen atoms.

If the position of the Ca^{2+} ion is vacant, the remaining framework is that of the ReO_3

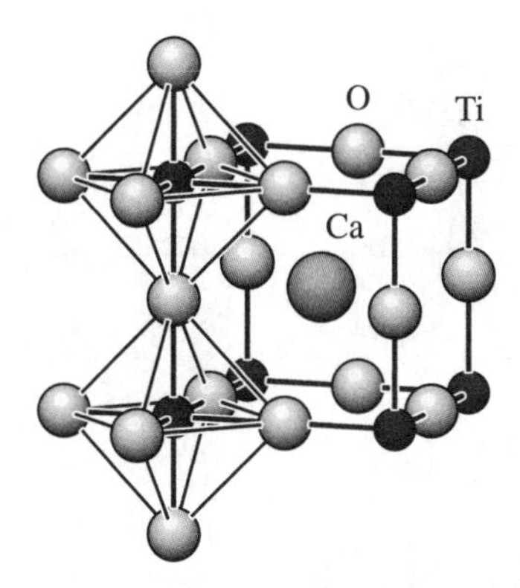

Fig. 17.10
The perovskite structure

type. The analogy between ReO_3 and $CaTiO_3$ is not a mere formalism, since the partial occupation of the Ca positions with varying amounts of metal ions can actually be achieved, specifically in the case of the cubic tungsten bronzes, A_xWO_3 (A = alkali metal, $x = 0.3$ to 0.93). Their color and the oxidation state of the tungsten depend on the value of x; they have metallic luster, and with $x \approx 1$ they are gold-colored, with $x \approx 0.6$ red and with $x \approx 0.3$ dark violet.

In normal, cubic perovskite the closest-packed hexagonal CaO_3 layers have the stacking sequence *ABC*... or *c*... and the occupied octahedra only share vertices. The structural family of perovskites also includes numerous other stacking variants, with *c* and *h* layers in different sequences. At an *h* layer the octahedra share faces. In a sequence like *chhc*, there is a group of three octahedra that share faces at the *h* layers; this group is connected with other octahedra by vertex-sharing at the *c* layers. The size of the groups of face-sharing octahedra depends on the nature of the metal atoms in the octahedra and especially on the ionic radius ratios. Fig. 17.11 shows some examples.

The ideal, cubic perovskite structure is not very common; even the mineral perovskite itself, $CaTiO_3$, is slightly distorted. $SrTiO_3$ is undistorted. As shown in Fig. 16.5 (p. 171), the ReO_3 type can be converted to a more dense packing by mutual rotation of the octahedra until a hexagonal closest-packing is obtained in the RhF_3 type. During the rotation the void in the center of the ReO_3 unit cell becomes smaller and finally becomes an octahedral interstice in the closest-packing of F atoms in the RhF_3 type. If this octahedral interstice is occupied, we have the ilmenite type ($FeTiO_3$). By an appropriate amount of rotation of the octahedra, the size of the hole can be adapted to the size of the A ion in a perovskite. In addition, some tilting of the octahedra allows a variation of the coordination number and coordination polyhedra. Distorted perovskites have reduced symmetry, which is important for the magnetic and electric properties of these compounds. Due to these properties perovskites have great industrial importance, especially the ferroelectric $BaTiO_3$. This is discussed in Chapter 19.

The *tolerance factor t* for perovskites AMX_3 is a value that allows us to estimate the degree of distortion. Its calculation is performed using ionic radii, *i.e.* purely ionic bonding is assumed:

$$t = \frac{r(\mathrm{A}) + r(\mathrm{X})}{\sqrt{2}[r(\mathrm{M}) + r(\mathrm{X})]}$$

Geometry requires a value of $t = 1$ for the ideal cubic structure. In fact, this structure occurs if $0.89 < t < 1$. Distorted perovskites occur if $0.8 < t < 0.89$. With values less than

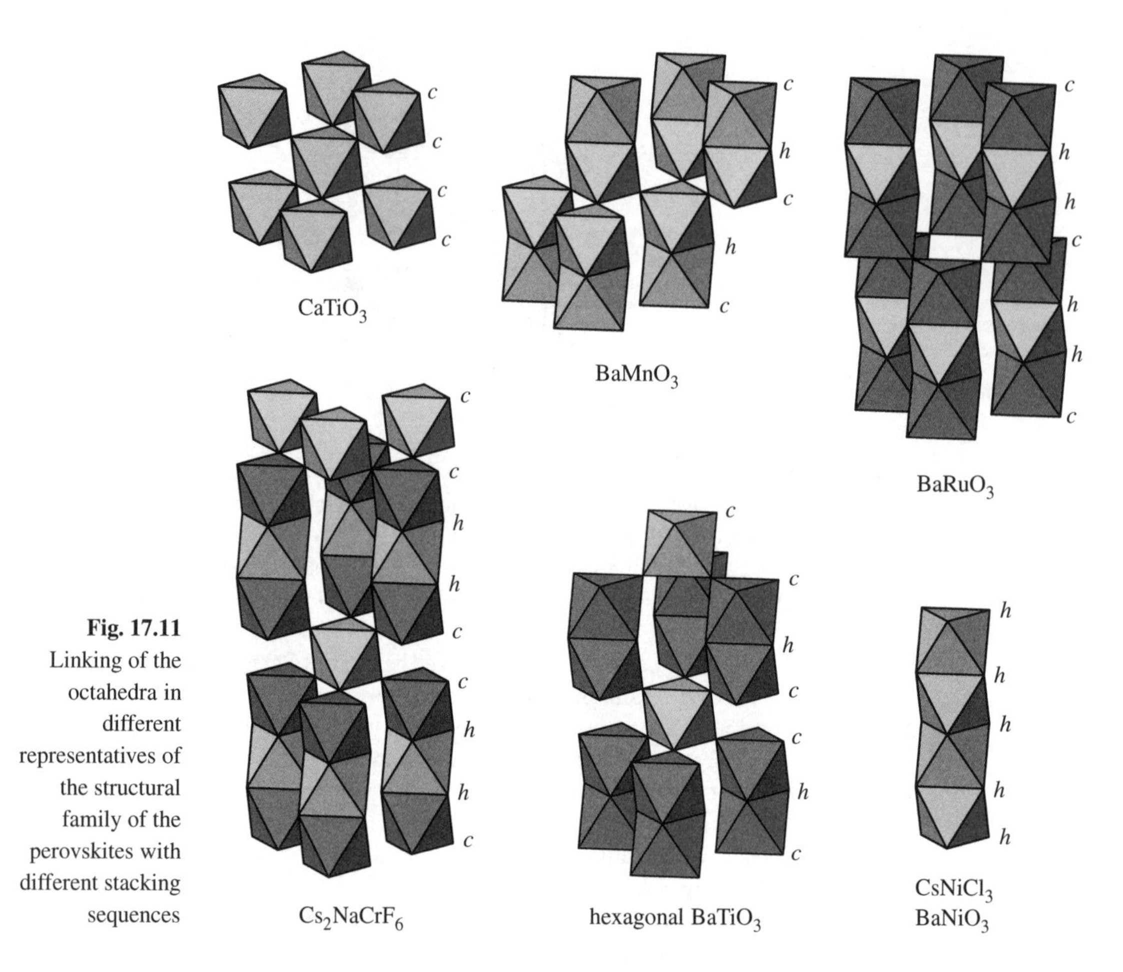

Fig. 17.11 Linking of the octahedra in different representatives of the structural family of the perovskites with different stacking sequences

0.8, the ilmenite type is more stable (Fig. 16.15, p. 179). The hexagonal stacking variants such as depicted in Fig. 17.11 usually have $t > 1$. Since perovskites are not truly ionic compounds and since the result also depends on which values are taken for the ionic radii, the tolerance factor is only a rough estimate.

Superstructures of the Perovskite Type

If we enlarge the unit cell of perovskite by doubling all three edges, it is possible to occupy equivalent positions with atoms of different elements. Fig. 17.12 shows some representatives of the elpasolite family. In elpasolite, K_2NaAlF_6, the potassium and the fluoride ions jointly form the cubic closest-packing, *i.e.* K^+ and F^- take the Ca and O positions of perovskite. The one-to-one relation can be recognized by comparison with the doubled formula of perovskite, $Ca_2Ti_2O_6$. The comparison also shows the partition of the octahedral Ti positions into two sites for Na and Al. In kryolite, Na_3AlF_6, the Na^+ ions occupy two different positions, namely those of Na^+ and K^+ in elpasolite, *i.e.* positions with coordination numbers of 6 and 12. Since this is not convenient for ions of the same size, the lattice experiences some distortion.

'High-temperature' superconductors show superconductivity at temperatures higher than the boiling point of liquid nitrogen (77 K). Their structures are superstructures of

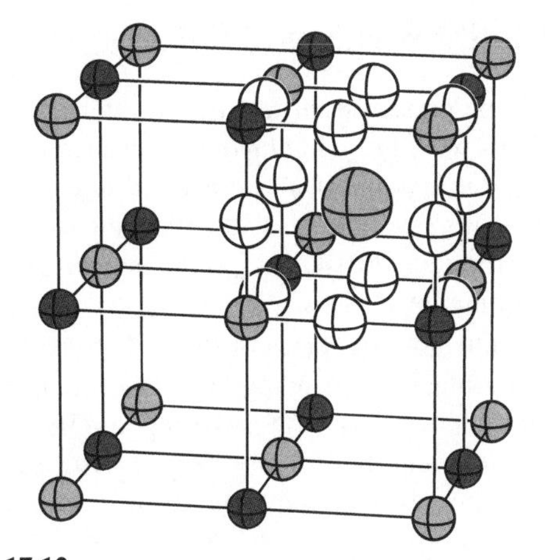

structure type	example	atomic positions			
perovskite	$SrTiO_3$	Sr	O	Ti	Ti
elpasolite	K_2NaAlF_6	K	F	Na	Al
kryolite	$(NH_4)_3AlF_6$	NH_4^+	F	NH_4^+	Al
K_2PtCl_6		K	Cl	□	Pt
CaF_2		F	□	□	Ca

Fig. 17.12
Superstructures of the perovskite type. Only in one octant have all atoms been plotted; the atoms on the edges and in the centers of all octants are the same

perovskite with copper atoms in the octahedral positions, and they are deficient in oxygen, $ACuO_{3-\delta}$. Alkaline earth metal ions and trivalent ions (Y^{3+}, lanthanoids, Bi^{3+}, Tl^{3+}) occupy the A positions. A typical composition is $YBa_2Cu_3O_{7-x}$ with $x \approx 0.04$. Approximately $\frac{7}{9}$ of the oxygen positions are vacant, in such a way that $\frac{2}{3}$ of the Cu atoms have square-pyramidal coordination and $\frac{1}{3}$ have square-planar coordination (Fig. 17.13). The structures of some other representatives of this compound class are considerably more complicated, and may exhibit disorder and other particularities.

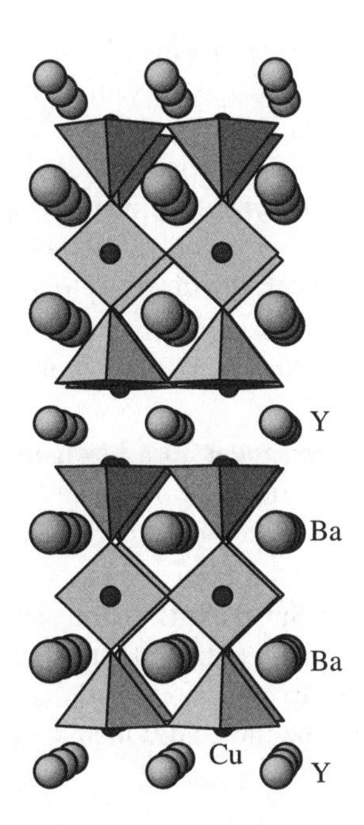

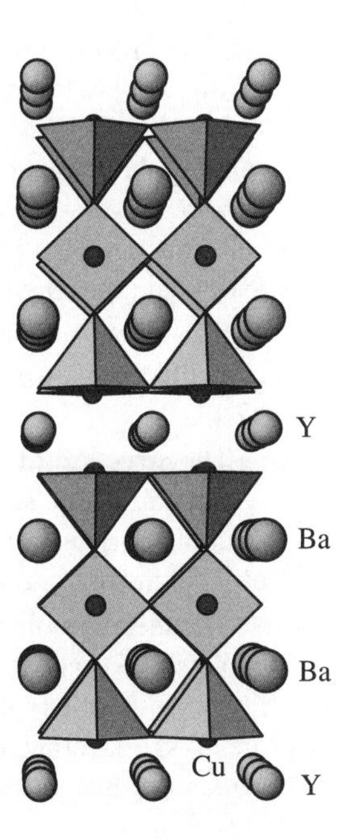

Fig. 17.13
Structure of $YBa_2Cu_3O_7$. The perovskite structure is attained by inserting O atoms between the strings of Y atoms and between the CuO_4 squares. Two unit cells are shown in each direction (stereo image)

17.5 Occupation of Tetrahedral Interstices in Closest-packings of Spheres

Coordination tetrahedra sharing faces would be present if all tetrahedral interstices in a hexagonal closest-packing were occupied. This would be an unfavorable arrangement for electrostatic reasons. On the other hand, the occupation of all tetrahedral interstices in cubic closest-packing results in an electrostatically favorable structure type: the CaF_2 type (F^- ions in the interstices), which also is the structure of Li_2O (Li^+ in the interstices). The tetrahedra are linked by common edges and common vertices.

Taking out half of the atoms from the tetrahedral interstices of the CaF_2 type leaves the composition MX. Several structure types can arise, depending on the selection of the vacated interstices: the zinc blende type with a network of vertex-sharing tetrahedra, the PbO type with layers of edge-sharing tetrahedra, and the PtS type (Fig. 17.14). In PbO and PtS the metal atoms form the packing of spheres. PbO only has tetrahedra occupied by O atoms at the height $z = \frac{1}{4}$, and those in $z = \frac{3}{4}$ are vacant; the Pb atoms at $z \approx 0$ and $z \approx \frac{1}{2}$ and the O atoms together form a layer in which every Pb atom has square-pyramidal coordination (*cf.* Fig. 16.28, p. 189). The distribution of the atoms in PtS results in a planar coordination around a Pt atom. The packing is a compromise between the requirements of tetrahedral coordination for sulfur and square coordination for platinum. With a ratio $c/a = 1.00$ the sulfur atoms would have an undistorted tetrahedral coordination, but it would be rectangular at Pt atoms; with $c/a = 1.41$ the bond angles would be 90° at Pt, but also at S. The actual ratio is $c/a = 1.24$.

HgI_2 and α-$ZnCl_2$ are examples of structures with cubic closest-packing of halogen atoms, having one-quarter of the tetrahedral interstices occupied. These tetrahedra share vertices, every tetrahedron vertex being common to two tetrahedra with bond angles of 109.5° at the bridging atoms. The HgI_2 structure corresponds to a PbO structure in which half of the O atoms have been removed and cations have been exchanged with anions (Fig. 17.14). There are layers, and all Hg atoms are at the same level within one layer (*cf.* also Fig. 16.23, p. 185).

If half of the atoms are removed from zinc blende, in the way shown in the right part of Fig. 17.14, the result is the α-$ZnCl_2$ structure. It has a framework of vertex-sharing tetrahedra. The zinc atoms form helices parallel to *c*. The *c* axis is doubled. By mutually rotating the tetrahedra, the lattice is widened and the bond angles at the bridging atoms become larger; the result is the cristobalite structure (Fig. 17.15). The face-centered unit cell shown in Fig. 12.9 (p. 125) is twice as big as the body-centered cell of Fig. 17.15; the axes **a** and **b** of the face-centered cell run diagonal to those of the body-centered cell.

SiS_2 offers another variant of the occupation of one-quarter of the tetrahedral interstices in a cubic closest-packing of S atoms. It contains strands of edge-sharing tetrahedra (Fig. 17.14).

The structure of wurtzite corresponds to a hexagonal closest-packing of S atoms in which half of the tetrahedral interstices are occupied by Zn atoms. In addition, any other stacking variant of closest-packings can have occupied tetrahedral interstices. Polytypes of this kind are known, for example, for SiC.

Tetrahedral molecules such as $SnCl_4$, $SnBr_4$, SnI_4, and $TiBr_4$ usually crystallize with a cubic closest-packing with $\frac{1}{8}$ of the tetrahedral interstices being occupied. Especially the lighter molecules like CCl_4 also exhibit modifications which have molecules rotating in the crystal; averaged over time, the molecules then appear as spheres, and adopt a body-centered cubic packing.

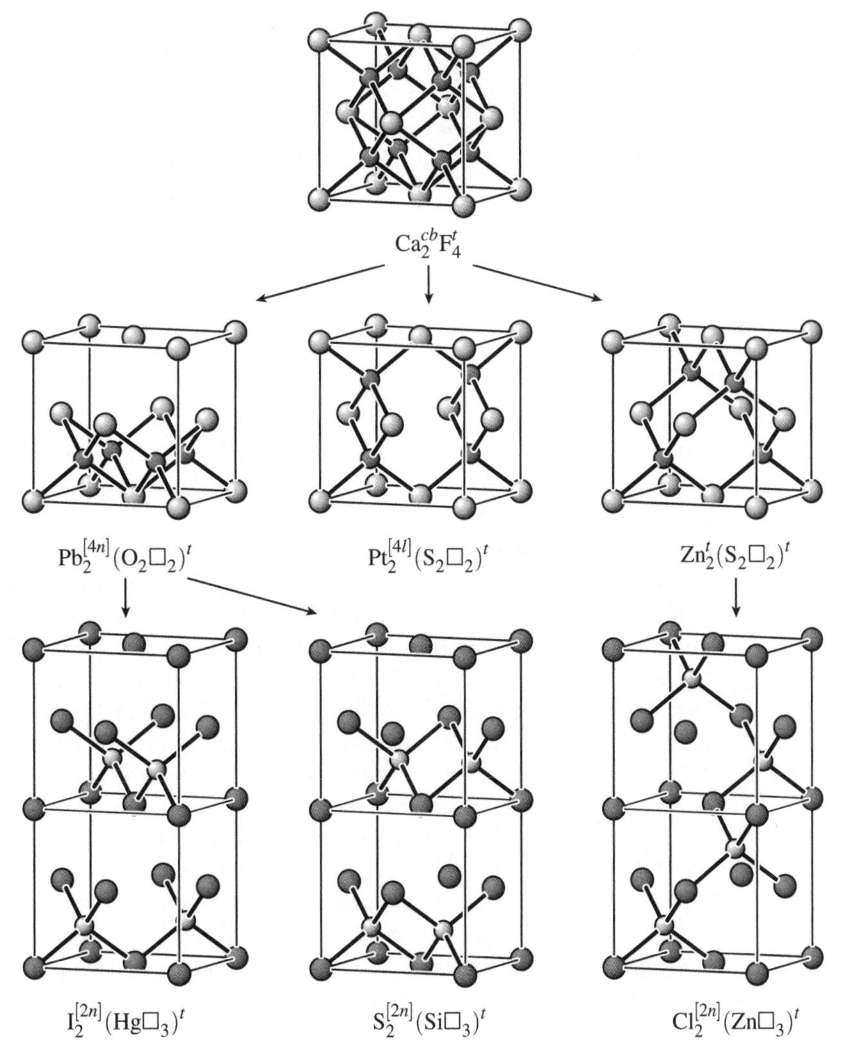

Fig. 17.14
Relationships among the structures of CaF_2, PbO, PtS, ZnS, HgI_2, SiS_2, and α-$ZnCl_2$. In the top row all tetrahedral interstices (= centers of the octants of the cube) are occupied. Every arrow designates a step in which the number of occupied tetrahedral interstices is halved; this includes a doubling of the unit cells in the bottom row. Light hatching = metal atoms, dark hatching = non-metal atoms. The atoms given first in the formulas form the cubic closest-packing

Whereas $AlCl_3$ and $FeCl_3$ have layer structures with octahedral coordination of the metal atoms in the solid state, they form dimeric molecules (two edge-sharing tetrahedra) in solution and in the gaseous state. Al_2Br_6, Al_2I_6 and the gallium trihalides, however, retain the dimeric structure even in the solid state. The halogen atoms form a hexagonal closest-packing in which $\frac{1}{6}$ of the tetrahedral interstices are occupied. Other molecules that consist of linked tetrahedra in many cases also are packed in the solid state according to the principle of closest-packings of spheres with occupied tetrahedral interstices, for example Cl_2O_7 or Re_2O_7.

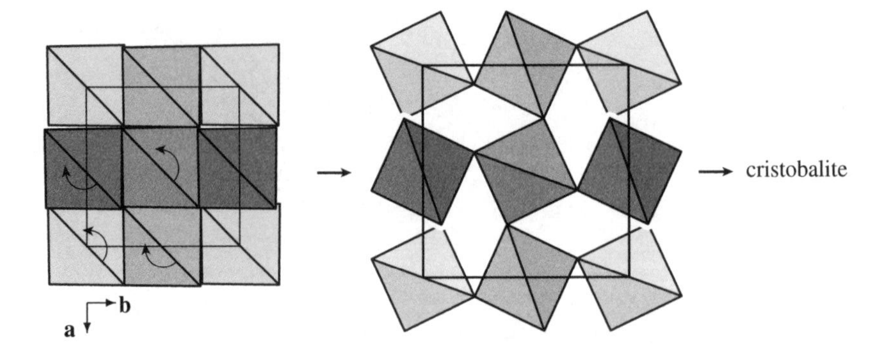

Fig. 17.15 By mutual rotation of the tetrahedra the α-$ZnCl_2$ structure is converted to the cristobalite structure

17.6 Spinels

Packings of spheres having occupied tetrahedral and octahedral interstices usually occur if atoms of two different elements are present, one of which prefers tetrahedral coordination, and the other octahedral coordination. This is a common feature among silicates (*cf.* Section 16.7). Another important structure type of this kind is the spinel type. Spinel is the mineral $MgAl_2O_4$, and generally spinels have the composition AM_2X_4. Most of them are oxides; in addition, there exist sulfides, selenides, halides and pseudohalides.

In the following, we start by assuming purely ionic structures. In spinel the oxide ions form a cubic closest-packing. Two-thirds of the metal ions occupy octahedral interstices, the rest tetrahedral ones. In a 'normal' spinel the A ions are found in the tetrahedral interstices and the M ions in the octahedral interstices; we express this by the subscripts T and O, for example $Mg_T[Al_2]_OO_4$. Since tetrahedral holes are smaller than octahedral holes, the A ions should be smaller than the M ions. Remarkably, this condition is not fulfilled in many spinels, and just as remarkable is the occurrence of 'inverse' spinels which have half of the M ions occupying tetrahedral sites and the other half occupying octahedral sites while the A ions occupy the remaining octahedral sites. Table 17.3 summarizes these facts and also includes a classification according to the oxidation states of the metal ions.

Arbitrary intermediate states also exist between normal and inverse spinels; they can be characterized by the *degree of inversion* λ:

$\lambda = 0$: normal spinel $\quad \lambda = 0.5$: inverse spinel

The distribution of the cations among the tetrahedral and octahedral sites is then expressed

Table 17.3: Summary of spinel types with examples

oxidation state combination	normal spinels $A_T[M_2]_OX_4$		inverse spinels $M_T[AM]_OX_4$			
II, III	$MgAl_2O_4$		$MgIn_2O_4$			
II, III	Co_3O_4		Fe_3O_4			
IV, II	$GeNi_2O_4$		$TiMg_2O_4$			
II, I	$ZnK_2(CN)_4$		$NiLi_2F_4$			
VI, I	WNa_2O_4					
Ionic radii:	Mg^{2+}	72 pm	Fe^{2+}	78 pm	Co^{2+}	75 pm
	Al^{3+}	54 pm	Fe^{3+}	65 pm	Co^{3+}	61 pm

in the following way: $(Mg_{1-2\lambda}Fe_{2\lambda})_T[Mg_{2\lambda}Fe_{2(1-\lambda)}]_O O_4$. The value of λ is temperature dependent. For example, at room temperature $MgFe_2O_4$ has $\lambda = 0.45$ and thus is essentially inverse.

The difficulties in understanding the cation distributions and in explaining the occurrence of inverse spinels on the basis of ionic radii show how insufficient this kind of approach is. A somewhat better approach considers the electrostatic part of the lattice energy, the calculated MADELUNG constant being a useful quantity. For a II,III-spinel which has an undistorted packing of anions, the MADELUNG constant of the normal spinel is 1.6 % smaller than that of the inverse one, *i.e.* the inverse distribution is slightly more favorable. However, small distortions that commonly occur in most spinels (widening of the tetrahedral interstices) can reverse this. In fact, spinels are not purely ionic compounds and the consideration of electrostatic interactions alone is hardly adequate, although it does work quite well for spinels of main group elements. With transition metals, in addition, the considerations of ligand field theory have to be taken into account. To illustrate this, we take as examples the spinels Mn_3O_4, Fe_3O_4 and Co_3O_4. Except for Co(III) these are made up of high-spin complexes. The relative ligand field stabilization energies are, expressed as multiples of Δ_O (*cf.* Table 9.1, p. 78):

Mn_O^{2+}	0	Fe_O^{2+}	$\frac{2}{5} = 0.4$	Co_O^{2+}	$\frac{4}{5} = 0.8$
Mn_T^{2+}	0	Fe_T^{2+}	$\frac{3}{5} \times \frac{4}{9} = 0.27$	Co_T^{2+}	$\frac{6}{5} \times \frac{4}{9} = 0.53$
Mn_O^{3+}	$\frac{3}{5} = 0.6$	Fe_O^{3+}	0	Co_O^{3+}	$\frac{12}{5} = 2.4$ (low-spin)
Mn_T^{3+}	$\frac{2}{5} \times \frac{4}{9} = 0.18$	Fe_T^{3+}	0	Co_T^{3+}	$\frac{3}{5} \times \frac{4}{9} = 0.27$

$\Delta_T = \frac{4}{9}\Delta_O$ was taken for tetrahedral ligand fields. Mn_3O_4 is a normal spinel, $Mn_T^{II}[Mn_2^{III}]_O O_4$. If it were to be converted to an inverse spinel, half of the Mn^{III} atoms would have to shift from the octahedral to the tetrahedral environment, and this would imply a decreased ligand field stabilization for these atoms (Table 17.4); for the Mn^{II} atoms the shifting would make no difference. Fe_3O_4 is an inverse spinel, $Fe_T^{III}[Fe^{II}Fe^{III}]_O O_4$. For

Table 17.4: Ligand field stabilization energies for Mn_3O_4, Fe_3O_4 and Co_3O_4. Values for high-spin complexes in all cases except for octahedral Co^{III}

	normal	inverse
	$Mn_T^{II}[Mn_2^{III}]_O O_4$	$Mn_T^{III}[Mn^{II}Mn^{III}]_O O_4$
Mn^{II}	0	0
Mn^{III}	$2 \times 0.6 = 1.2$	$0.18 + 0.6 = 0.78$
	$1.2\,\Delta_O$	$0.78\,\Delta_O$
	$Fe_T^{II}[Fe_2^{III}]_O O_4$	$Fe_T^{III}[Fe^{II}Fe^{III}]_O O_4$
Fe^{II}	0.27	0.40
Fe^{III}	0	0
	$0.27\,\Delta_O$	$0.40\,\Delta_O$
	$Co_T^{II}[Co_2^{III}]_O O_4$	$Co_T^{III}[Co^{II}Co^{III}]_O O_4$
Co^{II}	0.53	0.80
Co_T^{III} h.s.		0.27
Co_O^{III} l.s.	$2 \times 2.4 = 4.80$	2.40
	$5.33\,\Delta_O$	$3.47\,\Delta_O$

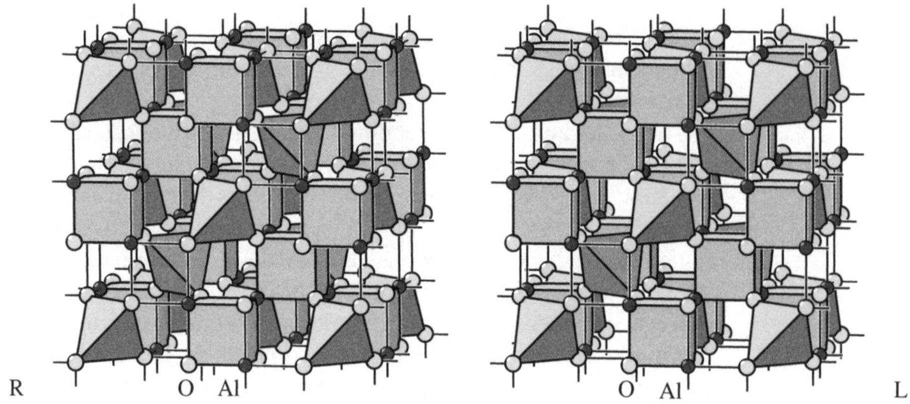

Fig. 17.16 The spinel structure (one unit cell). The Mg^{2+} ions are located in the centers of the tetrahedra (stereo image)

the Fe^{III} atoms the exchange of positions would make no difference; for the Fe^{II} atoms it would be unfavorable ($0.4\Delta_O \rightarrow 0.27\Delta_O$).

In the case of Co_3O_4, which is a normal spinel, $Co^{II}_T[Co^{III}_2]_O O_4$, the situation is different because octahedrally coordinated Co^{III} almost never occurs in high-spin complexes (its d^6 configuration corresponds to the maximum of ligand field stabilization energy in the low-spin state). If Co^{3+}_O were to adopt a high-spin state in Co_3O_4, it should be an inverse spinel. However, in the low-spin state the normal spinel is favored (Table 17.4). In addition, the ionic radius has also an effect; it decreases in the series Mn^{2+}–Fe^{2+}–Co^{2+}–Ni^{2+}–Cu^{2+}–Zn^{2+} and therefore favors the tetrahedral coordination towards the end of the series. Co^{2+} generally shows a tendency towards tetrahedral coordination in its compounds. In Fig. 9.4 (p. 79) this influence of the ionic size is taken into account by having the dashed line bent for octahedral coordination; this line corresponds to the notional reference state (spherical distribution of the *d* electrons), relative to which the ligand field stabilization energy is defined. According to Fig. 9.4, Co^{2+} is more stable in a tetrahedral environment.

Fig. 17.16 shows the spinel structure. There are four Al^{3+} and four O^{2-} ions in the vertices of an Al_4O_4 cube. Every Al^{3+} ion belongs to two such cubes, so that every cube is linked with four more cubes and every Al^{3+} ion has octahedral coordination. In addition, every O^{2-} ion belongs to an MgO_4 tetrahedron. Each of these tetrahedra shares vertices with four cubes. The cubic unit cell contains eight MgO_4 tetrahedra and eight Al_4O_4 cubes. The metal ions, taken by themselves, have the same arrangement as in the cubic LAVES phase $MgCu_2$ (*cf.* Fig. 15.4, p. 163).

The coordination of an O^{2-} ion is three Al^{3+} ions within an Al_4O_4 cube and one Mg^{2+} ion outside of this cube. This way it fulfills the electrostatic valence rule (PAULING's second rule, *cf.* p. 58), *i.e.* the sum of the electrostatic bond strengths of the cations corresponds exactly to the charge on an O^{2-} ion:

$$z(\mathrm{O}) = -(\underbrace{3\cdot\frac{3}{6}}_{3\,\mathrm{Al}^{3+}} + \underbrace{1\cdot\frac{2}{4}}_{1\,\mathrm{Mg}^{2+}}) = -2$$

The required local charge balance between cations and anions which is expressed in PAULING's rule causes the distribution of cations and anions among the octahedral and tetrahedral interstices of the sphere packing. Other distributions of the cations are not compatible with PAULING's rule.

The above-mentioned influence of the ligand field can be discerned when metal atoms with JAHN–TELLER distortions are present in a spinel. Mn_3O_4 is an example: its octahedra are elongated, and the structure is no longer cubic but tetragonal. Other examples with tetragonal distortions are the normal spinels $NiCr_2O_4$ and $CuCr_2O_4$ (Ni and Cu in tetrahedral interstices); in $NiCr_2O_4$ the tetrahedra are elongated, and in $CuCr_2O_4$ they are compressed.

Olivine $(Mg,Fe)_2SiO_4$ is the most abundant mineral of the upper Earth's mantle. Its oxygen atoms form a hexagonal closest-packing of spheres. One-eighth of the tetrahedral voids is occupied with Si atoms. Half of the octahedral voids are randomly occupied with Mg and Fe atoms. Therefore, the magnesium atoms occupy a different kind of void as compared to spinel. This accounts for the lower density of olivine, which is approximately 6 % less. Olivine is transformed to a spinel by the exertion of pressure. This transformation takes place in the Earth's mantle at a depth of 300 to 400 km, where a subduction zone is being forced under a continental plate. At the beginning, 'lenses' of spinel form; they have boundaries with the surrounding, yet untransformed olivine. Olivine and spinel can mutually slip at the boundaries. Therefore, the lenses are zones of instability which behave like cracks (they are called 'anticracks' because they have a higher density than the surrounding material). Such zones are the focuses of earthquakes.

17.7 Problems

17.1 Suppose the connection of tetrahedra shown in Fig. 17.2(a) were continued to form a layer. What would the composition be?

17.2 Why are the MX_3 strands shown in Fig. 16.10 compatible only with a hexagonal closest-packing of the X atoms?

17.3 What structure types would you expect for TiN, FeP, FeSb, CoS, and CoSb?

17.4 Why are CdI_2 and BiI_3 much more susceptible to stacking faults than $CaBr_2$ or RhF_3?

17.5 What fraction of the tetrahedral interstices are occupied in solid Cl_2O_7?

17.6 Decide whether the following compounds should form normal or inverse spinels using the ligand field stabilization energy as the criterion:
MgV_2O_4, VMg_2O_4, $NiGa_2O_4$, $ZnCr_2S_4$, $NiFe_2O_4$.

18 Symmetry as the Organizing Principle for Crystal Structures

18.1 Crystallographic Group–Subgroup Relations

A space group consists of a set of symmetry operations that always fulfills the conditions according to which a group is defined in mathematics. Group theory offers a mathematically clear-cut and very powerful tool for ordering the multitude of crystal structures according to their space groups. To this end we introduce some concepts without discussing group theory itself in detail. The following is rather an illustrative description of the facts than a strict mathematical treatment. However, this leads to no errors for our purpose. The exact mathematical treatment is rendered difficult due to the fact that space groups are infinite groups. For an exact treatment *cf.* [181].

A space group $\mathcal{G}_1$ consists of a set of symmetry operations. If another space group $\mathcal{G}_2$ consists of a subset of these symmetry operations, it is a *subgroup* of $\mathcal{G}_1$; at the same time $\mathcal{G}_1$ is a *supergroup* of $\mathcal{G}_2$. The symmetry operations present in the space group $\mathcal{G}_1$ multiply an atom which is placed in a general position by a factor n_1. A corresponding atom in the subgroup $\mathcal{G}_2$ is multiplied by analogy by a factor n_2. Since $\mathcal{G}_2$ possesses less symmetry operations than $\mathcal{G}_1$, $n_1 > n_2$ holds. The fraction n_1/n_2 is the *index* of the symmetry reduction from $\mathcal{G}_1$ to $\mathcal{G}_2$. It always is an integer and serves to order space groups hierarchically. For example, when passing from rutile to trirutile (*cf.* the end of Section 3.3, p. 21), there is a symmetry reduction of index 3.

$\mathcal{G}_2$ is a *maximal subgroup* of $\mathcal{G}_1$ if there exists no space group that can act as intermediate group between $\mathcal{G}_1$ and $\mathcal{G}_2$. $\mathcal{G}_1$ then is a *minimal supergroup* of $\mathcal{G}_2$. The index of symmetry reduction from a group to a maximal subgroup always is a prime number or a prime number power. According to the theorem of C. HERMANN a maximal subgroup is either a *translationengleiche* or *klassengleiche* subgroup.*

Translationengleiche subgroups have an unaltered translation lattice, *i.e.* the translation vectors and therefore the size of the primitive unit cells of group and subgroup coincide. The symmetry reduction in this case is accomplished by the loss of other symmetry operations, for example by the reduction of the multiplicity of symmetry axes. This implies a transition to a different crystal class. The example on the right in Fig. 18.1 shows how a fourfold rotation axis is converted to a twofold rotation axis when four symmetry-equivalent atoms are replaced by two pairs of different atoms; the translation vectors are not affected.

A space group and a *klassengleiche* subgroup belong to the same crystal class. The symmetry reduction takes place by the loss of translation symmetry, *i.e.* by enlargement of the unit cell or by the loss of centering of the unit cell. With the loss of translation symmetry the number of other symmetry operations is also reduced. Fig. 18.2 shows two examples. The example on the right side of Fig. 18.2 also shows how the enlargement of

*Terms taken from German: *translationengleiche* = with the same translations; *klassengleiche* = of the same class.

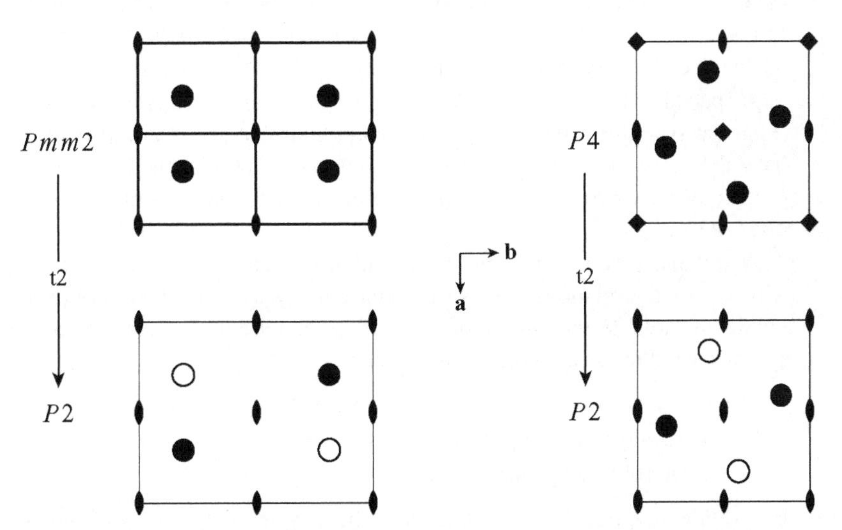

Fig. 18.1
Examples for *translationengleiche* group–subgroup relations: left, loss of reflection planes; right, reduction of the multiplicity of a rotation axis from 4 to 2. The circles of the same type, ○ and ●, designate symmetry-equivalent positions

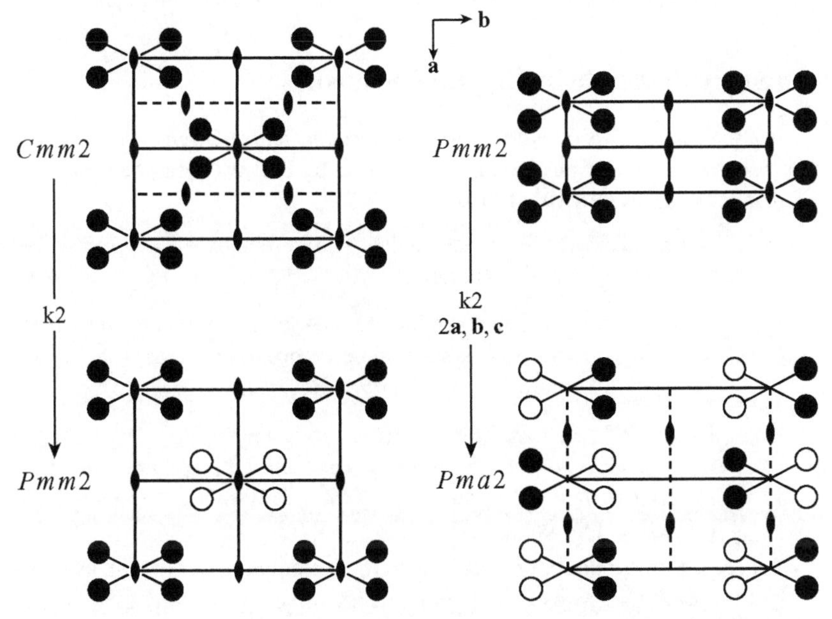

Fig. 18.2
Examples for *klassengleiche* group–subgroup relations: left, loss of centering including the loss of glide planes and twofold axes; right, enlargement of the unit cell including the symmetry reduction of reflection planes perpendicular to **b** to glide planes. Circles of the same type, ○ and ●, refer to symmetry-equivalent positions

the unit cell can cause the reduction of the symmetry of a reflection plane to a glide plane. In a similar way the symmetry of a rotation axis can be reduced to that of a screw axis when the cell enlargement is along the axis (*cf.* Problem 18.1, p. 225).

A special case of *klassengleiche* subgroups are the *isomorphic* subgroups. Group and subgroup belong to the same or the enantiomorphic space group type. Therefore, they have the same Hermann–Mauguin symbol or the symbol of the enantiomorphic space group type (*e.g.* $P3_1$ and $P3_2$). The subgroup has an enlarged unit cell. Rutile and trirutile offer an example (Fig. 3.10).

A suitable way to represent group–subgroup relations is by means of 'family trees' which show the relations from space groups to their maximal subgroups by arrows pointing downwards. In the middle of each arrow the kind of the relation and the index of the symmetry reduction are labeled, for example:

t2 = *translationengleiche* subgroup of index 2
k2 = *klassengleiche* subgroup of index 2
i3 = isomorphic subgroup of index 3

In addition, if applicable, it is stated how the new unit cell emerges from the old one (base vectors of the subgroup given as vectorial sums of the base vectors of the preceding group; see Fig. 18.2). Sometimes it is also necessary to state an origin shift. This applies when, according to the conventions of *International Tables for Crystallography* [48], the origin of the subgroup has to be placed in a different position than the origin of the preceding space group. The origin shift is given as a triplet of numbers, for example $0, \frac{1}{2}, -\frac{1}{4}$ = shift by $0, \frac{1}{2}\mathbf{b}, -\frac{1}{4}\mathbf{c}$; **b** and **c** being the base vectors of the preceding space group.

18.2 The Symmetry Principle in Crystal Chemistry

In crystalline solids a tendency to form arrangements of high symmetry is observable. The *symmetry principle*, put forward in this form by F. LAVES, has been stated in a more exact manner by H. BÄRNIGHAUSEN:

1. In the solid state the arrangement of atoms shows a pronounced tendency towards the highest possible symmetry.
2. Counteracting properties of the atoms or atom aggregates may prevent the attainment of the highest possible symmetry, but in most cases the deviations from the ideal symmetry are only small (keyword *pseudosymmetry*).
3. During a phase transition or a reaction in the solid state which results in one or more products of lower symmetry, very often the higher symmetry of the starting material is indirectly preserved by the orientation of domains formed within the crystalline matrix.

Aspect 1 corresponds approximately to the formulation of G. O. BRUNNER:

Atoms of the same kind tend to be in equivalent positions.

This formulation gives a clue to the physical background of the symmetry principle: Given certain conditions such as chemical composition, the kind of chemical bonding, electronic configuration of the atoms, relative size of the atoms, pressure, temperature etc., for every kind of atom there exists *one* energetically most favorable spatial arrangement which all atoms of this kind tend to adopt. Like atoms are indistinguishable particles according to quantum mechanics. However, in a crystal they are only indistinguishable if they are symmetry equivalent; only then do they have the same surrounding.

As an example of the prevalence of high-symmetry structures we can take the closest packings of spheres: only in the cubic and the hexagonal closest-packing of spheres are all atoms symmetry equivalent; in other stacking variants of closest-packings several non-equivalent atomic positions are present, and these packings only seldom occur.

The given conditions do not always allow for equivalent positions for all atoms. Take as an example the following conditions: composition MX_5, covalent M–X bonds, all X atoms bonded to M atoms. In this case all X atoms can only be equivalent if each set of five of them form a regular pentagon around an M atom (as for example in the XeF_5^- ion). If this is not possible for some reason, then there must be at least two non-equivalent positions for the X atoms. According to the symmetry principle the number of these non-equivalent positions will be as small as possible.

18.3 Structural Relationships by Group–Subgroup Relations

As was shown in various previous chapters, many structures of solids can be regarded as derivatives of simple, high-symmetry structure types. Let us recall some examples:

Body-centered cubic sphere packing ⇒ CsCl type ⇒ superstructures of the CsCl type (Section 15.3)

Diamond ⇒ zinc blende ⇒ chalcopyrite (Sections 12.2 and 12.4)

Closest-packings ⇒ closest-packings with occupied octahedral interstices (*e.g.* CdI_2 type, Section 17.3)

In all cases we start from a simple structure which has high symmetry. Every arrow (⇒) in the preceding examples marks a reduction of symmetry, *i.e.* a group–subgroup relation. Since these are well-defined mathematically, they are an ideal tool for revealing structural relationships in a systematic way. Changes that may be the reason for symmetry reductions include:

- Atoms of an element in symmetry-equivalent positions are substituted by several kinds of atoms. For example: CC (diamond) → ZnS (zinc blende).
- Atoms are replaced by voids or voids are occupied by atoms. For example: hexagonal closest-packing → CdI_2 type. If the voids are considered to be 'zero atoms', this can be considered as a 'substitution' of voids by atoms.
- Atoms of an element are substituted by atoms of another element that requires an altered kind of bonding. For example: $KMgF_3$ (perovskite type) → $CsGeCl_3$ (lone electron pair at the Ge atom, Ge atom shifted from the octahedron center towards an octahedron face so that the three covalent bonds of an $GeCl_3^-$ ion are formed).
- Distortions due to the JAHN–TELLER effect. For example: $CdBr_2$ (CdI_2 type) → $CuBr_2$ (distorted CdI_2 type).
- Emergence of new interactions. For example: iodine (high pressure, metallic, packing of spheres) → I_2 molecules (normal pressure).
- Distortions due to covalent bonds. For example: $RuCl_3$ (high temperature, hexagonal TiI_3 type) → $RuCl_3$ (low temperature, orthorhombic, Ru–Ru bonds).

- Phase transitions. Examples: $BaTiO_3$ (> 120 °C, cubic perovskite type) → $BaTiO_3$ (< 120 °C, tetragonal), *cf.* Fig. 19.5, p. 230; $CaCl_2$ (> 217 °C, rutile type) → $CaCl_2$ (< 217 °C), *cf.* Fig. 4.1, p. 33. For second-order phase transitions it is mandatory that there is a group–subgroup relation between the involved space groups (Section 18.4).

Structural relations can be presented in a clear and concise manner using family trees of group–subgroup relations as put forward by BÄRNIGHAUSEN. They can be set up with the aid of *International Tables for Crystallography*, Volume *A*1 [181], in which the maximal subgroups of every space group are completely listed. The top of the family tree (*Bärnighausen tree*) corresponds to the structure of highest symmetry, the *aristotype* (or basic structure), from which all other structures of a structure family are derived. The *hettotypes* (or derivative structures) are the structures that result by symmetry reduction.

Every space group listed in the family tree corresponds to a structure. Since the space group symbol itself states only symmetry, and gives no information about the atomic positions, additional information concerning these is necessary for every member of the family tree (Wyckoff symbol, site symmetry, atomic coordinates). The value of information of a tree is rather restricted without these data. In simple cases the data can be included in the family tree; in more complicated cases an additional table is convenient. The following examples show how specifications can be made for the site occupations. Because they are more informative, it is advisable to label the space groups with their full Hermann-Mauguin symbols.

Inspection of the atomic positions reveals how the symmetry is being reduced step by step. In the aristotype usually all atoms are situated in special positions, *i.e.* they have positions on certain symmetry elements, fixed values of the coordinates, and specific site symmetries. From group to subgroup at least one of the following changes occurs for each atomic position:

1. The site symmetry is reduced. Simultaneously, individual values of the coordinates x, y, z may become independent, *i.e.* the atom can shift away from the fixed values of a special position.
2. Symmetry-equivalent positions split into several positions that are independent of one another.

Diamond–Zinc Blende

The group–subgroup relation of the symmetry reduction from diamond to zinc blende is shown in Fig. 18.3. Some comments concerning the terminology have been included. In both structures the atoms have identical coordinates and site symmetries. The unit cell of diamond contains eight C atoms in symmetry-equivalent positions (Wyckoff position 8*a*). With the symmetry reduction the atomic positions split to two independent positions (4*a* and 4*c*) which are occupied in zinc blende by zinc and sulfur atoms. The space groups are *translationengleiche:* the dimensions of the unit cells correspond to each other. The index of the symmetry reduction is 2; exactly half of all symmetry operations is lost. This includes the inversion centers which in diamond are present in the centers of the C–C bonds.

The symmetry reduction can be continued. A (non-maximal) subgroup of $F\bar{4}3m$ is $I\bar{4}2d$ with doubled lattice parameter c. On the way $F\bar{4}3m \to I\bar{4}2d$ the Wyckoff position of the zinc atoms splits once more and can be occupied by atoms of two different elements.

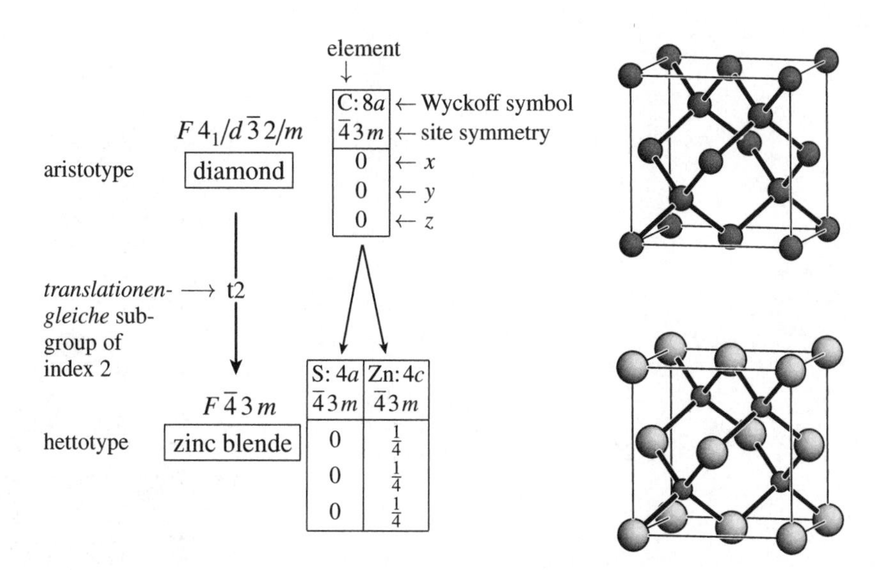

Fig. 18.3 Group–subgroup relation diamond–zinc blende

That corresponds to the structure of chalcopyrite, $CuFeS_2$. Another tetragonal subgroup of $F\bar{4}3m$ with doubled c axis is $I\bar{4}2m$. It has the position 4c of zinc blende split into three positions, 2a, 2b and 4d. Their occupation by atoms of three elements corresponds to the structure of stannite, $FeSnCu_2S_4$.

The Relation between NiAs and MnP

The symmetry reduction to the mentioned hettotypes of diamond is necessary to allow the substitution of the C atoms by atoms of different elements. No splitting of Wyckoff positions, but a reduction of site symmetries in necessary to account for distortions of a structure. Let us consider once more MnP as a distorted variant of the nickel arsenide type (Fig. 17.5, p. 197). Fig. 18.4 shows the relations together with images of the structures.

The first step entails a loss of the hexagonal symmetry; a slight distortion of the lattice would suffice to this end. To meet the conventions, a C-centered cell has to be chosen for the orthorhombic subgroup. Due to the centering it is a *translationengleiche* subgroup, even though the size of the cell has been doubled. The corresponding cell transformation is mentioned in the middle of the group–subgroup arrow. The centering is removed in the second step, which corresponds to a loss of half of the translations. Therefore, it is a *klassengleiche* symmetry reduction of index 2.

The images in Fig. 18.4 display which symmetry elements are being lost in the two steps of the symmetry reduction. Among others, half of the inversion centers are removed in the second step. The removed inversion centers of the space group $C2/m2/c2_1/m$ (for short $Cmcm$) are those of the Wyckoff positions 4a (0,0,0) and 4b ($\frac{1}{2}$,0,0), while those of the Wyckoff position 8d ($\frac{1}{4}$,$\frac{1}{4}$,0) are retained. Since the origin of the subgroup $P2_1/m2_1/c2_1/n$ (short $Pmcn$) is supposed to be situated in an inversion center, this requires an origin shift by $-\frac{1}{4},-\frac{1}{4},0$, as mentioned in the group–subgroup arrow. This shift means that $\frac{1}{4},\frac{1}{4},0$ have to be added to the coordinates. The necessary coordinate transformations are given between the boxes with the atomic coordinates.

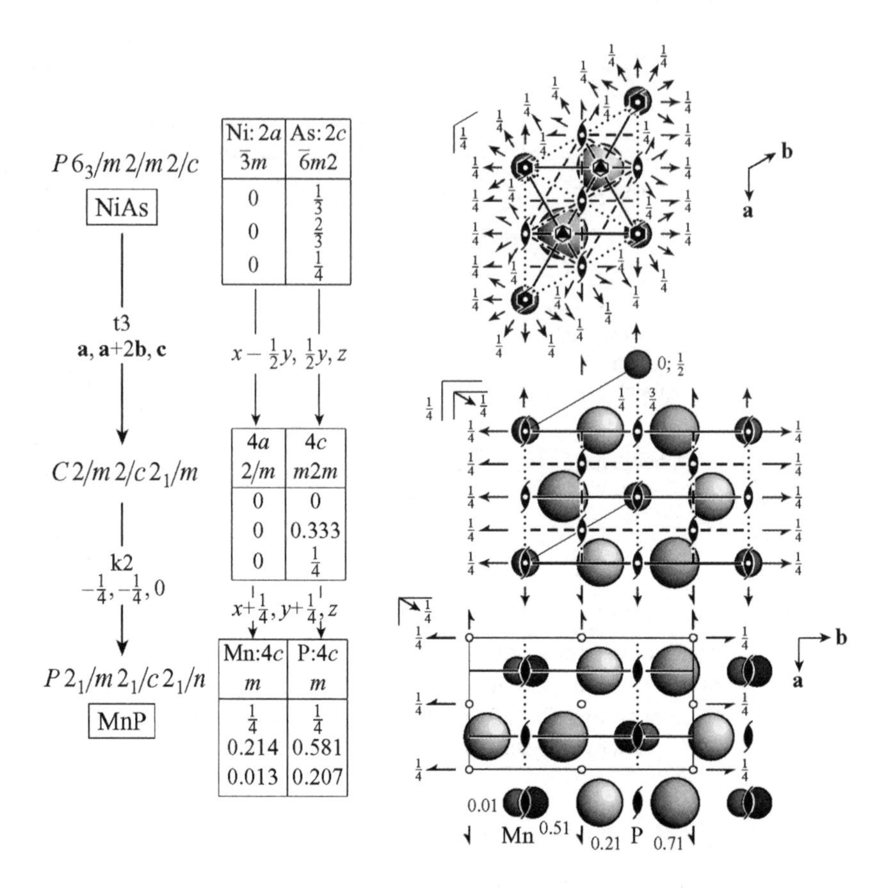

Fig. 18.4 The relation between NiAs and MnP. The numbers in the images are z coordinates

By the addition of $\frac{1}{4}, \frac{1}{4}, 0$ to the coordinates listed in Fig. 18.4 for the space group $Cmcm$, we obtain ideal values for an undistorted structure in $Pmcn$. However, due to the missing distortion the symmetry would still be $Cmcm$. The space group $Pmcn$ is only attained by the shift of the atoms from the ideal positions. First of all, the deviations concern the y coordinate of the Mn atom (0.214 instead of $\frac{1}{4}$) and the z coordinate of the P atom (0.207 instead of $\frac{1}{4}$). These are rather small deviations, so we have good reasons to consider MnP as being a distorted variant of the NiAs type.

The relation between diamond and zinc blende shown above is a formal view. The substitution of carbon atoms by zinc and sulfur atoms cannot be performed in reality. The distortion of the NiAs structure according to Fig. 18.4, however, can actually be performed. This happens during phase transitions (Section 18.4). For example, MnAs exhibits this kind of phase transition at 125 °C (NiAs type above 125 °C, second-order phase transition; another transition takes place at 45 °C, *cf.* p. 238).

Occupation of Octahedral Interstices in Hexagonal Closest-packing

According to the discussion in Section 17.3, many structures can be derived from the hexagonal closest-packing of spheres by occupying a fraction of the octahedral interstices with other atoms. If the X atoms of a compound MX_n form the packing of spheres, then the fraction $1/n$ of the octahedral interstices must be occupied. The unit cell of the

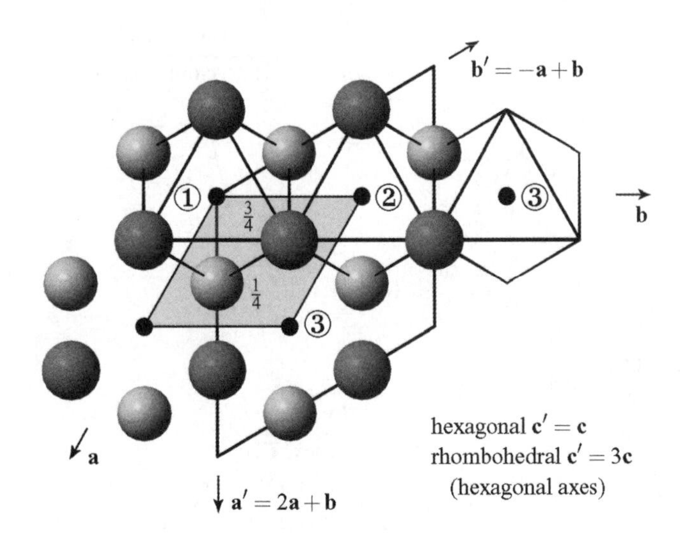

Fig. 18.5 Section of the hexagonal closest-packing. Gray area: Unit cell, space group $P6_3/m2/m2/c$. Large cell: Base of the triple cell. The z coordinates of the spheres refer to $\mathbf{c}' = \mathbf{c}$. The dots labeled ①, ② and ③ mark six octahedral voids at $z = 0$ and $z = \frac{1}{2}$ (for $\mathbf{c}' = \mathbf{c}$) and at $z = 0$ and $z = \frac{1}{6}$ ($\mathbf{c}' = 3\mathbf{c}$)

hexagonal closest-packing contains two octahedral interstices, so that only the fractions $\frac{1}{1}$ or $\frac{1}{2}$ of the octahedral interstices can be occupied without enlargement of the unit cell. Occupation of any other fraction requires an enlargement of the unit cell. In other words, only the compositions MX and MX_2 allow for structures without cell enlargement. Cell enlargement is equivalent to loss of translation symmetry, therefore *klassengleiche* group–subgroup relations must occur.

The aristotype can be considered to be either the packing of spheres itself, or the NiAs type which corresponds to the packing in which all octahedral interstices are occupied by Ni atoms (Wyckoff position 2*a*). In the aristotype these interstices are symmetry equivalent; subgroups result if the interstices are occupied only partially or by different kinds of atoms (or if the Ni atoms of NiAs are partially removed or substituted). By this procedure the sites of the interstices become non-equivalent.

$P6_3/m2/m2/c$, which is the space group of the hexagonal closest-packing of spheres, has only one maximal subgroup in which the position 2*a* is split into two independent positions, namely $P\overline{3}2/m1$. If one of these positions is occupied and the other one remains vacant, this corresponds to the CdI_2 type.

Fig. 18.5 shows how the unit cell of the hexagonal closest-packing can be enlarged by a factor of three. The enlarged cell remains hexagonal or trigonal if the base vector **c** (viewing direction) is not changed. A rhombohedral cell results if **c** is also triplicated, together with a centering in $\frac{2}{3},\frac{1}{3},\frac{1}{3}$ and $\frac{1}{3},\frac{2}{3},\frac{2}{3}$. The volume of the cell itself then is nine times larger, but due to the centering the primitive cell is only three times larger. In both cases the primitive, enlarged cell contains six spheres (X atoms) and six octahedral holes. The composition M_2X_6 or MX_3 is the result of the occupation of two of the octahedral holes, the other four remaining vacant.

The Bärnighausen tree shown in Fig. 18.6 shows how the structures of some compounds MX_3 and M_2X_3 can be derived from a hexagonal closest-packing of spheres, taking the mentioned triplicated cells. The octahedral positions for each space group are represented by two or six small boxes instead of specifying numerical values. The boxes refer to the octahedral voids in the unit cell. The corresponding coordinates are given in the image at the top left. The positions of the octahedron centers are labeled by their Wyckoff letters;

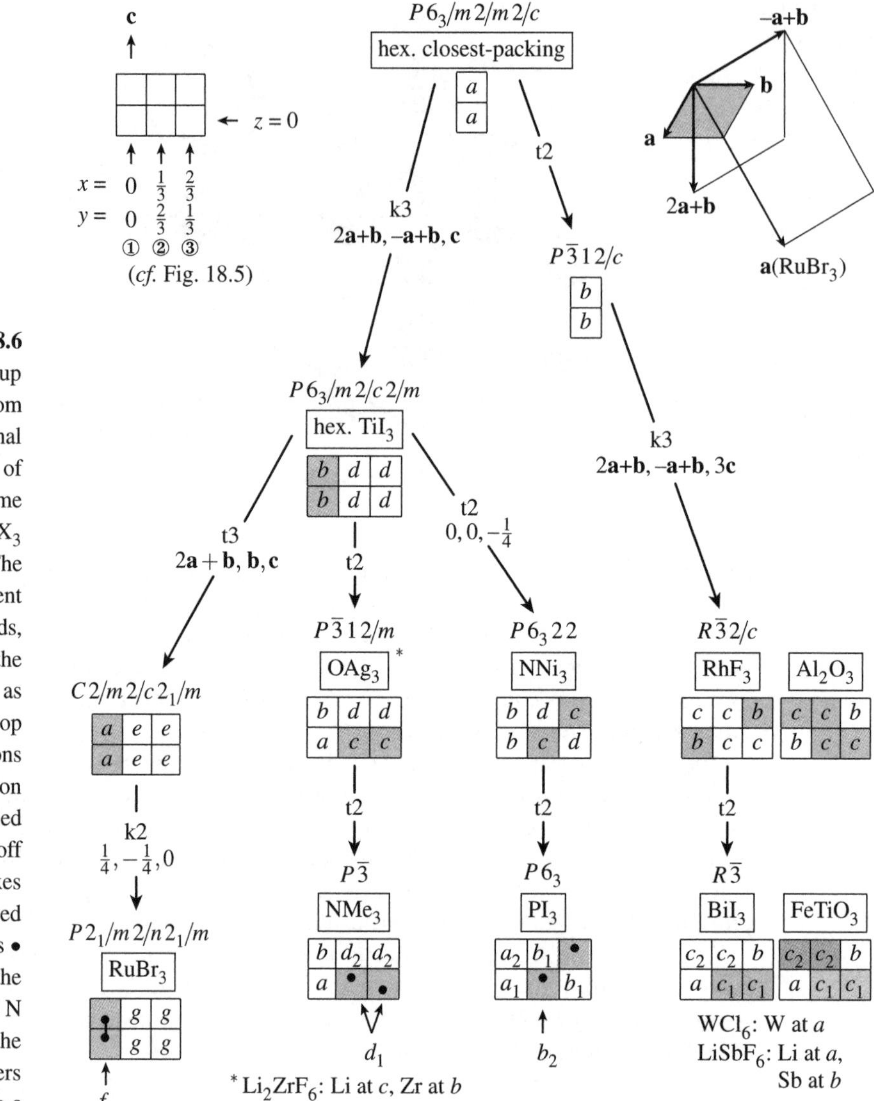

Fig. 18.6 Group–subgroup relations from hexagonal closest-packing of spheres to some MX_3 and M_2X_3 structures. The boxes represent octahedral voids, with the coordinates as given at the top left. The positions of the octahedron centers are labeled by their Wyckoff letters. Gray boxes refer to occupied voids. The dots • indicate how the atoms Ru, P and N are shifted from the octahedron centers parallel to **c**

symmetry-equivalent octahedra have the same letter. The symmetry reduction from top to bottom can be recognized by the increasing number of different letters.

The right branch of the tree lists rhombohedral subgroups. The triplication of the unit cell happens at the *klassengleiche* reduction $P\bar{3}12/c$ —k3→ $R\bar{3}2/c$. The two octahedral voids in the position 2*b* of $P\bar{3}12/c$ result in six octahedral voids in the positions 2*b* and 4*c* of $R\bar{3}2/c$. Occupation of the position 2*b*, leaving vacant 4*c*, corresponds to the RhF_3 type. The opposite, occupation of 4*c*, leaving vacant 2*b*, corresponds to corundum (α-Al_2O_3). Further splitting of the positions takes place at the next step, $R\bar{3}2/c$ —t2→ $R\bar{3}$. Depending on which one of them is occupied, we obtain the structure types of BiI_3, ilmenite ($FeTiO_3$), WCl_6 and $LiSbF_6$.

Three structure types of composition MX_3 are mentioned in the left branch of Fig. 18.6. Hexagonal TiI_3 has strands of occupied face-sharing octahedra in the direction of **c** (gray boxes one on top of the other; *cf.* also Fig. 16.10, p. 175). OAg_3 has edge-sharing octahedra joined to layers as in the BiI_3 type (gray boxes side-by-side; *cf.* Fig. 16.8, p. 174). The occupied octahedra of NNi_3 share vertices.

The structures of TiI_3, OAg_3 and NNi_3 are followed by distorted variants with the same occupation patterns (lower part of the left branch). The symmetries are further reduced because in all three cases the atoms are shifted away from the octahedron centers; this is marked by the dots •. The Ru atoms of $RuBr_3$ have shifted toward each other in pairs (Ru–Ru bonds). The P atoms of PI_3 have lone electron pairs. Each P atoms has shifted toward an octahedron face parallel to +**c**; in this way the P atoms obtain three short P–I bonds and three long P···I contacts. A similar situation arises in crystalline trimethylamine, whose methyl groups form a hexagonal closet-packing. The N atoms are alternately shifted in the directions +**c** and −**c** out of the midpoints of the octahedra.

The crystal data compared to expected values assuming no distortions are summarized in Table 18.1. Inspection of the atomic coordinates reveals that the distortions of the packing of spheres are only marginal. As expected, the greatest deviations are observed for the molecular compounds PI_3 and NMe_3.

18.4 Symmetry Relations at Phase Transitions. Twinned Crystals

As discussed in Section 4.4 (p. 32), reconstructive phase transitions are always first-order transitions. The transformation begins at a nucleation site, which may be a site with a vacancy in the crystal. There, the movement of the atoms sets in and is followed by the growth of the nucleus and the reconstruction of the structure. Bonds of the old structure are broken and new bonds of the growing new structure are joined at the boundary between the two phases, while atoms perform the necessary diffusion. Group–subgroup relations between the space groups are of no importance in this kind of phase transition. Occasional speculations that assume a first-order phase transition to proceed via an intermediate common subgroup of the space groups concerned have no physical foundation whatsoever; it is not possible to assign a space group to the atomic arrangement along a phase boundary.

The situation is different for second-order phase transitions. These can only occur when the space group of the one phase is a (not necessarily maximal) subgroup of the space group of the other phase. Many displacive structure transformations are second-order transitions, in which atomic groups experience only small mutual motions. As an example, Fig. 4.1 (p. 33) shows the second-order phase transition of calcium chloride from the rutile type to the $CaCl_2$ type. All that happens is a mutual rotation of the strands of edge-sharing octahedra. This implies that the tetragonal symmetry cannot be retained; the symmetry is 'broken'. The space group of the $CaCl_2$ type necessarily is a subgroup of the space group $P4_2/m\,2_1/n\,2/m$ of the rutile type. Since the crystal class changes from tetragonal to orthorhombic, the subgroup is *translationengleiche*. The corresponding relation is depicted in Fig. 18.7.

The lattice parameters a and b become different during the transition from the tetragonal to the orthorhombic structure. Either $a > b$ or $a < b$ results, depending on the direction of the rotation of the octahedra strands. During the phase transition both directions of rotation occur at random; domains appear, in which either $a > b$ or $a < b$. The obtained crystal is a *twinned crystal* or *twin*, consisting of intergrown domains. The twin domains are

Table 18.1: Crystal data of structures mentioned in Fig. 18.6. The ideal coordinates would apply to an undistorted packing of spheres. Coordinate values fixed by symmetry are stated as 0 or fractional numbers, otherwise as decimal numbers

	space group	$\frac{a}{\text{pm}}$	$\frac{c}{\text{pm}}$	Wyckoff position		x	y	z	ideal coordinates x	y	z
TiI_3-hex.	$P6_3/mmc$	715	650	Ti	$2b$	0	0	0	0	0	0
				I	$6g$	0.313	0	0	0.333	0	0
$RuBr_3$	$Pmnm$	1126	650	Ru	$4f$	$\frac{1}{4}$	0.746	0.015	$\frac{1}{4}$	0.75	0.0
		$b = 587$		Br	$2a$	$\frac{1}{4}$	0.431	$\frac{1}{4}$	$\frac{1}{4}$	0.417	$\frac{1}{4}$
				Br	$2b$	$\frac{1}{4}$	0.052	$\frac{3}{4}$	$\frac{1}{4}$	0.083	$\frac{3}{4}$
				Br	$4e$	0.597	0.407	$\frac{1}{4}$	0.583	0.417	$\frac{1}{4}$
				Br	$4e$	0.408	0.903	$\frac{1}{4}$	0.417	0.917	$\frac{1}{4}$
OAg_3	$P\bar{3}1m$	532	495	O	$2c$	0	0	0	0	0	0
				Ag	$6k$	0.699	0	0.276	0.667	0	0.25
Li_2ZrF_6	$P\bar{3}1m$	497	466	Zr	$1b$	0	0	$\frac{1}{2}$	0	0	$\frac{1}{2}$
				Li	$2c$	$\frac{1}{3}$	$\frac{2}{3}$	0	$\frac{1}{3}$	$\frac{2}{3}$	0
				F	$6k$	0.672	0	0.245	0.667	0	0.25
NMe_3	$P\bar{3}$	614	685	N	$2c$	$\frac{1}{3}$	$\frac{2}{3}$	0.160	$\frac{1}{3}$	$\frac{2}{3}$	0.0
				C	$6g$	0.576	−0.132	0.227	0.667	0.0	0.25
NNi_3	$P6_322$	463	431	N	$2c$	$\frac{1}{3}$	$\frac{2}{3}$	$\frac{1}{4}$	$\frac{1}{3}$	$\frac{2}{3}$	$\frac{1}{4}$
				Ni	$6g$	0.328	0	0	0.333	0	0
PI_3	$P6_3$	713	741	P	$2b$	$\frac{1}{3}$	$\frac{2}{3}$	0.146	$\frac{1}{3}$	$\frac{2}{3}$	0.25
				I	$6c$	0.686	0.034	0	0.667	0.0	0
RhF_3	$R\bar{3}c$	487	1355	Rh	$6b$	0	0	0	0	0	0
				F	$18e$	0.652	0	$\frac{1}{4}$	0.667	0	$\frac{1}{4}$
α-Al_2O_3	$R\bar{3}c$	476	1300	Al	$12c$	$\frac{1}{3}$	$\frac{2}{3}$	0.019	$\frac{1}{3}$	$\frac{2}{3}$	0.0
				O	$18e$	0.694	0	$\frac{1}{4}$	0.667	0	$\frac{1}{4}$
BiI_3	$R\bar{3}$	752	2070	Bi	$6c_1$	$\frac{1}{3}$	$\frac{2}{3}$	−0.002	$\frac{1}{3}$	$\frac{2}{3}$	0.0
				I	$18f$	0.669	0.000	0.246	0.667	0.0	0.25
$FeTiO_3$	$R\bar{3}$	509	1409	Ti	$6c_1$	$\frac{1}{3}$	$\frac{2}{3}$	0.020	$\frac{1}{3}$	$\frac{2}{3}$	0.0
				Fe	$6c_2$	0	0	0.145	0	0	0.167
				O	$18f$	0.683	−0.023	0.255	0.667	0.0	0.25
α-WCl_6	$R\bar{3}$	609	1668	W	$3a$	0	0	0	0	0	0
				Cl	$18f$	0.628	−0.038	0.247	0.667	0.0	0.25
$LiSbF_6$	$R\bar{3}$	518	1360	Li	$3a$	0	0	0	0	0	0
				Sb	$3b$	$\frac{1}{3}$	$\frac{2}{3}$	$\frac{1}{6}$	$\frac{1}{3}$	$\frac{2}{3}$	$\frac{1}{6}$
				F	$18f$	0.598	−0.014	0.246	0.667	0.0	0.25

interrelated by symmetry: their mutual orientation corresponds to a symmetry operation that has been lost during the symmetry reduction. *Cf.* aspect 3 of the symmetry principle mentioned on page 214.

Definition: An intergrowth of two or more macroscopic individuals of the same crystal species is a twin, if the orientation relations between the individuals conform to crystallographic laws. The individuals are called twin partners, twin components or twin domains.

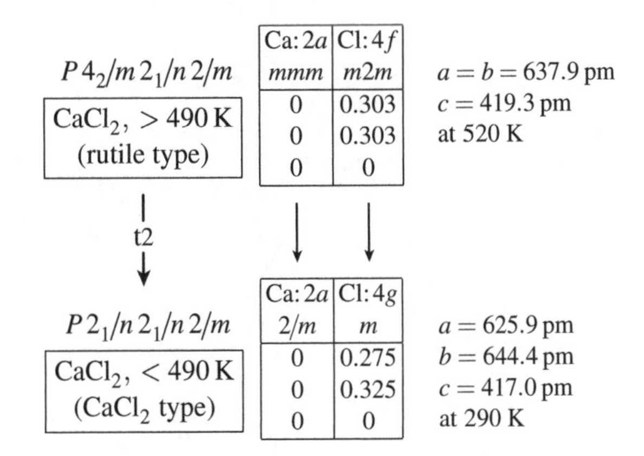

Fig. 18.7 Group–subgroup relation between the modifications of calcium chloride (*cf.* Fig. 4.1, p. 33)

The occurrence of twinned crystals is a widespread phenomenon. They may consist of individuals that can be depicted macroscopically as in the case of the 'dovetail twins' of gypsum, where the two components are mirror-inverted (Fig. 18.8). There may also be numerous alternating components which sometimes cause a streaky appearance of the crystals (polysynthetic twin). One of the twin components is converted to the other by some symmetry operation (twinning operation), for example by a reflection in the case of the dovetail twins. Another example is the 'Dauphiné twins' of quartz which are interconverted by a twofold rotation axis (Fig. 18.8). Threefold or fourfold axes can also occur as symmetry elements between the components; the domains then have three or four orientations. The twinning operation is *not* a symmetry operation of the space group of the structure, but it must be compatible with the given structural facts.

The formation of twins is to be expected if a phase transition takes place from a space group to another less symmetrical space group and if there occurs a *translationengleiche* group–subgroup relation. If it is a *translationengleiche* subgroup of index 2, the twin will consist of two kinds of domains, with index 3 three kinds and with index 4 four kinds of domains (indices higher than 4 do not occur among *translationengleiche* maximal subgroups). If the symmetry reduction entails several steps and there occur two *translationengleiche* steps, twins of twins may emerge. Among phase transitions induced by a change of temperature, as a rule, the high-temperature modification has the higher symmetry.

The Dauphiné twins of quartz are formed when quartz is transformed from its high-temperature form (β or high quartz, stable above of 573 °C) to the low-temperature form

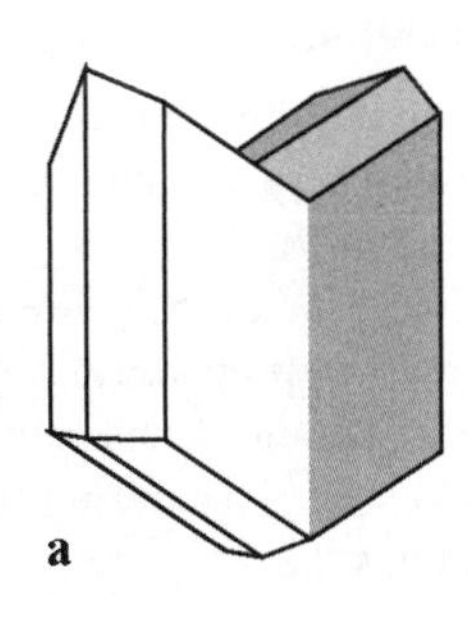

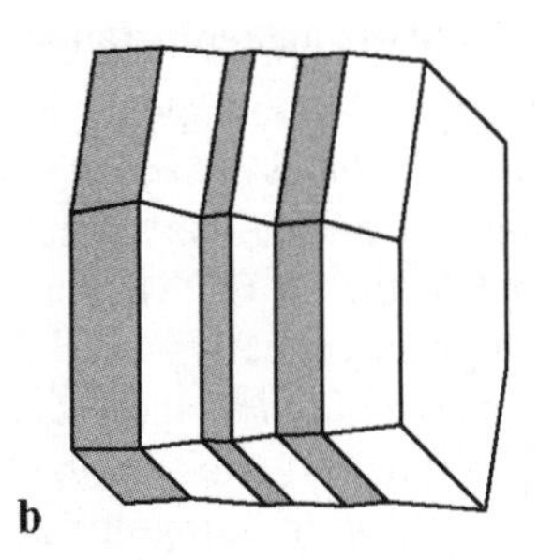

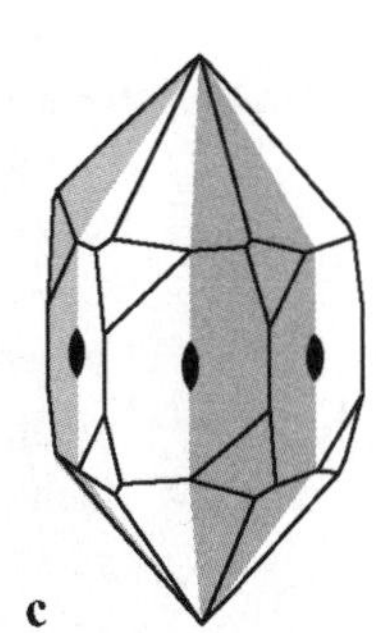

Fig. 18.8 (**a**) Dovetail twin (gypsum). (**b**) Polysynthetic twin (feldspar). (**c**) Dauphiné twin (quartz)

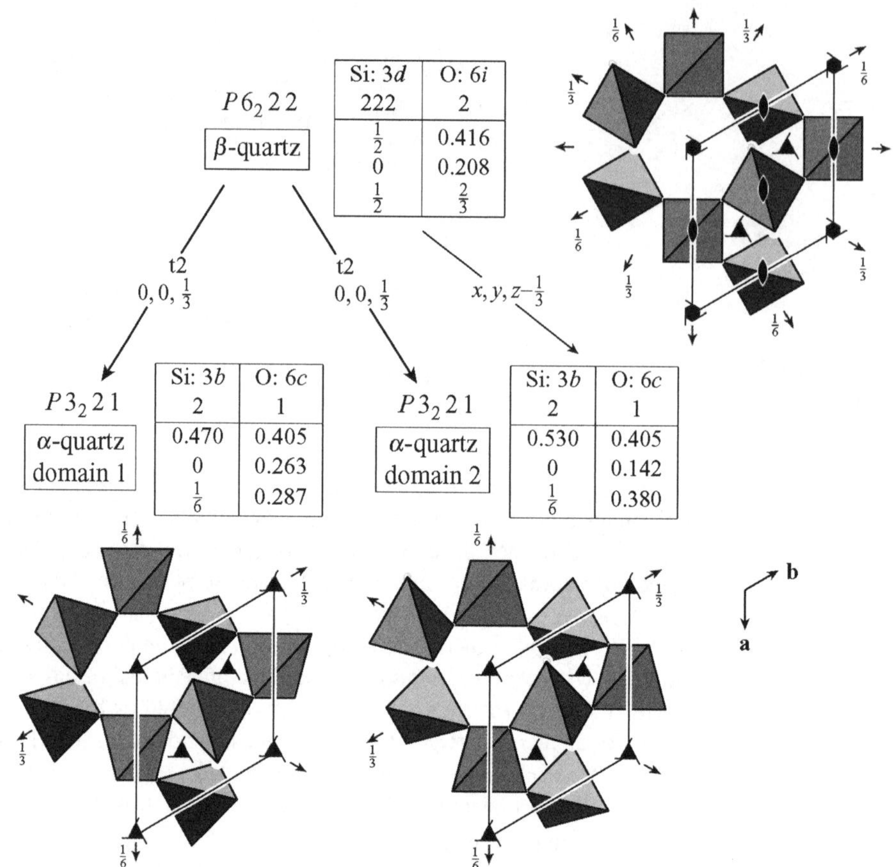

Fig. 18.9 Group–subgroup relations and the emergence of twins in the phase transition β-quartz → α-quartz (Dauphiné twins)

(α or low quartz). The space group of α-quartz is a *translationengleiche* subgroup of index 2 of the space group of β-quartz, and twins are the result. During the symmetry reduction certain twofold axes are lost (the last 2 in the space group symbol $P6_222$). These axes, present only in the higher symmetry space group, produce the twinning operation. If they were present, they would map an atom from the site x, y, z to the sites $x, x-y, \frac{2}{3}-z$; $y-x, y, \frac{1}{3}-z$ and $-y, -x, -z$ in terms of the coordinate system of $P3_221$. Compare this with the atomic coordinates of the two twin components given in Fig. 18.9. Quartz crystals twinned in this manner are unsuitable as piezoelectric components for electronic devices, as the polar directions of the twins compensate each other. During the production of piezoelectric quartz the temperature must therefore never exceed 573 °C.

The symmetry reduction shown in Fig. 18.6 resulting in the space group of $RuBr_3$ ($P2_1/m2/n2_1/m$) includes a *translationengleiche* step of index 3. The structure deviates only slightly from the hexagonal TiI_3 type in the higher symmetrical space group $P6_3/m2/c2/m$ (shifting of the Ru atoms from the octahedron centers), and at higher temperatures the higher symmetry is true. $RuBr_3$ that was prepared at higher temperatures and then cooled consists of twinned crystals with three components being mutually rotated by 120°, corresponding to the loss of the threefold symmetry during the step $P6_3/m2/c2/m$ —t3→ $C2/m2/c2_1/m$. In X-ray diffraction studies the X-ray reflections of the tree components appear superimposed, so that they give the impression of hexagonal symmetry and the resulting Ru atom positions seem to be exactly in the octahedron centers. Unrecognized twins that feign a wrong symmetry can be a treacherous problem during structure determination and may cause erroneous results.

In addition to the formation of twins during phase transformations in the solid state, twins can also develop during the growth of crystals. The emergence of crystallization nuclei controls how growth twins will be intergrown. Group–subgroup relations are irrelevant in this case.

18.5 Problems

18.1 To solve the following problems it is useful to draw images of the symmetry axes in the way as in Fig. 3.4, p. 16.

(a) Let a space group (*e.g.* $P3$) have threefold rotation axes parallel to **c**. What kind of screw axes can remain when c triplicated?
(b) The space group $P3_1$ has threefold screw axes parallel to **c**. Which screw axes remain in the maximal subgroup having c doubled?
(c) Let a space group (*e.g.* $P2_1$) have twofold screw axes parallel to **b**. Can this space group have *klassengleiche* or isomorphic, maximal subgroups in which b has been increased by a factor of two or three?

18.2 Determine whether the following group–subgroup relations are *translationengleiche, klassengleiche* or isomorphic. If the unit cell of the subgroup is enlarged, this is stated within the arrow.
(a) $Cmcm \rightarrow Pmcm$;   (b) $P2_1/c \rightarrow P\bar{1}$; (c) $Pbcm$ —2**a**,**b**,**c**→ $Pbca$;
(d) $C12/m1$ —**a**,3**b**,**c**→ $C12/m1$;   (e) $P6_3/mcm \rightarrow P6_322$;
(f) $P2_1/m2_1/m2/n \rightarrow Pmm2$;   (g) $P2_1/m2_1/m2/n \rightarrow P12_1/m1$.

18.3 Set up the Bärnighausen tree for the relation between disordered and ordered $AuCu_3$, including the relations between the Wyckoff positions (Fig. 15.1, p. 158). You will need *International Tables for Crystallography* [48], Volume A, and advantageously also Volume $A1$ [181]. Will ordered $AuCu_3$ form twins?

18.4 Set up the Bärnighausen tree for the relation between perovskite and elpasolite, including the relations between the Wyckoff positions (Fig. 17.10, p. 203 and Fig. 17.12, p. 205). Make use of *International Tables*, Volumes A and $A1$ [48, 181].

18.5 Set up the Bärnighausen tree for the relation between cubic $BaTiO_3$ (perovskite type, Fig. 17.10) and tetragonal $BaTiO_3$ (Fig. 19.5, p. 230). Remark: the subgroup is not maximal. Is it to be expected that the phase transition from the cubic to the tetragonal form will yield twins? Atomic coordinates for tetragonal $BaTiO_3$: Ba, $\frac{1}{2}\ \frac{1}{2}\ \frac{1}{2}$; Ti, 0 0 0.020; O1, 0 0 0.474; O2, $\frac{1}{2}$ 0 −0.012. Make use of *International Tables*, Volumes A and $A1$ [48, 181].

18.6 The phase transition from α-tin to β-tin involves a change of space group $F4_1/d\bar{3}2/m$ —t3→ $I4_1/a2/m2/d$. Would you expect β-tin to form twins?

18.7 The phase transition of $NaNO_2$ at 164 °C from the paraelectric to the ferroelectric form involves a change of space group from $I2/m2/m2/m$ to $Imm2$. Will the ferroelectric phase be twinned?

19 Physical Properties of Solids

The majority of the materials we use and handle every day are solid. We take advantage of their physical properties in manifold ways. The properties are intimately related to the structures. In the following we will deal only briefly with a few properties that are directly connected with some structural aspects. Many other properties such as electrical and thermal conductivity, optical transparency and reflectivity, color, luminescence etc. require the discussion of sophisticated theories that are beyond the scope of this book.

19.1 Mechanical Properties

In addition to specific properties of interest for a particular application of a material, its elasticity, compressive and tensile strength, deformability, hardness, wear-resistance, brittleness and cleavability also determine whether an application is possible. No matter how good the electric, magnetic, chemical or other properties are, a material is of no use if it does not fulfill mechanical requirements. These depend to a large extent on the structure and on the kind of chemical bonding. Mechanical properties usually are anisotropic, *i.e.* they depend on the direction of the applied force.

A framework of strong covalent bonds as in diamond results in high hardness and compressive strength. It also accounts for a high tensile strength; in this case it is sufficient if the covalent bonds are present in the direction of the tensile stress. Hardness can be determined in a qualitative manner by the MOHS scratch test: a material capable of scratching another is the harder of the two. Standard materials on the MOHS scale are talc at the lower end (hardness 1) and diamond (hardness 10) at the upper end. Talc is soft because its structure consists of electrically uncharged layers; only VAN DER WAALS forces act between the layers (*cf.* Fig. 16.21), and the layers can easily slide over each other. The same applies to graphite and MoS_2, which are used as lubricants. Crystals consisting of stiff parallel chain molecules have strong bonding forces in the chain direction and weak ones in perpendicular directions. They can be cleaved to form fiber bunches.

Ionic crystals have moderate to medium hardness, those incorporating highly charged ions being harder (*e.g.* NaCl hardness 2, CaF_2 hardness 4). Quartz with its network of polar covalent bonds is harder (hardness 7). The surfaces of materials with hardness below 7 become lusterless in everyday use because they continually suffer scratching from quartz particles in dust. The differing cohesion due to covalent and to ionic bonds in different directions is apparent in micas. Micas consist of anionic layers and intercalated cations. Within the layers atoms are held together by (polar) covalent bonds. Micas can easily be cleaved parallel to the layers, which allows the manufacture of plates with sizes of several square decimeters and a thickness of less than 0.01 mm.

Ionic crystals can be cleaved in certain directions. Fig. 19.1 shows why the exertion of a force results in cleavage: if two parts of a crystal experience a mutual displacement by a shearing force, ions of like charges come to lie side by side and repel each other. The displacement is easiest along planes which have the fewest cation–anion contacts. In

Inorganic Structural Chemistry, Second Edition Ulrich Müller

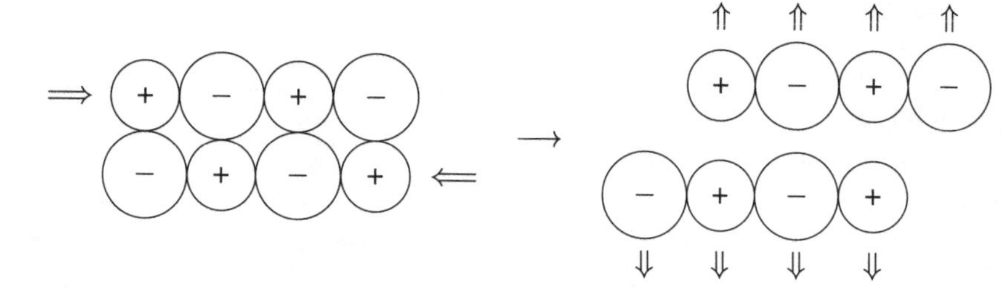

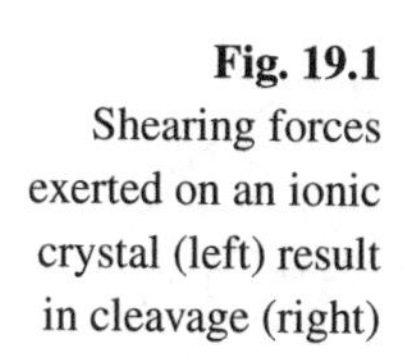

Fig. 19.1 Shearing forces exerted on an ionic crystal (left) result in cleavage (right)

rock salt, for example, one encounters one Cl^- neighbor when looking from an Na^+ ion in the direction parallel to an edge of the unit cell, two Cl^- neighbors if one looks in the direction diagonal to a face, and three Cl^- neighbors in the direction of a body diagonal. An NaCl crystal is most easily cleaved perpendicular to a cell edge.

Metals behave differently since the metal atoms are embedded in an electron gas. The attractive forces remain active even after mutual displacement of parts of a crystal has occurred. Metals therefore can be deformed without fracture.

Most ceramic materials are oxides (MgO, Al_2O_3, ZrO_2, silicates), though some are nitrides (BN, AlN, Si_3N_4) or carbides (B_4C, SiC, WC). Because of the short range of action of the chemical bonds, the material suffers a substantial loss of strength once a rupture has begun. The resulting brittleness is one of the most severe drawbacks of ceramic materials. This problem has been largely solved for one material, zirconium dioxide. ZrO_2 adopts several modifications: at temperatures above 2370 °C it has the cubic CaF_2 structure (Zr atoms with c.n. 8), between 1170 and 2370 °C it has a slightly distorted, tetragonal CaF_2 structure (Zr coordination 4 + 4), and below 1170 °C baddeleyite is the stable form; this is a more distorted variant of the CaF_2 type in which a Zr atom only has the coordination number 7. The tetragonal form can be stabilized down to room temperature by doping with a few percent of Y_2O_3. Compared to the tetragonal modification, baddeleyite requires 7 % more volume. For this reason *pure* ZrO_2 is not appropriate as a high-temperature ceramic: it cracks during heating when the transition temperature of 1170 °C is reached. But it is precisely this volume effect which is taken advantage of in order to reduce the brittleness, thus rendering ZrO_2 a high-performance ceramic material. Such material consists of 'partially stabilized' tetragonal ZrO_2, *i.e.* it is maintained metastable in this modification by doping. The mechanical forces have their maximum at the tip of a crack, and this is where the crack propagates in common materials. In metastable tetragonal ZrO_2, however, mechanical strain at the tip of a crack induces a transition to the baddeleyite form at this site, and by the volume expansion the crack is sealed.

19.2 Piezoelectric and Ferroelectric Properties

The Piezoelectric Effect

Within a crystal, consider an atom with a positive partial charge that is surrounded tetrahedrally by atoms with negative partial charges. The center of gravity of the negative charges is at the center of the tetrahedron. By exerting pressure on the crystal in an appropriate direction, the tetrahedron will experience a distortion, and the center of gravity of the negative charges will no longer coincide with the position of the positive central atom (Fig. 19.2); an electric dipole has developed. If there are inversion centers in the

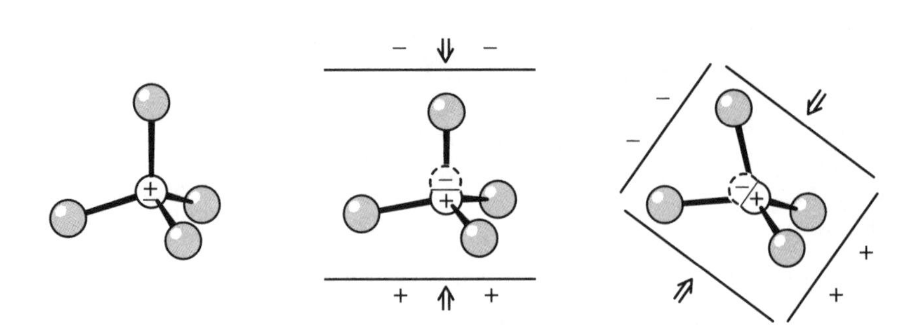

Fig. 19.2 Explanation of the piezoelectric effect: external pressure causes the deformation of a coordination tetrahedron, resulting in a shift of the centers of gravity of the electric charges

structure, then for every tetrahedron there is another tetrahedron which has the exact opposite orientation; the electric fields of the dipoles compensate each other. If, however, all tetrahedra have the same orientation or some other mutual orientation that does not allow for a compensation, then the action of all dipoles adds up and the whole crystal becomes a dipole. Two opposite faces of the crystal develop opposite electric charges. Depending on the direction of the acting force, the faces being charged are either the two faces experiencing the pressure (longitudinal effect) or two other faces in a perpendicular or an inclined direction (transversal effect).

This described *piezoelectric effect* is reversible. If the crystal is introduced into an external electric field, it experiences a contraction or an elongation. Zinc blende, turmaline, ammonium chloride and quartz are examples. The effect is used in the quartz resonators that beat time in electronic watches and computers. The quartz resonator is a sheet cut from a quartz crystal in an appropriate direction. Metal coatings act as electrical contacts. Mechanical vibrations are induced in the quartz with the aid of electric pulses; these vibrations have an exactly defined frequency and produce a corresponding alternating electric field. Aside from quartz, $Pb(Ti,Zr)O_3$ (PZT) is mainly used; its properties can be controlled by the Ti/Zr ratio. It has a distorted perovskite structure (space group $P4mm$, Fig. 19.5, or $R3c$, depending on composition). Piezoelectric crystals serve whenever mechanical pulses are to be converted to electrical signals or vice versa, for example in seismometers, vacuum meters, acceleration meters, press keys, microphones or in the production of ultrasound.

Crystals can only be piezoelectric when they are non-centrosymmetric. In addition, they may not belong to the crystal class 4 3 2. The effect is thus restricted to 20 out of the 32 crystal classes.

Ferroelectricity

In some crystalline substances the centers of gravity of positive and negative charges do not coincide in the first place, *i.e.* permanent dipoles are present. Concerning the electrical properties, the following cases can be distinguished.

A *paraelectric* substance is not polarized macroscopically because the dipoles are oriented randomly. However, they can be oriented by an external electric field (orientation polarization). The orientation is counteracted by thermal motion, *i.e.* the degree of polarization decreases with increasing temperature.

An *electret* is a crystal which has dipoles oriented permanently in one direction. The crystal therefore is a macroscopic dipole.

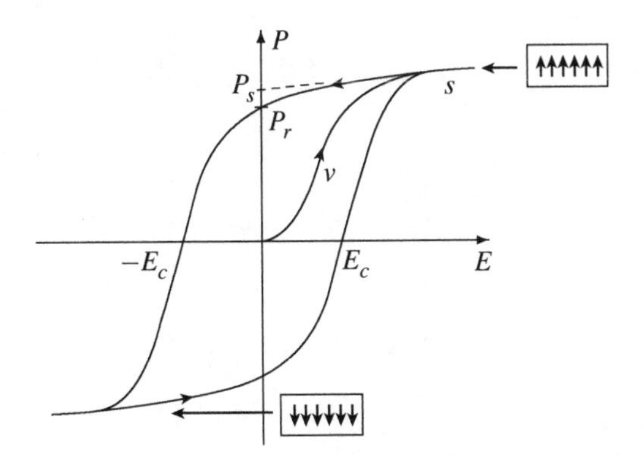

Fig. 19.3
Hysteresis curve of a ferroelectric crystal. v = initial (virginal) curve, P_r = remanent polarization, P_s = spontaneous polarization, E_c = coercive field

In a *ferroelectric* substance all dipoles are also oriented uniformly, but they can be reoriented by an external electric field. A previously untreated crystal can be made up of domains, and the uniform orientation of the dipoles is fulfilled within a domain. It differs from domain to domain. As a whole, the dipole moments of different domains can compensate each other. If an increasing external electric field acts on the sample, those domains whose polarization corresponds to the direction of the electric field will grow at the expense of the remaining domains. The total polarization of the crystal increases (curve v in Fig. 19.3). Finally, if the external field is strong enough, the whole crystal is one large domain, and the polarization continues to increase only slightly with increasing electric field (curve s; the continuing increase is due to the normal dielectric polarization which takes place in any substance by polarization of the electrons). If the external electric field is removed, a *remanent polarization* P_r remains, *i.e.* the crystal now is a macroscopic dipole. In order to remove the remanent polarization, an electric field $-E_c$ with the opposite orientation to has to be applied; this is the coercive field. The value of P_s, the spontaneous polarization, corresponds to the polarization within a domain.

Above a specific temperature, the CURIE temperature, a ferroelectric substance becomes paraelectric since the thermal vibrations counteract the orientation of the dipoles. The coordinated orientation of the dipoles taking place during the ferroelectric polarization is a *cooperative phenomenon*. This behavior is similar to that of ferromagnetic substances, which is the reason for its name; the effect has to do nothing with iron (it is also called seignette or rochelle electricity).

The polarization induced by the electric field is considerably larger than in a nonferroelectric substance, and therefore the dielectric constants are much larger. $BaTiO_3$ in particular has practical applications in the manufacture of capacitors with large capacitance. Further examples include SbSI, KH_2PO_4, and $NaNO_2$, as well as certain substances which have a distorted perovskite structure such as $LiNbO_3$ and $KNbO_3$. Fig. 19.4 shows how all nitrite ions in sodium nitrite are oriented in one direction below 164 °C, thus producing a macroscopic dipole moment. It also shows how the differently oriented domains alternate as long as there has been no electric field to shift all the NO_2^- ions into the same orientation. Above the CURIE temperature of 164 °C all NO_2^- ions are randomly oriented and $NaNO_2$ is paraelectric.

In sodium nitrite the ferroelectric polarization only occurs in one direction. In $BaTiO_3$ it is not restricted to one direction. $BaTiO_3$ has the structure of a distorted perovskite between 5 and 120 °C. Due to the size of the Ba^{2+} ions, which form a closest packing of spheres together with the oxygen atoms, the octahedral interstices are rather too large for

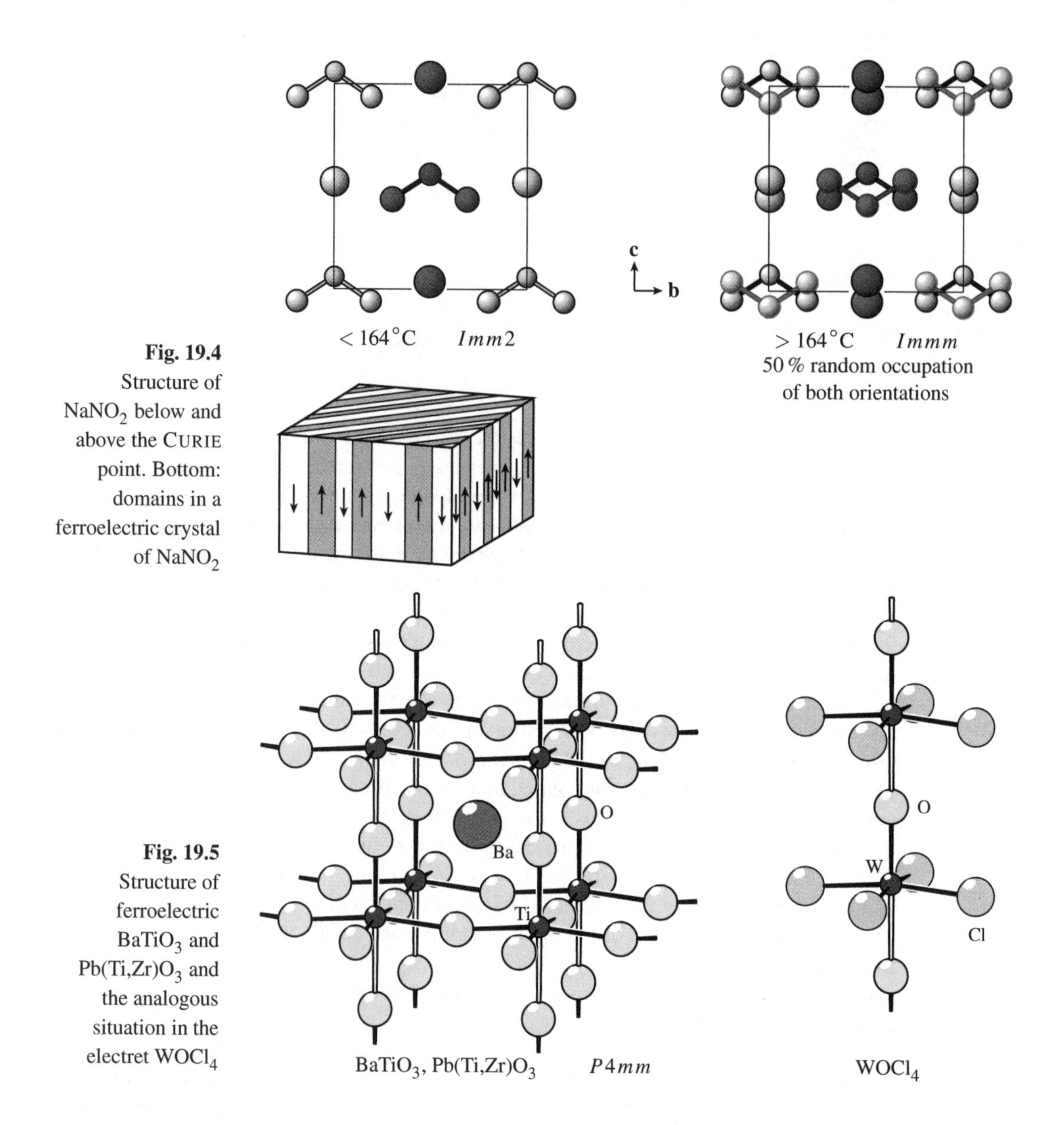

Fig. 19.4 Structure of $NaNO_2$ below and above the CURIE point. Bottom: domains in a ferroelectric crystal of $NaNO_2$

Fig. 19.5 Structure of ferroelectric $BaTiO_3$ and $Pb(Ti,Zr)O_3$ and the analogous situation in the electret $WOCl_4$

the titanium atoms, and these consequently do not occupy exactly the octahedron centers. The titanium atom in an octahedron is shifted towards one of the O atoms, the direction of the shift being the same for all octahedra of one domain (Fig. 19.5). The result is a polarization in the domain. The shift is similar for the W atoms in $WOCl_4$, which has square-pyramidal molecules associated to form a strand with alternating short and long W–O distances. Above the CURIE temperature of 120 °C, $BaTiO_3$ has the cubic perovskite structure, with all titanium atoms occupying the octahedron centers. A considerably higher CURIE temperature (1470 °C) and also a much larger polarization have been found for $LiNbO_3$.

No ferroelectricity is possible when the dipoles in the crystal compensate each other due to the crystal symmetry. All centrosymmetric, all cubic and a few other crystal classes are

Table 19.1: Crystal classes permitting ferroelectric crystals

crystal class	direction of polarization
1	arbitrary
2	parallel to the monoclinic axis
m	perpendicular to the monoclinic axis
$mm2$	parallel to the c axis
4, $4mm$	parallel to the c axis
3, $3m$	parallel to the c axis
6, $6mm$	parallel to the c axis

excluded. The allowed crystal classes are listed in Table 19.1. All ferroelectric materials also are piezoelectric.

Ferroelasticity is the mechanical analogon to ferroelectricity. A crystal is ferroelastic if it exhibits two (or more) differently oriented states in the absence of mechanical strain, and if one of these states can be shifted to the other one by mechanical strain. $CaCl_2$ offers an example (Fig. 4.1, p. 33). During the phase transition from the rutile type to the $CaCl_2$ type, the octahedra can be rotated in one or the other direction. If either rotation takes place in different regions of the crystal, the crystal will consist of domains having the one or the other orientation. By exerting pressure all domains can be forced to adopt only one orientation.

19.3 Magnetic Properties

An unpaired electron executes a spin about its own axis. The mechanical spin momentum is related to a spin vector which specifies the direction of the rotation axis and the magnitude of the momentum. The spin vector **s** of an electron has an exactly defined magnitude:

$$|\mathbf{s}| = \frac{h}{2\pi}\sqrt{s(s+1)} = \frac{h}{4\pi}\sqrt{3}$$

The spin quantum number s is used to characterize the spin. It can have only the one numerical value of $s = \frac{1}{2}$. $h = 6.6262 \cdot 10^{-34}$ J s = PLANCK's constant.

The spin is associated with a magnetic moment, *i.e.* an electron behaves like a tiny bar magnet. An external magnetic field exerts a force on an electron, resulting in a precession of the electron about the direction of the magnetic field which is similar to the precession of a top; the rotation axis of the electron is thus inclined relative to the magnetic field. Quantum theory permits only two values for the inclination; they are expressed by the magnetic spin quantum number m_s, which has the values of $m_s = +s = +\frac{1}{2}$ or $m_s = -s = -\frac{1}{2}$. The two inclinations are also called 'parallel' and 'antiparallel', although the spin vectors are not really exactly parallel or antiparallel to the magnetic field.

The magnetic moment of an isolated electron has a definite values of

$$\mu_s = 2\mu_B\sqrt{s(s+1)} = 2\mu_B\sqrt{3} \qquad (19.1)$$

$$\text{with} \quad \mu_B = \frac{eh}{4\pi m_e} = 9.274 \cdot 10^{-24}\ \mathrm{J\,T^{-1}} \qquad (19.2)$$

e = unit charge, h = PLANCK's constant, m_e = mass of the electron; 1 tesla is the unit of the magnetic flux density, 1 T = 1 V s m^{-2}

μ_B is termed the BOHR *magneton*. Magnetic moments are given as multiples of μ_B.

An electron orbiting in an atom is a circular electric current that is surrounded by a magnetic field. This also can adopt only certain orientations in an external magnetic field according to quantum mechanics. The state of an electron in an atom is characterized by four quantum numbers:

Principal quantum number, $n = 1, 2, 3, \ldots$
Orbital quantum number, $l = 0, 1, 2, \ldots, n-1$
Magnetic quantum number, $m_l = -l, \ldots, 0, \cdots + l$
Magnetic spin quantum number, $m_s = -\frac{1}{2}, +\frac{1}{2}$

Two electrons in an atom exert an influence on each other, *i.e.* their spins and their orbital angular momenta are coupled. Two electrons are termed paired if they coincide in all of their quantum numbers except the magnetic spin quantum number. In such an electron pair the magnetic moments of the electrons compensate each other. Unpaired electrons in different orbitals tend to orient their spins parallel and thus produce an accordingly larger magnetic field (HUND's rule); they have the same magnetic spin quantum number and differ in some other quantum number.

Substances having only paired electrons are *diamagnetic*. When they are introduced into an external magnetic field, a force acts on the electrons, *i.e.* an electric current is induced. The magnetic field of this current is opposed to the external field (LENZ's rule). As a result, the substance experiences a repulsion by the external magnetic field; the repulsive force is rather weak, but ever present.

In a *paramagnetic* substance unpaired electrons are present. Frequently the unpaired electrons can be assigned to certain atoms. When an external magnetic field acts on a paramagnetic substance, the magnetic moments of the electrons adopt the orientation of this field, the sample is magnetized and a force pulls the substance into the field. The magnetization can be determined quantitatively by measuring this force. Thermal motion of the atoms prevents a complete orientation, and higher temperatures therefore cause a smaller degree of magnetization.

The additional magnetic field produced by the orientation is a measure of the magnetization M. It is proportional to the external magnetic field H:

$$M = \chi H$$

The dimensionless proportionality factor χ is the *magnetic susceptibility*. The magnetization and consequently also the susceptibility depend on the number of orientable particles in a given volume. A volume-independent, material-specific magnitude is the *molar susceptibility* χ_{mol}:

$$\chi_{\text{mol}} = \chi V_{\text{mol}} = \chi_g M_{\text{mol}}$$

V_{mol} is the molar volume, M_{mol} the molar mass and $\chi_g = \chi / \rho$ the commonly specified mass susceptibility (ρ = density).

Taking the susceptibility, we can classify materials according to their magnetic properties in the following way:

$\chi_{\text{mol}} < 0$ diamagnetic
$\chi_{\text{mol}} > 0$ paramagnetic
$\chi_{\text{mol}} \gg 0$ ferromagnetic

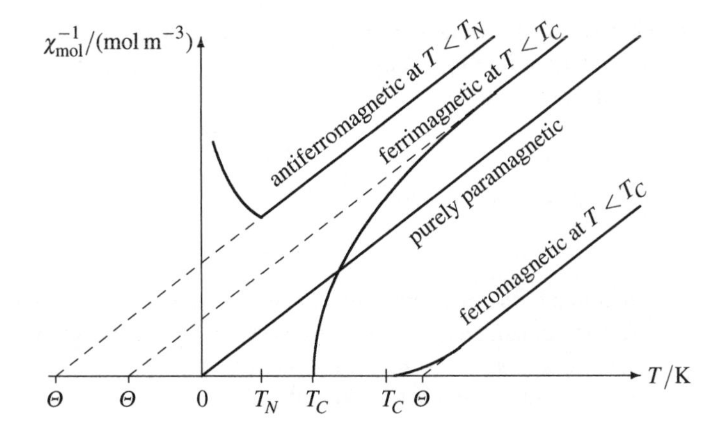

Fig. 19.6 Schematic plot of the reciprocal molar susceptibility *vs.* temperature

Paramagnetism

The temperature dependence of the molar susceptibility of a paramagnetic substance follows the CURIE–WEISS law (if the magnetic field is not too strong):

$$\chi_{mol} = \frac{C}{T - \Theta} \tag{19.3}$$

where T = absolute temperature, C = CURIE constant, and Θ = WEISS constant.

A plot of the reciprocal of the measured susceptibility $1/\chi_{mol}$ *vs.* T is a straight line with slope $1/C$, and which crosses the abscissa at $T = \Theta$ (Fig. 19.6). For $\Theta = 0$ the equation is simplified to the classic CURIE law $\chi_{mol} = C/T$. Generally, values of $\Theta \neq 0$ are found when cooperative effects arise at low temperatures (ferro-, ferri- or antiferromagnetism). The straight line then has to be extrapolated from high to low temperatures (dashed lines in Fig. 19.6).

The following discussion is restricted to the case of a substance containing only one species of paramagnetic atoms (atoms with unpaired electrons). The *magnetic moment* μ is used to specify how magnetic an atom is. An increasing magnetic moment results in an increasing susceptibility; the quantitative relation is given by means of the CURIE constant:

$$C = \mu_0 \frac{N_A^2 \mu^2}{3R} \tag{19.4}$$

μ_0 = magnetic field constant = $4\pi \times 10^{-7}$ V s A^{-1}m^{-1}; N_A = AVOGADRO's number, R = gas constant.

The solution of equation (19.4) and division of μ by μ_B yields the experimentally determined magnetic moment μ_{eff} of a sample, expressed as a multiple of μ_B:

$$\mu_{eff} = \frac{\mu}{\mu_B} = \frac{1}{\mu_B}\sqrt{\frac{3R}{\mu_0 N_A^2}\chi_{mol}(T-\Theta)} = 800\sqrt{\frac{\chi_{mol}}{\text{mol}^{-1}\text{m}^3}\,\frac{(T-\Theta)}{\text{K}}} \tag{19.5}$$

The coupling of the spins of the electrons in an atom is accounted for by adding their magnetic spin quantum numbers. Since they add up to zero for paired electrons, it is sufficient to consider only the unpaired electrons. The spins of n unpaired electrons add up according to HUND's rule to a total spin quantum number $S = \frac{1}{2}n$. The magnetic moment of these n electrons, however, is not the scalar sum of the single magnetic moments. The spin momenta must be added vectorially, taking into account the particular directions they may adopt according to quantum theory. The addition of the spin momenta yields a total spin momentum with the magnitude:

$$|S| = \frac{h}{2\pi}\sqrt{S(S+1)}$$

which is related to a magnetic moment:

$$\mu = 2\mu_B\sqrt{S(S+1)} \qquad (19.6)$$

In addition to the magnetism due to the electron spin, the magnetism of the orbital motion has to be considered. For this purpose the magnetic quantum numbers of the electrons are added to a *resultant orbital quantum number* $L = \sum m_l$, beginning with the highest magnetic quantum number. For example:

Cr^{2+}:	d^4	↑ | ↑ | ↑ | ↑ |	$l = 2$; $L = +2+1+0-1 = 2$; $S = 4 \times \frac{1}{2} = 2$
Cu^{2+}:	d^9	↑↓ | ↑↓ | ↑↓ | ↑↓ | ↑	$l = 2$; $L = +2+2+1+1+0+0-1-1-2 = 2$; $S = \frac{1}{2}$

Finally, the magnetic moments resulting from the spin and the orbital motion interact. This spin–orbit coupling is taken into account by the *total angular momentum quantum number* J (RUSSEL-SAUNDERS coupling):

$J = L - S$ if the subshell is less than half-occupied, otherwise $J = L + S$.

The corresponding magnetic moment is:

$$\mu = g\mu_B\sqrt{J(J+1)} \qquad (19.7)$$

$$\text{with} \quad g = 1 + \frac{J(J+1)+S(S+1)-L(L+1)}{2J(J+1)} \quad \text{(LANDÉ factor)}$$

$L = 0$ and $S = 0$ always hold for a fully occupied subshell. Therefore, the core electrons of an atom do not contribute to magnetism. $L = 0$ and $g = 2$ hold for half-filled subshells, resulting in a pure spin paramagnetism according to equation (19.6).

However, compounds of the lighter elements including the $3d$ transition elements also show only a spin magnetism with nearly no contribution from orbital motion. Their orbits are strongly influenced by the ligand field and cannot orient themselves freely in a magnetic field. The ligand field quenches the contributions of the orbital angular momentum completely or partially. As shown in Table 19.2, the spin-only approximation according to equation (19.6) is fulfilled quite well for $3d$ ions having electron configurations $3d^1$ to $3d^5$; a small amount of spin–orbit coupling is observed for the configurations $3d^6$ to $3d^9$.

The spin–orbit coupling generally cannot be neglected for compounds of the heavier $4d$ transition metals and especially of the $5d$ metals. This leads to large discrepancies between the spin-only and the observed values of the magnetic moments. The coupling depends on the electron configuration and for certain configurations also on temperature. Usually, it leads to reduced magnetic moments for d^1 to d^4 and to enhanced moments for d^5 to d^9 as compared to the spin-only value. For more details the specialist literature should be consulted.

The $4f$ electrons of lanthanoid ions, being shielded by the fully occupied, spherical shells $5s$ and $5p$, experience almost no influence from the ligand field. In this case the orbit magnetism is fully effective. Good agreement is observed between experimental magnetic moments and values calculated from the RUSSELL–SAUNDERS coupling according to equation (19.7) (Table 19.2).

Table 19.2: Calculated values of $2\sqrt{S(S+1)}$ and $g\sqrt{J(J+1)}$ for some high-spin ions and comparison with experimental values $\mu_{\text{eff}} = \mu/\mu_B$

		spin-only $2\sqrt{S(S+1)}$	μ_{eff}			spin-orbit $g\sqrt{J(J+1)}$	μ_{eff}
Sc^{3+}	$3d^0$	0	0	Ce^{3+}	$4f^1$	2.54	2.3 – 2.5
Ti^{3+}	$3d^1$	1.73	1.7 – 1.8	Pr^{3+}	$4f^2$	3.58	3.4 – 3.6
V^{3+}	$3d^2$	2.83	2.8 – 2.9	Nd^{3+}	$4f^3$	3.62	3.5 – 3.6
Cr^{3+}	$3d^3$	3.87	3.7 – 3.9	Sm^{3+}	$4f^5$	0.85	1.6*
Cr^{2+}, Mn^{3+}	$3d^4$	4.90	4.8 – 5.0	Eu^{3+}	$4f^6$	0	3.3 – 3.5*
Mn^{2+}, Fe^{3+}	$3d^5$	5.92	5.7 – 6.1	Gd^{3+}	$4f^7$	7.94	7.9 – 8.0
Fe^{2+}, Co^{3+}	$3d^6$	4.90	5.1 – 5.7	Tb^{3+}	$4f^8$	9.72	9.7 – 9.8
Co^{2+}	$3d^7$	3.87	4.3 – 5.2	Dy^{3+}	$4f^9$	10.65	10.2 – 10.6
Ni^{2+}	$3d^8$	2.83	2.8 – 3.0	Ho^{3+}	$4f^{10}$	10.61	10.3 – 10.5
Cu^{2+}	$3d^9$	1.73	1.7 – 2.0	Er^{3+}	$4f^{11}$	9.58	9.4 – 9.5
Zn^{2+}	$3d^{10}$	0	0	Yb^{3+}	$4f^{13}$	4.54	4.5

* Deviation because the first excited state is only slightly above the ground state and a fraction of the atoms is always excited; $\mu_{\text{eff}} \to 0$ when $T \to 0$

Ferro-, Ferri- and Antiferromagnetism

The term ferromagnetism reflects the fact that iron shows this effect, but it is by no means restricted to iron or iron compounds. Ferromagnetism is a *cooperative phenomenon, i.e.* many particles in a solid behave in a coupled manner. Paramagnetic atoms or ions exert influence on each other over extended regions.

In a ferromagnetic substance the magnetic moments of adjacent atoms orient themselves mutually parallel, and their action is added up. However, most materials that have not been treated magnetically exhibit no macroscopic magnetic moment. This is due to the presence of numerous domains (WEISS domains). In each domain the orientation of all spins is parallel, but it differs from domain to domain. An external magnetic field causes the growth of those domains oriented parallel with the magnetic field at the expense of the other domains. When the spins of all domains have been oriented, saturation has taken place. To achieve this state, a magnetic field with some minimum field strength is required; its magnitude is dependent on the material.

A hysteresis curve shows this kind of behavior; it is like the hysteresis of a ferroelectric material (Fig. 19.7). Starting with an untreated sample, an increasing magnetic field causes an increasing magnetization until saturation has been reached. After turning off the magnetic field, there is some loss of magnetization, but a remanent magnetization R is retained. By reversing the magnetic field, the spins experience a reorientation. The minimum magnetic field required for this is the coercive force or coercive field K. Depending on the application, magnetic materials with different magnetic 'hardness' are required. For example, a permanent magnet in a direct-current electric motor must have a high coercive force in order to maintain its magnetization. Small coercive forces are required whenever frequent and fast remagnetizations are required, as for example in recording heads.

Above a critical temperature T_C, the CURIE temperature, a ferromagnetic material becomes paramagnetic, since thermal motion inhibits the parallel orientation of the magnetic moments. The susceptibility then follows the CURIE–WEISS law with a positive value of the WEISS constant, $\Theta > 0$ (Fig. 19.6).

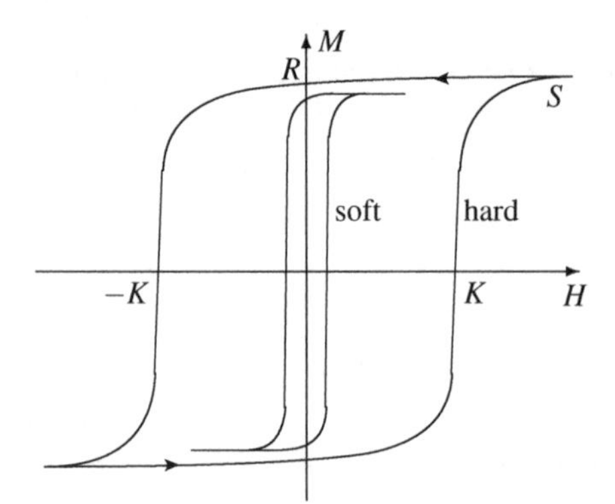

Fig. 19.7
Hysteresis curves for a magnetically 'hard' and a magnetically 'soft' ferromagnetic material. S = saturation magnetization, R = remanent magnetization, K = coercive force

The coupling of the magnetic moments of the atoms can also result in spins having opposite orientations; in this case the material is *antiferromagnetic* (Fig. 19.8). At very low temperatures its total magnetic moment is zero. With increasing temperature the thermal motion interferes with the antiparallel orientation of all particles and the magnetic susceptibility increases; at even higher temperatures, thermal motion increasingly causes a random distribution of the spin orientations and the magnetic susceptibility decreases again, as in a paramagnetic material. Therefore, an antiferromagnetic material exhibits a maximum susceptibility at a certain temperature, the NÉEL temperature.

The symmetry of antiferromagnetic crystals with the inclusion of the spin orientation can be described with the aid of *black–white space groups* (SHUBNIKOV groups, antisymmetry space groups). They are an extension of the space groups. Every point in space is distinguished to be either black or white. The common symmetry operations then can be coupled with a change of the color. In the Hermann–Mauguin symbol this is expressed by a prime $'$. For example, a twofold rotation that is coupled with a color change (or with a spin inversion) obtains the symbol $2'$. There exist 1191 black–white space group types.

In a *ferrimagnetic* material the situation is the same as in an antiferromagnetic material, but the particles bearing opposite magnetic moments occur in different quantities and/or their magnetic moments differ in magnitude. As a consequence, they do not compensate each other even at very low temperatures. The behavior in a magnetic field is like that of ferromagnetic materials.

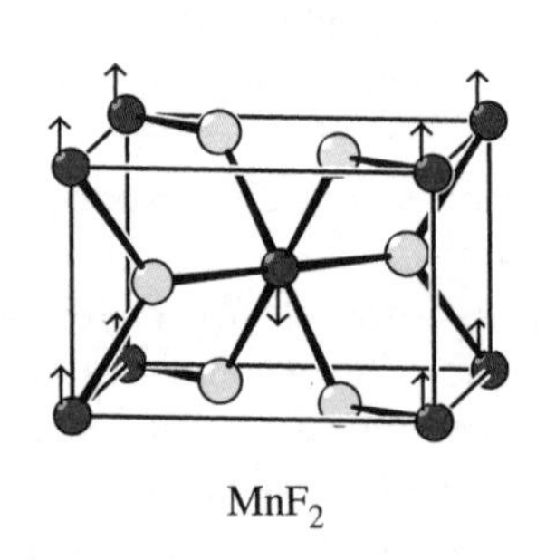

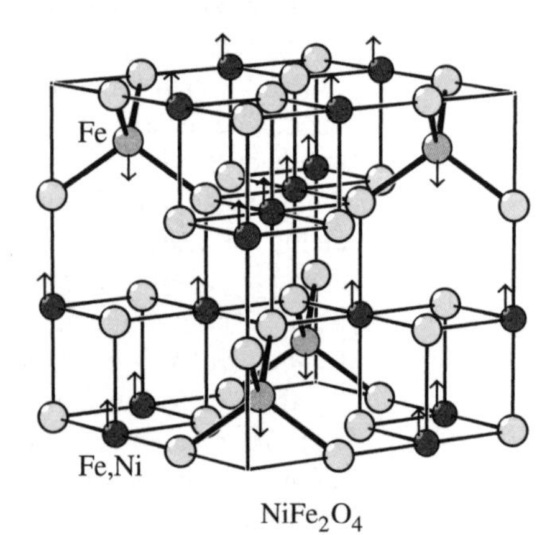

Fig. 19.8
Orientation of the spins in antiferromagnetic MnF_2 (rutile type) and in the ferrimagnetic inverse spinel $NiFe_2O_4$ ($\frac{1}{8}$ of the unit cell); the sites with octahedral coordination are randomly occupied by Fe and Ni

Table 19.3: Coupling of the spin vectors related to cooperative magnetic effects

	spin orientation within a domain	examples
ferromagnetism	↑↑	α-Fe, Ni, Gd; EuO (NaCl type)
antiferromagnetism	↑↓	MnF_2, FeF_2 (rutile type)
ferrimagnetism	↑↓	Fe_3O_4, $NiFe_2O_4$ (inverse spinels)
	↑↑↑↓↓	$Y_3Fe_5O_{12}$ (garnet)

The order occurring among the spins of the atoms in the unit cell can be determined experimentally by neutron diffraction. Since a neutron itself has a spin and a magnetic moment, it is diffracted by an atom to an extent which depends on the orientation of the magnetic moment.

Different kinds of spin coupling are listed in Table 19.3 .

What determines the way in which the spins couple? Parallel orientation always occurs when the corresponding atoms act directly on one another. This is the case in pure metals like iron or nickel, but also in EuO (NaCl type). Antiparallel orientation usually occurs when two paramagnetic particles interact indirectly by means of the electrons of an intermediate particle which itself is not paramagnetic; this is called *superexchange* mechanism. That is the case in the commercially important spinels and garnets.

In $NiFe_2O_4$, an inverse spinel $Fe_T^{3+}[Ni^{2+}Fe^{3+}]_O O_4$, the spins of the octahedral sites are parallel with one another; the same applies to the tetrahedral sites (Fig. 19.8). The interaction between the two kinds of sites is mediated by superexchange via the oxygen atoms. High-spin states being involved, Fe^{3+} (d^5) has five unpaired electrons, and Ni^{2+} (d^8) has two unpaired electrons. The coupled parallel spins at the octahedral sites add up to a spin of $S = \frac{5}{2} + \frac{2}{2} = \frac{7}{2}$. It is opposed to the spin of $S = \frac{5}{2}$ of the Fe^{3+} particles at the tetrahedral sites. A total spin of $S = 1$ remains which is equivalent to two unpaired electrons per formula unit.

Garnet is an orthosilicate, $Mg_3Al_2(SiO_4)_3$, which has a complicated cubic structure. The structure is retained when all metal atoms are trivalent according to the following substitution:

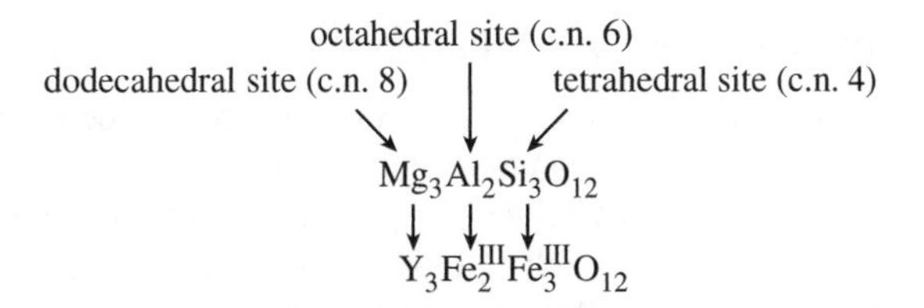

In yttrium iron garnet $Y_3Fe_5O_{12}$ ('YIG') a ferrimagnetic coupling (superexchange) is active between the octahedral and the tetrahedral sites. Since the tetrahedral sites are in excess, the magnetic moments do not compensate each other. The magnetic properties can be varied by substitution of yttrium by lanthanoids.

Magnetic Materials of Practical Importance

Iron is a material whose ferromagnetic properties have been applied for a long time. It becomes paramagnetic when heated above the CURIE temperature of 766 °C; this does not involve a phase transition of the body-centered cubic structure (a phase transition to cubic closest-packing occurs at 906 °C). Being magnetically soft, iron and alloys of iron with silicon, cobalt or nickel are applied in electric motors and in transformers. Finely dispersed iron powder ('iron pigment') serves as magnetic material in data storage devices (since it is pyrophoric it is stabilized by plating it by evaporation with a Co–Cr alloy). A disadvantage of iron is its electrical conductivity; an alternating magnetic field induces electric eddy currents which cause energy losses by heating the iron. Using stacks of mutually insulated iron plates can decrease but not fully avoid the eddy currents.

Wherever there are no alternating magnetic fields, the electric conductivity causes no trouble, for example in certain applications of permanent magnets. Permanent magnets with especially high magnetization and coercivity are made of $SmCo_5$ or Sm_2Co_{17}.

Nearly no eddy current losses occur in electrically insulating magnetic materials. This is one of the reasons for the importance of oxidic materials, especially of spinels and garnets. Another reason is the large variability of the magnetic properties that can be achieved with spinels and garnets of different compositions. The tolerance of the spinel structure to substitution at the metal atom sites and the interplay between normal and inverse spinels allow the adaptation of the properties to given requirements.

Spinel ferrites are iron-containing spinels $M^{II}Fe_2O_4$. They are magnetically soft to medium hard. γ-Fe_2O_3 is a spinel with defective structure, $Fe^{III}_T[Fe^{III}_{1.67}\square_{0.33}]_OO_4$; it is used as a storage medium (diskettes, recording tapes). Fe_3O_4 is applied in 'magnetic liquids' that are used to seal bearings against vacuum. These are suspensions of magnetic pigments in oil; in a magnetic field the pigment collects in the region with the most intense field and causes an increase in the density and viscosity of the liquid. Magnetically soft materials are needed in high-frequency electronics; manganese–zinc ferrites are most important for this purpose.

Hexagonal ferrites of the magnetoplumbite type serve as magnetic hard materials. They have high coercivities combined with low remanent magnetizations. They are used as nonconducting permanent magnets, for example in electric motors, generators, and closet locks. Structurally, they are related to spinels, but with some of the oxygen atoms substituted by larger cations like Ba^{2+} or Pb^{2+}. The two main types are $BaFe_{12}O_{19}$ ('*M* phase') and $Ba_2Zn_2Fe_{12}O_{22}$ ('*Y* phase').

The Magnetocaloric Effect

A phase transition of a magnetic material can be connected with a change of the magnetic properties (magnetostructural phase transition). In some cases the phase transition can be induced by an external magnetic field. If the transformation is of first order and thus involves the exchange of an enthalpy of transformation with the surroundings, this phenomenon is called *magnetocaloric effect*.

MnAs exhibits this behavior. It has the NiAs structure at temperatures exceeding 125 °C. When cooled, a second-order phase transition takes place at 125 °C, resulting in the MnP type (*cf.* Fig. 18.4, p. 218). This is a normal behavior, as shown by many other substances. Unusual, however, is the reappearance of the higher symmetrical NiAs structure at lower temperatures after a second phase transition has taken place at 45 °C. This second transformation is of first order, with a discontinuous volume change ΔV and with enthalpy of transformation ΔH. In addition, a reorientation of the electronic spins occurs from a low-spin to a high-spin state. The high-spin structure ($< 45\,°C$) is ferromagnetic, the low-spin structure is paramagnetic. The ferromagnetic structure can be stabilized by a magnetic field above 45 °C.

Refrigeration devices that take advantage of the magnetocaloric effect are being developed. For this purpose, there is a need for materials exhibiting a strong magnetocaloric effect at some desired temperature. $Gd_5Si_xGe_{4-x}$ may be a candidate. Gd_5Si_4 and Gd_5Ge_4 have very similar structures, but they differ in a crucial minor detail. They consist of slabs in which cubes and pairs of trigonal prisms of Gd atoms are joined (Fig. 19.9). There is another Gd atom inside of every cube, and an Si_2 or Ge_2 dumbbell with a covalent Si–Si or Ge–Ge bond is placed inside of every pair of prisms. The slabs are stacked with a mutual shift. Further Si_2 or Ge_2 dumbbells are located at an angle in between the slabs. The last-mentioned dumbbells make the difference. Their Si–Si bond length measures 247 pm in Gd_5Si_4, which is consistent with a covalent bond. Taking a calculation according to the ZINTL rules, this corresponds to the formula $Gd_5^{3+}(Si_2^{6-})_2e_3^-$. In Gd_5Ge_4, however, the Ge···Ge distance in these dumbbells amounts to 363 pm, which is too long for a bond. The ZINTL formula would be $Gd_5^{3+}(Ge_2^{6-})(Ge^{4-})_2e^-$, albeit the assumption of Ge^{4-} ions is not realistic. The difference entails marked differences in the properties: Gd_5Si_4 is fer-

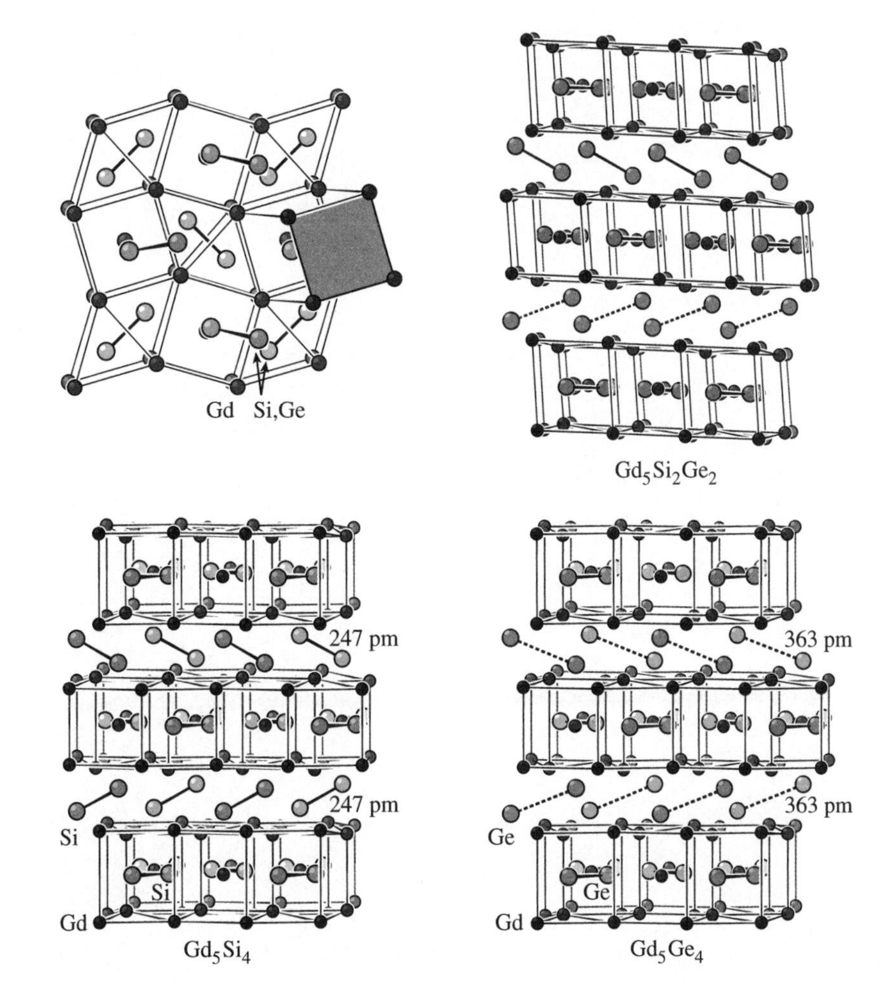

Fig. 19.9 Top left: Plan view of a slab consisting of cubes and double trigonal prisms in $Gd_5(Si,Ge)_4$ with $(Si,Ge)_2$ dumbbells on top; the gray square is part of the next slab and shows the mutual stacking of the slabs. Remaining images: Side views parallel to the slabs

romagnetic below 6 °C; Gd_5Ge_4 becomes antiferromagnetic below the NÉEL temperature of −148 °C and ferromagnetic below 20 K.

$Gd_5Si_2Ge_2$ has a structure with short and long dumbbells. It crystallizes in the monoclinic space group $P112_1/a$ (Fig. 19.9), which is a subgroup of $Pnma$, the orthorhombic space group of Gd_5Si_4 and Gd_5Ge_4. Si and Ge atoms have a random distribution in both kinds of dumbbells. $Gd_5Si_2Ge_2$ is ferrimagnetic below the Curie temperature of 26 °C, with a small magnetic moment. Upon cooling, a first-order phase transition takes place at 3 °C. Below this temperature, the atoms of the long dumbbells have moved up, such that all dumbbells now have short bonds as in Gd_5Si_4. The low-temperature form is ferromagnetic, with the full magnetic moment of the Gd^{3+} ions. This transition exhibits a strong magnetocaloric effect. Mixed crystals $Gd_5Si_xGe_{4-x}$ with $0.2 < x < 0.5$ show the same structure as $Gd_5Si_2Ge_2$ and the same properties; the transition temperature can be controlled by the value of x (the higher the Ge content, the lower is the transition temperature).

20 Nanostructures

Materials that consist of particles or that have structural attributes (such as pores) with sizes of 2 to 1000 nanometers are called nanostructured or nanocrystalline materials. Since the diameter of an atom is approximately 0.2 to 0.3 nm, we are dealing with distances that comprise 8 to 5000 atoms. The properties of a material that consists of particles of this magnitude differ from those of the same material consisting of larger aggregates. A crystal measuring $1 \times 1 \times 1\ mm^3$ has a fraction of about 10^{-6} atoms at its surface; but it amounts to approximately 1 % of the atoms if the size of the crystal is only $100 \times 100 \times 100\ nm^3$. The surface is the most severe disturbance of the periodic order of a crystal. Atoms at the surface have a different bonding and differ electronically as compared to interior atoms. The properties of a nanostructured material are influenced to a large extent by the atoms at the surface. Mechanical, electric, magnetic, optical and chemical properties depend on the particle sizes and shapes. At even smaller sizes, in addition, quantum-mechanical effects become effective. If the color of luminescence of a semiconductor is red at a particle size of 8 nm, it becomes green at 2.5 nm. If one wants to achieve certain properties, the particles have to have a uniform size, shape and orientation.

The terminology is not yet homogeneous. The use of the prefix 'nano' spread out in the 1990s. Until then, the common term used to be *mesoscopic structures*, which continues to be used. According to a definition by IUPAC of 1985, the following classification applies to porous materials: microporous, < 2 nm pore diameter; mesoporous, 2–50 nm; macroporous, > 50 nm.

Nanostructured materials are nothing new. Chrysotile fibers are an example (Fig. 16.22), as are bones, teeth and shells. The latter are *composite materials* made up of proteins and embedded hard, nanocrystalline, inorganic substances like apatite. Just as with the imitated artificial composite materials, the mechanical strength is accomplished by the combination of the components.

Chemists have been working for a long time with particles having sizes of nanometers. The novelty of recent developments concerns the ability to make nanostructured substances with *uniform* particle sizes and in regular arrays. In this way it becomes feasible to produce materials that have definite and reproducible properties that depend on the particle size. The development began with the discovery of carbon nanotubes by IJIMA in 1991 (Fig. 11.15, p. 116).

Aside from the methods for the production of carbon nanotubes mentioned on page 115, a number of methods to make nanostructured materials have been developed. In the following we mention a selection.

In the LAMER process particles are grown from solution. The formation of crystallization nuclei and their growth are strictly separated. First, a large number of crystal nuclei are produced in a short interval of time from a supersaturated solution; then the crystals are allowed to grow slowly with avoidance of further nucleation from a solution which is only slightly supersaturated. Surfactants such as thiols or amines with long-chain alkyl groups can be used to influence the crystal growth. In this way, one can obtain spherical particles with uniform diameters of 3 to 15 nm.

Inorganic Structural Chemistry, Second Edition Ulrich Müller

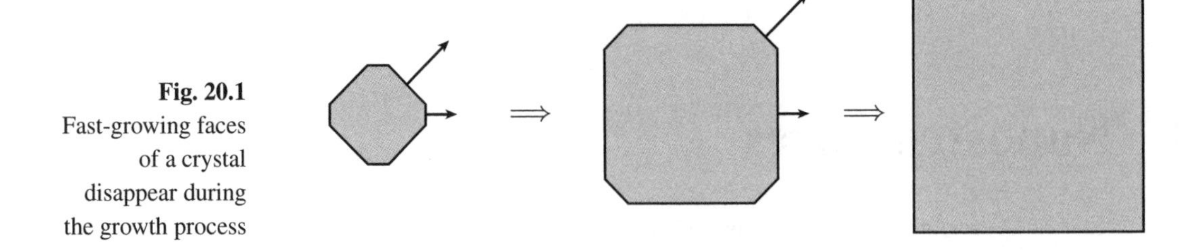

Fig. 20.1 Fast-growing faces of a crystal disappear during the growth process

In order to control the shape of the particles, the rate of growth of the crystal faces has to be controlled. Fast-growing crystal faces disappear after a while (Fig. 20.1). Ligands such as long-chain carbonic acids are deposited in a selective manner on certain crystal faces and suppress their growth, with the consequence that only these faces remain at the end. If the growth in the direction of the cube vertices of a cubic crystal is faster, one obtains a cube. If the growth of the faces of the cube is faster, one obtains an octahedron. If the rate of growth of both is the same, the result is a cuboctahedron. For example, nanocubes of silver can be made by the reduction of silver nitrate in alcoholic solution. These cubes can be used as a template to produce hollow gold boxes. This is done by the reaction with tetrachloroauric acid, $HAuCl_4$; gold precipitates on the surface of the cubes while the silver dissolves. One obtains gold nanoboxes in the form of empty cuboctahedra.

Nanowires of hexagonal cobalt can be grown by the selective adsorption of ligands on all crystal faces except the faces that become the sides of the wires.

Another method takes advantage of an oriented solidification from eutectic melts. The two components solidify simultaneously when a eutectic mixture is cooled (p. 36). A BRIDGMAN furnace serves to produce large single crystals. This is done by slowly drawing downward a crucible from the furnace (Fig. 20.2). The lower apex of the crucible cools first, giving rise to the formation of a crystallization nucleus, followed by the slow crystallization of the melt in the crucible. There is a narrow zone of supercooled melt between the crystal and the remaining melt on top of it. In the case of a eutectic mixture, the separation of the phases takes place by horizontal diffusion in the supercooled melt. Continuous single crystals form in the drawing direction. If one of the components has a small volume fraction, it solidifies in the form of parallel nanowires with uniform thickness, embedded in the other phase. For example, rhenium wires embedded in a single-crystal matrix of NiAl can be obtained from a eutectic melt of NiAl/Re. The result can be influenced by the temperature gradient and the drawing speed. The wires can be separated by complete or partial dissolution of the NiAl matrix with an acid. Electrochemical oxidation of the rhenium to perrhenate permits the removal of the rhenium, leaving behind a nanofilter of NiAl.

The anodic oxidation of sheet aluminum has been used for a long time to protect aluminum against corrosion by a well-adhering oxide layer. Porous oxide layers are formed if acid electrolytes are used that can redissolve the aluminum oxide (mostly sulfuric or phosphoric acid). A compact oxide layer is formed at the beginning of the electrolysis (Fig. 20.3). Simultaneously, the current decreases, due to the electric resistance of the oxide. Subsequently follows a process in which the oxide is redissolved by the acid, and the current increases until it reaches a steady state. The electrochemical oxidation continues to take place with formation of pores. At the end of a pore, where it has the largest curvature, the electric field has its largest gradient and the process of redisolution is fastest.

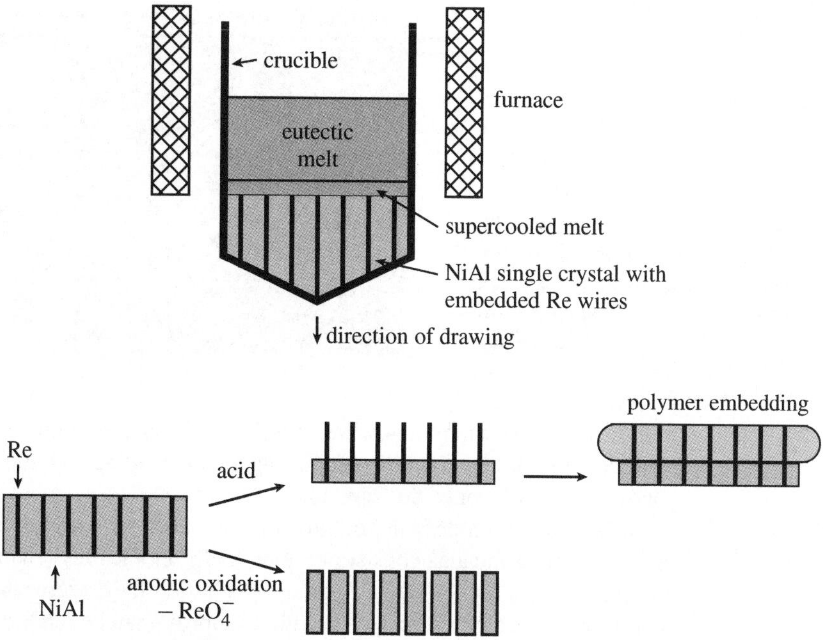

Fig. 20.2 Top: BRIDGMAN procedure for the production of large single crystals. From a eutectic melt one can obtain a single crystal with embedded wires. Bottom: possible subsequent steps of processing

Therefore, there is no growth of the oxide layer at the base of a pore. Instead, the walls between the pores grow and become higher and higher. Depending on the electric tension and the kind of acid, a certain curvature results at the base of a pore. Finally, as a consequence, all pores have the same diameter and form a regular array. Pore diameters of 25 to 400 nm and pore depths of 0.1 mm can be achieved. The pore walls have the approximate composition AlOOH; they still contain anions of the electrolyte, and they are amorphous.

Porous aluminum oxide can be used as a template for the production of nanowires and nanotubes. For example, metals can be deposited on the pore walls by the following procedures: deposition from the gas phase, precipitation from solution by electrochemical reduction or with chemical reducing agents, or by pyrolysis of substances that have previously been introduced into the pores. Wires are obtained when the pore diameters are 25 nm, and tubes from larger pores; the walls of the tubes can be as thin as 3 nm. For example, nanowires and nanotubes of nickel, cobalt, copper or silver can be made by electrochemical deposition. Finally, the aluminum oxide template can be removed by dissolution with a base.

One can also take advantage of reactions with the pore walls. For example, if the pore walls of aluminum oxide are coated with $\mathrm{Sn(SePh)_4}$, then they react with each other at

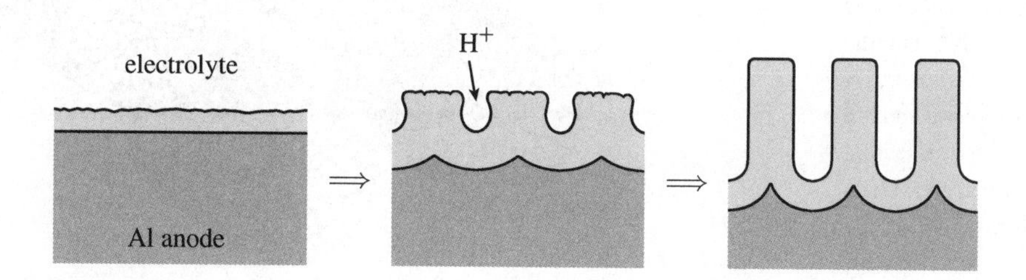

Fig. 20.3 Formation of mesoporous aluminum oxide by anodic oxidation of aluminum in acid electrolytes

Fig. 20.4
Nanotubes of tin
(Scanning electron microscope image by S. SCHLECHT, Freie Universität Berlin. Reprinted form *Angewandte Chemie* 118 (2006) 317, with permission from Wiley-VCH)

650 °C, and one obtains nanowires of SnO_2. If the same is done in a mesoporous template of silicon, the silicon acts as a reducing agent. Depending on temperature, the products are nanowires of SnSe or of tin (Fig. 20.4).

Surfactants, cyclodextrins or proteins can also be used as template materials. The molecules of surfactants consist of a long hydrophobic alkyl chain and a hydrophilic end group ($-SO_3^-$, $-CO_2^-$, $-NR_3^+$). In aqueous solution they aggregate to *micelles* if the concentration exceeds a critical value. First, approximately spherical micelles are formed, which have the hydrophilic groups on the surface and the alkyl groups pointing inwards. At higher concentrations the micelles become rod-like and adopt a liquid-crystalline order similar to a hexagonal rod-packing (Fig. 20.5). When a solid product is prepared by a precipitation reaction from such a solution, it encloses the micelles. If the solid is cross-linked and sufficiently stable, the micelle molecules can be removed, for example by calcination.

As an example, take the manufacture of mesoporous silica (amorphous SiO_2) with uniformly sized pores. Rod-like micelles of alkyltrimethylammonium ions form in an aqueous solution of an alkyltrimethylammonium halide. The micelles have a positive electric

Fig. 20.5
Rod-like micelles of surfactant molecules with a liquid-crystalline order. The aqueous solution is located between the micelles. Spheres = hydrophilic ends of the surfactant molecules, black zigzag lines = long alkyl groups

charge at their surfaces and arrange themselves parallel. The counter-ions are located in the aqueous solution between the micelles. Precipitation of silica is induced by slow hydrolysis of tetraethylorthosilicate, $Si(OC_2H_5)_4$, at pH $\approx$ 11 at 100 to 150 °C (hydrothermal conditions, *i.e.* under pressure). The enclosed micelles can be removed by calcination at 540 °C, and mesoporous silica ('MCM-41') is left. The pore size can be controlled by the length of the alkyl groups.

In mixtures of nonpolar solvents with little water, surfactants form spherical reverse micelles. They have a reversed orientation of the molecules with the hydrophilic groups in the interior and a drop of enclosed water in the middle. Starting from a precursor material, metal oxides in the form of uniform nanosized spheres can be obtained by hydrolysis under controlled conditions (pH, concentration, temperature). For example, titanium oxide spheres are obtained from a titanium alkoxide, $Ti(OR)_4 + 2\,H_2O \rightarrow TiO_2 + 4\,ROH$.

Nanostructured materials have found several applications, and more are to be expected, for example:

Encapsulation in nanocontainers, which crack under pressure and set free an enclosed substance. This includes adhesives that become effective under pressure, or perfume that is freed upon rubbing.

Nanoparticles of TiO_2 in sun-ray filter cream remain adhered to the skin and do not migrate into wrinkles.

Water- and dirt-repelling coatings have nanoparticles pointing outwards. Water is in touch with the surface at only a few points; due to the surface tension it contracts to droplets which roll off ('lotus effect').

SiO_2 nanospheres are fused onto the surface of glass at 650 °C, resulting in a a surface that reflects nearly no light.

Ferroelectric $Pb(Ti,Zr)O_3$ nanoparticles for electronic data storage.

21 Pitfalls and Linguistic Aberrations

In the literature and speech it is common to find incorrect, inaccurate, misleading or inane expressions. Below, some are listed which it is hoped you will never use.

Do not confuse *crystal structure* and *crystal lattice*. The crystal structure designates a regular array of atoms, the crystal lattice corresponds to an infinity of translation vectors (Section 2.2). The terms should not be mixed up either. There exists no 'lattice structure' and no 'diamond lattice', but a diamond structure.

Caesium chloride is *not body-centered* cubic, but cubic primitive. A structure is body centered only if for every atom in the position x, y, z there is another *symmetry-equivalent* atom in the position $x+\frac{1}{2}, y+\frac{1}{2}, z+\frac{1}{2}$ in the unit cell. The atoms therefore must be of the same kind. It is unfortunate to call a cluster with an interstitial atom a 'centered cluster' because this causes a confusion of the well-defined term 'centered' with a rather blurred term. Do not say, the O_4 tetrahedron of the sulfate ion is 'centered' by the sulfur atom.

Do not confuse *symmetry operation* and *symmetry element* (p. 12).

Do not confuse *space group* and *space-group type* (p. 20).

Do not call an achiral *Sohncke space group* (or space-group type) a 'chiral space group'. Most chiral molecules *do not* crystallize in a chiral space group (p. 83).

Identical means 'the very same'. Avoid the expression 'identical atoms' when you mean symmetry-equivalent or translationally equivalent atoms. It is impossible to lessen or to increase the term identical; 'nearly identical' is nonsense.

A crystal that consists of only one kind of atoms is not a 'monoatomic crystal'; it consists of more than one atom.

A molecule does not contain 'hydrogens', 'oxygens' or 'borons', but hydrogen atoms, oxygen atoms or boron atoms.

Do not say 'octahedrally coordinated complex' if you mean an octahedrally coordinated *atom* in a complex compound. An 'octahedrally coordinated complex' could at best be a complex *molecule* that is surrounded by six other molecules.

Molecules with a planar coordination figure do not contain 'planar atoms'. Further, 'tetrahedral atoms', 'chiral atoms' etc. are nonsense. A minimum of four atoms is required for a chiral structure.

'One-dimensional structures' do not exist. Structures are always three-dimensional. The term one-dimensional is acceptable to express specifically named, highly anisotropic properties, such as 'one-dimensional conductivity', 'one-dimensional connection' or 'one-dimensional disorder'. The 'dimensionality' of a structure is a meaningless phrase.

A structure can be tetragonal or not. That is a clear yes-or-no case. When the symmetry of an orthorhombic structure is close to tetragonal, it is nonsense to talk of a 'tertagonality' of the structure.

The image of a *single molecule*, the structure of which has been determined by crystal structure analysis, shows a *molecular structure*, but not a 'crystal structure', an 'X-ray structure', or even worse an 'X-ray'. The packing and the spatial arrangement of the molecules in a crystal are always an indispensable part of a crystal structure.

Inorganic Structural Chemistry, Second Edition Ulrich Müller

Stoichiometry is the *discipline* that deals with the amount relationships of the elements in compounds and during chemical reactions. It is not the amount relationship itself. A specific compound has no stoichiometry, but a composition.

Geometry is the science of bodies in space. It is a mathematical discipline, subdivided into subdisciplines such as Euclidean geometry, spherical geometry, analytical geometry etc., but it is never the *property* of an object. Analogous considerations are valid for *architecture*. A molecule has a structure, a constitution, a configuration, an atomic arrangement, a spatial arrangement, a shape, but no geometry and no architecture. 'Molecular geometry' or 'coordination geometry' could perhaps be used to designate a discipline that deals with the geometric conditions in molecules, but it should not be used to designate the spatial arrangement of atoms in a molecule. There exists neither a 'tetrahedral geometry' nor a 'chiral geometry', and the quantum-chemical calculation of a molecular structure is by no means a 'geometry optimization'.

The idiomatic changes taking place in everyday life also affect the language used by chemists. This includes the spreading of certain fatuous expressions. One of them is the 'center' instead of atom. According to this, 10^{23} 'centers' are located in one gram of water, and a C_{60} molecule has 60 'centers', but none of them is in the center. Advocates in favor of the 'center' (if they happen to meditate on this in the first place) argue that only this is an expression which comprises atoms as well as monoatomic ions. That is wrong: ions (monoatomic ones) are atoms! They only have the special property of bearing an electric charge. 'Excited centers', 'peripheral centers', 'tetrahedral zinc centers' etc. are especially absurd terms.

Another superfluous vogue word is 'self organization' or 'self assembly', meaning a more or less ordered atomic or molecular association, crystal nucleation, or crystal growth. Whenever molecules become associated, this happens by itself. 'Self organization' has become one of the favorite expressions in the science of nanostructures. The terms 'self organization' or 'self assembled' can almost always be completely deleted without changing the sense of a text or making it even slightly less understandable or readable. A 'self-assembled' monolayer of molecules is just a monolayer of molecules; it goes without saying that the molecules have assembled and that they have done so by themselves.

Correct Use of Units of Measurement

Science cannot be performed without an accurate system of measurement, which is globally standardized and compulsory. Units and standards of measurement are agreed upon and harmonized on an international basis by the *Bureau International des Poids et Mesures* in Sèvres, France, and by the *International Organization for Standardization* in Geneva, Switzerland. The units and standards are then laid down in national laws. Nearly all countries have accepted the *Système International d'Unités* (SI units) as their system of measurement. This also applies to countries that had been accustomed to use British units like Australia, Canada, South Africa and the United States. In Britain, SI units are official from January 2010. The valid standards are available from the competent bureaus, for example:

Bureau International des Poids et Mesures, www.bip.fr

National Measurement Institute (Australia), www.measurement.gov.au

National Institute of Standards and Technology (NIST, USA), www.physics.nist.gov/Pubs/SP811/

The specification of a value like '$d = 235$ pm' means: the distance d amounts to 235 times 1 picometer. Arithmetically, this is a multiplication of the numerical value with

the unit. In tables and diagrams, it is common not to repeat the unit with every listed number. That means that the actual values have been divided by the unit. Therefore, the correct heading of a table or marking at a diagram axis reads: d/pm or $\frac{d}{\text{pm}}$ or $d\ \text{pm}^{-1}$. The frequently encountered notation 'd [pm]' is not in accordance with the SI standards. Brackets have a special meaning in the SI system, namely:
'$[d]$ = pm' means, 'the selected unit for d is picometer'.

The use of non-SI units is strongly discouraged. For these units there often do not exist standards, and for historical reasons the same denomination may mean sundry units. For example, it is common practice in theoretical chemistry to state energy values in kilocalories. However, to convert a calorie to the SI unit Joule, there exist different conversion factors:
1 cal = 4.1868 J ('international calorie'), 1 cal = 4.1855 J ('15° calorie') or 1 cal = 4.184 J ('thermochemical calorie'). Which one is applicable?

Some non-SI units are explicitly permitted. They include for crystallographic statements: 1 Å = 10^{-10} m = 100 pm and 1° = $\pi/180$ rad (plane angle). The liter is also permitted (both abbreviations are official, L or l); 1 L = 1 dm^3.

The much used term 'molarity' instead of *amount-of-substance concentration* (for short *amount concentration* or just *concentration*) is obsolete and should not be used. The same applies to specifications such as '0.5*M*' or '0.5 M'; the correct statement is $c = 0.5$ mol/L.

Hints for the Publication of Data on Crystal Structures

Crystal structure analysis by X-ray diffraction has become the most powerful tool for structure determination in chemistry. Certain rules should be observed for the documentation of the results, taking into account the standards common in crystallography. The *most important information* are the *lattice parameters*, the *space group* and **all atomic coordinates**. If the parameters of the thermal displacement of all atoms are not listed, at least one figure should show the corresponding ellipsoids, because unusual ellipsoids indicate flaws in the structure determination or the presence of special structural circumstances. Those who do not depict the ellipsoids are suspect of trying to conceal something. Even with scrupulous work, crystal structure determinations are not always free of flaws (wrong space groups are a frequent error, resulting in unreliable atomic coordinates).

Nowadays, most data are deposited in databases. Good-working databases are very helpful, but in a way treacherous. Due to the rapid development of computer and data storage technologies, data need uninterrupted attention in order to be saved reliably. The financial breakdown of a database, computer viruses and other events can lead to a total loss of the data. Data that have been stored on other media (like CDs) remain legible only for the few years of existence of devices that can process them. Therefore, results of structure determinations, especially the atomic coordinates, should be published on a durable material in a way that can be read without the need for machines, and they should remain generally accessible without the need for a license for a database. It is an absurdity that scientific results are accessible only against the annually repeated payment of license fees even though they have been elaborated nearly exclusively by the employment of public funds. Another problem is the incomplete inclusion of the data in the databases. The *Inorganic Crystal Structure Database* (ICSD) certainly is a very useful tool, but on the one side it is incomplete, and on the other side it contains many duplicates and unfortunately also many errors.

References

General Literature, Text Books

[1] A. F. Wells, *Structural Inorganic Chemistry*, 5th ed. Clarendon, 1984.

[2] A.R. West, *Basic Solid State Chemistry*, 2nd ed. John Wiley & Sons, Ltd, 1999.

[3] L. Smart, E. Moore, *Solid State Chemistry, an Introduction*, Chapman & Hall, 1992.

[4] R. C. Ropp, *Solid State Chemistry*. Elsevier, 2003.

[5] G. M. Clark, *The Structures of Non-molecular Solids*. Applied Science Publishers, 1972.

[6] D. M. Adams, *Inorganic Solids*. John Wiley & Sons, Ltd, 1974.

[7] H. Krebs, *Fundamentals of Inorganic Crystal Chemistry*. McGraw-Hill, 1968.

[8] A. K. Cheetham, *Solid State Chemistry and its Applications*, John Wiley & Sons, Ltd, 1984.

[9] A. K. Cheetham, P. Day, *Solid State Chemistry: Compounds*. Oxford University Press, 1992.

[10] B. K. Vainshtein, V. M. Fridkin, V. L. Indenbom, *Modern Crystallography II: Structure of Crystals*, 3rd ed. Springer, 2000.

[11] K. Schubert, *Kristallstrukturen zweikomponentiger Phasen*. Springer, 1964.

[12] B. G. Hyde, S. Andersson, *Inorganic Crystal Structures*. John Wiley & Sons, Inc., 1989.

[13] M. O'Keeffe, B. G. Hyde, *Crystal Structures. I. Patterns and Symmetry*. Mineralogical Society of America, 1996.

[14] H. D. Megaw, *Crystal Structures, a Working Approach*. Saunders, 1973.

[15] D. L. Kepert, *Inorganic Stereochemistry*. Springer, 1982.

[16] E. Parthé, *Elements of Inorganic Structural Chemistry*. Sutter-Parthé, Petit-Lancy (Switzerland), 1990.

[17] D. M. P. Mingos, *Essential Trends in Inorganic Chemistry*. Oxford University Press, 1998.

[18] H. F. Franzen, *Physical Chemistry of Solids. Basic Principles of Symmetry and Stability of Crystalline Solids*. World Scientific, 1994.

[19] J. I. Gersten, F. W. Smith, *The Physics and Chemistry of Materials*. John Wiley & Sons, Inc., 2001.

Collections of Molecular and Crystal Structure Data

[20] R. W. G. Wyckoff, *Crystal Structures*, Vols. 1–6. John Wiley & Sons, Inc., 1962–1971.

[21] J. Donohue, *The Structures of the Elements*. John Wiley & Sons, Inc., 1974.

[22] *Strukturbericht* 1–7. Akademische Verlagsgesellschaft, 1931–1943. *Structure Reports* 8–58. Kluwer, 1956–1990. Collection of crystal structure data of one year, published annually.

[23] *Molecular Structures and Dimensions* (O. Kennard and others, eds.). Reidel, 1970ff.

[24] Landolt-Börnstein, *Numerical Data and Functional Relationships in Science and Technology*, New Series. Springer:
Group II, Vols. 7, 15, 21, 23, 25 A–D, Structural data of free polyatomic molecules (W. Martiensen, ed.), 1976–2003.
Group III, Vol. 6, Structural data of the elements and intermetallic phases (P. Eckerlin, H. Kandler, K. Hellwege, A. M. Hellwege, eds.), 1971.
Group III, Vols. 7; 43 A1, A2, Crystal structures of inorganic compounds (K. H. Hellwege, ed.; P. Villars, K. Cenzual, eds.), 1973–1978; 2004, 2005.
Group IV, Vol. 14, Zeolite-type crystal structures and their chemistry (W. H. Baur, R. X. Fischer, eds.), 2000, 2002, 2006.

[25] P. Villars, L. D. Calvert, *Pearson's Handbook of Crystallographic Data for Intermetallic Phases*, 2nd ed., Vols. 1–4. ASM International, 1991.
P. Villars, *Pearson's Handbook: Desk Edition of Crystallographic Data for Intermetallic Phases*. ASM International, 1998.

[26] J. L. C. Daams, P. Villars, J. H. N. van Vucht, *Atlas of Crystal Structure Types for Intermetallic Phases*, Vols. 1–4. ASM International, 1991.

[27] *Molecular Gas Phase Documentation* (MOGADOC). Chemieinformationssysteme, Universität Ulm, Germany. Electronic database of molecular structures in the gas phase. www.uni-ulm.de/strudo/mogadoc/.

[28] *Inorganic Crystal Structure Database* (ICSD). Fachinformationszentrum Karlsruhe, Germany, and National Institute of Standards and Technology, Gaithersburg, MD, USA. Electronic database of crystal structures of inorganic compounds. www.fiz-karlsruhe.de/ecid/Internet/en/DB/icsd/.

[29] *Cambridge Structural Database* (CSD). Cambridge Crystallographic Data Centre, University Chemical Laboratory, Cambridge, UK. Electronic database of crystal structures of organic and metallorganic compounds. www.ccdc.cam.ac.uk.

[30] *Metals Crystallographic Data File* (CRYSTMET). Toth Information Systems Inc., Ottawa, Canada. Electronic database of crystal structures of metals, intermetallic compounds and minerals. www.Tothcanada.com.

Literature to Individual Chapters

Chapter 2

[31] G. O. Brunner, D. Schwarzenbach, Zur Abgrenzung der Koordinationssphäre und Ermittlung der Koordinationszahl in Kristallstrukturen. *Z. Kristallogr.* **133** (1971) 127.

[32] F. L. Carter, Quantifying the concept of coordination number. *Acta Crystallogr.* **B 34** (1978) 2962.

[33] R. Hoppe, Effective coordination numbers and mean fictive ionic radii. *Z. Kristallogr.* **150** (1979) 23.

[34] R. Hoppe, The coordination number — an inorganic chameleon. *Angew. Chem. Int. Ed.* **9** (1970) 25.

[35] J. Lima-de-Faria, E. Hellner, F. Liebau, E. Makovicky, E. Parthé, Nomenclature of inorganic structure types. *Acta Crystallogr.* **A 46** (1990) 1.

[36] E. Parthé, L. M. Gelato, The standardization of inorganic crystal-structure data. *Acta Crystallogr.* **A 40** (1984) 169.

[37] S. W. Bailey, V. A. Frank-Kamentski, S. Goldsztaub, H. Schulz, H. F. W. Taylor, M. Fleischer, A. J. C. Wilson, Report of the International Mineralogical Association – International Union of Crystallography joint committee on nomenclature. *Acta Crystallogr.* **A 33** (1977) 681.

[38] J. Lima-de-Faria, Crystal chemical formulae for inorganic structure types. In: *Modern Perspectives in Inorganic Crystal Chemistry* (E. Parthé, ed.), p. 117. Kluwer, 1992.

[39] S. Andersson, A description of complex inorganic crystal structures. *Angew. Chem. Int. Ed.* **22** (1983) 69.

[40] S. Alvarez, Polyhedra in (inorganic) chemistry. *J. Chem. Soc. Dalton* **2005**, 2209.

Chapter 3

[41] S. F. A. Kettle, *Symmetry and Structure*, 2nd ed. John Wiley & Sons, Ltd, 1995.

[42] C. Hammond, *The Basics of Crystallography and Diffraction*, 2nd ed. Oxford University Press, 2001.

[43] W. Borchardt-Ott, *Crystallography*, 2nd ed. Springer, 1995.

[44] D. Schwarzenbach, *Crystallography.* John Wiley & Sons, Ltd, 1996.

[45] H. L. Monaco, D. Viterbo, F. Sordari, G. Grilli, G. Zanotti, M. Catti, *Fundamentals of Crystallography,* 2nd ed. Oxford University Press, 2002.

[46] B. K. Vainshtein, *Modern Crystallography I: Fundamentals of Crystals: Symmetry and Methods of Structural Crystallography.* 2nd ed. Springer, 1994.

[47] G. Burns, A. M. Glazer, *Space Groups for Solid State Scientists*, 2nd ed. Academic Press, 1990.

[48] *International Tables for Crystallography*, Vol. A: *Space-group Symmetry* (T. Hahn, ed.), 5th ed. Kluwer, 2002.

[49] *International Tables for Crystallography*, Vol. E: *Subperiodic Groups* (V. Kopský, D. B. Litvin, eds.). Kluwer, 2002.

[50] B. Grünbaum, G. C. Shephard, *Tilings and Patterns.* Freeman, 1987.

[51] A. I. Kitaigorodsky, *Molecular Crystals and Molecules.* Academic Press, 1973.

[52] A. Yamamoto, Crystallography of quasiperiodic crystals. *Acta Crystallogr.* **A 52** (1996) 509.

[53] S. v. Smaalen, An elementary introduction to superspace crystallography. *Z. Kristallogr.* **219** (2004) 681.

[54] W. Steurer, Twenty years of research on quasicrystals. *Z. Kristallogr.* **219** (2004) 391.

Chapter 4

[55] D. A. Porter, K. E. Easterling, *Phase Transformations in Metals and Alloys.* Chapman & Hall, 1992.

[56] *International Tables for Crystallography*, Vol. D: *Physical Properties of Crystals* (A. Authier, ed.), Chapter 3.1 (J.-C. Tolédano, V. Janovec, V. Kospký, J. F. Scott, P. Boček): Structural phase transitions. Kluwer, 2003.

[57] *Binary Alloy Phase Diagrams* (T. B. Massalski, H. Okamoto, P. R. Subramanian, L. Kapczak, eds.). ASM International, 1990.

[58] U. F. Petrenko, R. W. Whitworth, *Physics of Ice.* Oxford University Press, 1999.

Chapter 5

[59] A. J. Pertsin, A. I. Kitaigorodskii, *The Atom–Atom Potential Method.* Springer, 1987.

[60] T. C. Waddington, Lattice energies and their significance in inorganic chemistry. *Adv. Inorg. Chem. Radiochem.* **1** (1959) 157.

[61] M. F. C. Ladd, W. H. Lee, Lattice energies and related topics, *Prog. Solid State Chem.* **1** (1964) 37; **2** (1965) 378; **3** (1967) 265.

[62] R. Hoppe, Madelung constants as a new guide to the structural chemistry of solids. *Adv. Fluorine Chem.* **6** (1970) 387.

[63] A. Gavezzoti, The crystal packing of organic molecules. *Crystallogr. Rev.* **7** (1998) 5.

[64] A. Gavezzoti, Calculation of lattice energies of organic crystals. *Z. Kristallogr.* **220** (2005) 499.

Chapter 6

[65] A. Bondi, Van der Waals volumes and radii. *J. Phys. Chem.* **68** (1964) 441.

[66] S. C. Nyborg, C. H. Faerman, A review of van der Waals atomic radii for molecular crystals. I: N, O, F, S, Se, Cl, Br and I bonded to carbon. *Acta Crystallogr.* **B 41** (1985) 274. II: Hydrogen bonded to carbon. *Acta Crystallogr.* **B 43** (1987) 106.

[67] R. S. Rowland, R. Taylor, Intermolecular nonbonded contact distances in organic crystal structures. *J. Chem. Phys.* **100** (1996) 7384.

[68] S. Israel, R. Saravanan, N. Srinivasam, R. K. Rajaram, High resolution electron density mapping for LiF and NaF. *J. Phys. Chem. Solids* **64** (2003) 43.

[69] R. D. Shannon, Revised effective ionic radii and systematic studies of interatomic distances in halides and chalcogenides. *Acta Crystallogr.* **A 32** (1976) 751.

[70] R. D. Shannon, Bond distances in sulfides and a preliminary table of sulfide crystal radii. In: *Structure and Bonding in Crystals*, Vol. II (M. O'Keeffe, A. Navrotsky, eds.). Academic Press, 1981.

[71] G. P. Shields, P. R. Raithby, F. H. Allen, W. D. S. Motherwell, The assignment and validation of metal oxidation states in the Cambridge Structural Database. *Acta Crystallogr.* **B 46** (2000) 244.

[72] A. Simon, Intermetallic compounds and the application of atomic radii for their description. *Angew. Chem. Int. Ed.* **22** (1983) 95.

Chapter 7

[73] L. Pauling, *The Nature of the Chemical Bond*, 3rd ed. Cornell University Press, 1960.

[74] W. H. Baur, Bond length variation and distorted coordination polyhedra in inorganic crystals. *Trans. Am. Crystallogr. Assoc.* **6** (1970) 129.

[75] W. H. Baur, Interatomic distance predictions for computer simulation of crystal structures, in: *Structure and Bonding in Crystals*, Vol. II, S. 31 (M. O'Keeffe, M. Navrotsky, eds.). Academic Press, 1981.

[76] V. S. Urusov, I. P. Orlov, State-of-the-art and perspectives of the bond-valence model in inorganic chemistry. *Crystallogr. Rep.* **44** (1999) 686.

Chapter 8

[77] R. J. Gillespie, I. Hargittai, *VSEPR Model of Molecular Geometry.* Allyn & Bacon, 1991.

[78] R. J. Gillespie, P. L. P. Popelier, *Chemical Bonding and Molecular Geometry.* Oxford University Press, 2001.

[79] R. J. Gillespie, The VSEPR model revisited. *Chem. Soc. Rev.* **21** (1992) 59.

[80] R. J. Gillespie, E. A. Robinson, Electron domains and the VSEPR model of molecular geometry. *Angew. Chem. Int. Ed.* **35** (1996) 495.

[81] J. R. Edmundson, On the distribution of point charges on the surface of a sphere. *Acta Crystallogr.* **A 48** (1991) 60.

[82] M. Hargittai, Molecular structure of the metal halides. *Chem. Rev.* **100** (2000) 2233.

Chapter 9

[83] C. J. Ballhausen, *Ligand Field Theory.* McGraw-Hill, 1962.

[84] B. N. Figgis, M. A. Hitchman, *Ligand Field Theory and its Applications.* Wiley-VCH, 2000.

[85] I. B. Bersuker, *Electronic Structure and Properties of Transition Metal Compounds.* John Wiley and Sons, Inc., 1996.

[86] H. D. Flack, Chiral and achiral structures. *Helv. Chim. Acta* **86** (2003) 905.

Chapter 10

[87] R. Hoffmann, *Solids and Surfaces. A Chemist's View of Bonding in Extended Structures.* VCH, 1988.

[88] D. Pettifor, *Bonding and Structure of Molecules and Solids.* Clarendon, 1995.

[89] J. M. Burdett, *Chemical Bonding in Solids.* Oxford University Press, 1995.

[90] R. V. Dronskowski, *Computational Chemistry of Solid State Materials.* Wiley-VCH, 2005.

[91] J. A. Duffy, *Bonding, Energy Levels and Bands in Inorganic Chemistry.* Longman Scientific & Technical, 1990.

[92] H. Jones, *The Theory of Brillouin Zones and Electronic States in Crystals.* North Holland, 1962.

[93] P. A. Cox, *The Electronic Structure and Chemistry of Solids.* Clarendon, 1987.

[94] S. L. Altmann, *Band Theory of Solids. An Introduction from the Point of View of Symmetry.* Clarendon, 1991.

[95] I. D. Brown, *The Chemical Bond in Inorganic Chemistry.* Oxford University Press, 2002.

[96] J. M. Burdett, Perspectives in structural chemistry. *Chem. Rev.* **88** (1988) 3.

[97] A. Savin, R. Nesper, S. Wengert, T. F. Fässler, ELF: The electron localization function. *Angew. Chem. Int. Ed.* **36** (1997) 1808.

[98] T. F. Fässler, The role of non-bonding electron pairs in intermetallic compounds. *Chem. Soc. Rev.* **32** (2003) 80.

[99] G. A. Landrum, R. Dronskowki, The orbital origins of magnetism: from atoms to molecules to ferromagnetic alloys. *Angew. Chem. Int. Ed.* **39** (2000) 1560.

Chapters 11, 12

[100] H. Zabel, S. A. Solin, *Graphite Intercalation Compounds I; Structure and Dynamics.* Springer, 1990.

[101] H. W. Kroto, Buckminsterfullerene, a celestial sphere that fell to Earth. *Angew. Chem. Int. Ed.* **31** (1992) 111.

[102] C, N. R. Rao, A. Govindaraj, Carbon nanotubes from organometallic precursors. *Acc. Chem. Res.* **35** (2002) 998.

[103] H. Selig, L.B. Ebert, Graphite intercalation compounds. *Advan. Inorg. Chem. Radiochem.* **23** (1980) 281.

[104] M. Makha, A. Purich, C. L. Raston, A. N. Sbolev, Strucutral diversity of host–guest and intercalation complexes of fullerene C_{60}. *Eur. J. Inorg. Chem.* **2006**, 507.

[105] W. B. Pearson, The crystal structures of semiconductors and a general valence rule. *Acta Crystallogr.* **17** (1964) 1.

[106] R. Steudel, B. Eckert, Solid sulfur allotropes. *Top. Curr. Chem.* **230** (2003) 1.

[107] U. Schwarz, Metallic high-pressure modifications of main group elements. *Z. Kristallogr.* **219** (2004) 376.

[108] U. Häusermann, High-pressure structural trends of group 15 elements. *Chem. Eur. J.* **9** (2003) 1471.

[109] M. McMahon, R. Nelmes, Incommensurate crystal structures in the elements at high pressures. *Z. Kristallogr.* **219** (2004) 742.

[110] J. S. Tse, Crystallography of selected high pressure elemental solids. *Z. Kristallogr.* **220** (2005) 521.

Chapter 13

[111] H. G. v. Schnering, W. Hönle, Bridging chasms with polyphosphides, *Chem. Rev.* **88** (1988) 243.

[112] H. G. v. Schnering, Homonuclear bonds with main group elements. *Angew. Chem. Int. Ed.* **20** (1981) 33.

[113] W. S. Sheldrick, Network self-assembly patterns in main group metal chalcogenides. *J. Chem. Soc. Dalton* **2000** 3041.

[114] P. Böttcher, Tellurium-rich tellurides. *Angew. Chem. Int. Ed.* **27** (1988) 759.

[115] M. G. Kanatzidis, From cyclo-Te_8 to Te_x^{n-} sheets: are nonclassical polytellurides more classiscal than we thought? *Angew. Chem. Int. Ed.* **34** (1995) 2109.

[116] H. Schäfer, B. Eisenmann, W. Müller, Zintl-phases: intermediate forms between metallic and ionic bonding. *Angew. Chem. Int. Ed.* **12** (1973) 694.

[117] H. Schäfer, Semimetal clustering in intermatallic phases. *J. Solid State Chem.* **57** (1985) 97.

[118] R. Nesper, Structure and chemical bonding in Zintl phases containing lithium. *Progr. Solid State Chem.* **20** (1990) 1.

[119] R. Nesper, Chemical bonds — intermetallic compounds. *Angew. Chem. Int. Ed.* **30** (1991) 789.

[120] S. M. Kauzlarich (ed.), *Chemistry, Structure and Bonding of Zintl Phases and Ions*. Wiley-VCH, 1996.

[121] G. A. Papoian, R. Hoffmann, Hypervalent bonding in one, two and three dimensions: extending the Zintl–Klemm concept to nonclassical electron-rich networks. *Angew. Chem. Int. Ed.* **39** (2000) 2408.

[122] T. Fässler, S. D. Hoffmann, Endohedral Zintl ions: intermetallic clusters. *Angew. Chem. Int. Ed.* **43** (2004) 6242.

[123] R. J. Gillespie, Ring, cage, and cluster compounds of the main group elements. *Chem. Soc. Rev.* **1979**, 315.

[124] J. Beck, Rings, cages and chains – the rich structural chemitry of the polycations of the chalcogens. *Coord. Chem. Rev.* **163** (1997) 55.

[125] K. Wade, Structural and bonding patterns in cluster chemistry. *Adv. Inorg. Chem. Radiochem.* **18** (1976) 1.

[126] S. M. Owen, Electron counting in clusters: a view of the concepts. *Polyhedron* **7** (1988) 253.

[127] J. W. Lauher, The bonding capabilites of transition metals clusters. *J. Am. Chem. Soc.* **100** (1978) 5305.

[128] B. K. Teo, New topological electron-counting theory. *Inorg. Chem.* **23** (1984) 1251.

[129] D. M. P. Mingos, D. J. Wals, *Introduction to Cluster Chemistry*. Prentice-Hall, 1990.

[130] C. E. Housecroft, *Cluster Molecules of the p-Block Elements*. Oxford University Press, 1994.

[131] G. González-Moraga, *Cluster Chemistry*. Springer, 1993.

[132] D. M. P. Mingos (ed.), Structural and electronic paradigms in cluster chemistry. *Struct. and Bonding* **93** (1999).

[133] D. M. P. Mingos, T. Slee, L. Zhenyang, Bonding models for ligated and bare clusters. *Chem. Rev.* **90** (1990) 383.

[134] J. D. Corbett, Polyatomic Zintl anions of the post-transition elements. *Chem. Rev.* **85** (1985) 383.

[135] R. Chevrel, in: *Superconductor Materials Sciences Metallurgy, Fabrication and Applications* (S. Foner, B. B. Schwarz, eds.), Chap. 10. Plenum Press, 1981.

[136] R. Chevrel, Chemistry and structure of ternary molybdenum chalcogenides. *Top. Curr. Phys.* **32** (1982) 5.

[137] R. Chevrel, Cluster solid state chemistry. In: *Modern Perspectives in Inorganic Crystal Chemistry* (E. Parthé, ed.), p. 17. Kluwer, 1992.

[138] J. D. Corbett, Polyanionic clusters and networks of the early *p* metals in solids: beyond the Zintl frontier. *Angew. Chem. Int. Ed.* **39** (2000) 670.

[139] J.-C. P. Gabriel, K. Boubekeur, S. Uriel, O. Batail, Chemistry of hexanuclear rhenium chalcogenide clusters. *Chem. Rev.* **101** (2001) 2037.

[140] A. Simon, Condensed metal clusters. *Angew. Chem. Int. Ed.* **20** (1981) 1.

[141] A. Simon, Clusters of metals poor in valence electrons – structures, bonding, properties. *Angew. Chem. Int. Ed.* **27** (1988) 159.

[142] J.D. Corbett, Extended metal metal bonding in halides of the early transition metals. *Acc. Chem. Res.* **14** (1981) 239.

[143] M. Ruck, From metals to molecules – ternary subhalides of bismuth. *Angew. Chem. Int. Ed.* **40** (2001) 1182.

[144] J. D. Corbett, Structural and bonding principles in metal halide cluster chemistry. In: *Modern Perspectives in Inorganic Crystal Chemistry* (E. Parthé, ed.), p. 27. Kluwer, 1992.

[145] T. Hughbanks, Bonding in clusters and condesed cluster compounds that extend in one, two and three dimensions. *Prog. Solid State Chem.* **19** (1990) 329.

[146] A. Schnepf, H. Schnöckel, Metalloid aluminium and gallium clusters: element modifications on a molecular scale? *Angew. Chem. Int. Ed.* **41** (2002) 3532.

Chapters 14, 15

[147] W. Hume-Rothery, R. E. Sallmann, C. W. Haworth, *The Structures of Metals and Alloys*, 5th ed. Institute of Metals, 1969.

[148] W. B. Pearson, *The Crystal Chemistry and Physics of Metals and Alloys.* John Wiley and Sons, Inc., 1972.

[149] F. C. Frank, J. S. Kasper, Complex alloy structures regarded as sphere packings. I: Definitions and basic principles, *Acta Crystallogr.* **11** (1958) 184. II: Analysis and classification of representative structures, *Acta Crystallogr.* **12** (1959) 483.

[150] W. B. Holzapfel, Physics of solids under strong compression. *Rep. Prog. Phys.* **59** (1996) 29.

[151] R. L. Johnston, R. Hoffmann, Structure bonding relationships in the Laves phases. *Z. Anorg. Allg. Chem.* **616** (1992) 105.

Chapters 16, 17

[152] O. Muller, R. Roy, *The Mayor Ternary Structural Families.* Springer, 1974.

[153] D. J. M. Bevan, P. Hagenmuller, *Nonstoichiometric Compounds: Tungsten Bronzes, Vanadium Bronzes and Related Compounds.* Pergamon, 1975.

[154] K. Wold, K. Dwight, *Solid State Chemistry – Synthesis, Structure and Properties of Selected Oxides and Sulfides.* Chapman & Hall, 1993.

[155] F. Liebau, *Structural Chemistry of Silicates.* Springer, 1985.

[156] J. Lima-de-Faria. *Structural Mineralogy.* Kluwer, 1994.

[157] D. W. Breck, *Zeolite Molecular Sieves.* John Wiley and Sons, Inc., 1974.

[158] T. Lundström, Preparation and crystal chemistry of some refractory borides and phosphides. *Ark. Kemi* **31** (1969) 227.

[159] P. Hagenmuller, Les bronzes oxygénés, *Prog. Solid State Chem.* **5** (1971) 71.

[160] M. Greenblatt, Molybdenum oxide bronzes with quasi low-dimensional properties. *Chem. Rev.* **88** (1988) 31.

[161] M. Figlharz, New oxides in the MoO_3–WO_3 system. *Prog. Solid State Chem.* **19** (1989) 1.

[162] F. Hulliger, Crystal chemistry of the chalcogenides and pnictides of the transition metals. *Struct. Bonding* **4** (1968) 82.

[163] S. C. Lee and R. H. Holm, Nonmolecular metal chalcogenide/halide solids and their molecular cluster analogues. *Angew. Chem. Int. Ed.* **29** (1990) 840.

[164] A. Kjeskhus, W. B. Pearson, Phases with the nickel arsenide and closely-related structures, *Prog. Solid State Chem.* **1** (1964) 83.

[165] D. Babel, A. Tressaud, Crystal chemistry of fluorides. In: *Inorganic Solid Fluorides* (P. Hagenmuller, ed.). Academic Press, 1985.

[166] G. Meyer, The syntheses and structures of complex rare-earth halides, *Progr. Solid State Chem.* **14** (1982) 141.

[167] D. M. P. Mingos (ed.), Bonding and charge distribution in polyoxometalates. *Struct. Bonding* **87** (1997).

[168] D. G. Evans, R. C. T. Slade, Structural aspects of layered double hydroxides. *Struct. Bonding* **119** (2000) 1.

[169] M.T. Pope, A. Müller, Chemistry of polyoxometallates: variations of an old theme with interdisciplinary relations. *Angew. Chem. Int. Ed.* **30** (1991) 34.

[170] A. Müller, F. Peters, M. T. Pope, D. Gatteschini, Polyoxometallates: very large clusters – nanoscale magnets. *Chem. Rev.* **98** (1998) 239.

[171] H. Müller-Buschbaum, The crystal chemistry of copper oxometallates. *Angew. Chem. Int. Ed.* **30** (1991) 723.

[172] H. Müller-Buschbaum, Zur Kristallchemie der Oxoargentate und Silberoxometallate, *Z. Anorg. Allg. Chem.* **630** (2004) 2175; — Oxoplatinate, **630** (2004) 3; — Oxopalladate, **630** (2004) 339; — Oxoiridate, **631** (2005) 1005.

[173] W. Schnick, Solid state chemistry of nonmetal nitrides. *Angew. Chem. Int. Ed.* **32** (1993) 806.

[174] W. Bronger, Complex transition metal hydrides, *Angew. Chem. Int. Ed.* **30** (1991) 759.

[175] B. Krebs, Thio and seleno compounds of main group elements – new inorganic oligomers and polymers. *Angew. Chem. Int. Ed.* **22** (1983) 113.

[176] B. Krebs and G. Henkel, Transition metal thiolates — from molecular fragments of sulfidic solids to models of active centers in biomolecules. *Angew. Chem. Int. Ed.* **30** (1991) 769.

[177] K. Mitchell, J. A. Ibers, Rare-erath transition-metal chalcogenides. *Chem. Rev.* **102** (2002) 1929.

[178] J. V. Smith, Topochemistry of zeolite and related materials. *Chem. Rev.* **88** (1988) 149.

[179] M. T. Telly, Where zeolites and oxides merge: semi-condensed tetrahedral frameworks. *J. Chem. Soc. Dalton Trans.* **2000**, 4227.

[180] A. K. Cheetham, G. Férey, L. Loiseau, Inorganic materials with open frameworks. *Angew. Chem. Int. Ed.* **38** (1999) 3268.

Chapter 18

[181] *International Tables for Crystallography*, Vol. A1: *Symmetry Relations between Space Groups.* (H. Wondratschek, U. Müller, eds.). Kluwer, 2004.

[182] H. Bärnighausen, Group–subgroup relations between space groups: a useful tool in crystal chemistry. *MATCH, Commun. Math. Chem.* **9** (1980) 139.

[183] G. C. Chapuis, Symmetry relationships between crystal structures and their practical applications. In: *Modern Perspectives in Inorganic Crystal Chemistry* (E. Parthé, ed.), p. 1. Kluwer, 1992.

[184] U. Müller, Kristallographische Gruppe-Untergruppe-Beziehungen und ihre Anwendung in der Kristallchemie. *Z. Anorg. Allg. Chem.* **630** (2004) 1519.

[185] G. O. Brunner, An unconventional view of the closest sphere packings. *Acta Crystallogr.* **A 27** (1971) 388.

[186] *International Tables for Crystallography*, Vol. D: *Physical Properties of Crystals* (A. Authier, ed.), Chap. 3.1 (J.-C. Tolédano, V. Janovec, V. Kospký, J. F. Scott, P. Boěk): Structural phase transitions; Chap. 3.2 (V. Janovec, Th. Hahn, H. Klapper): Twinning and domain structures. Kluwer, 2003.

Chapter 19

[187] C. N. R. Rao, B. Raveau, *Transition Metal Oxides; Structures, Properies, and Synthesis of Ceramic Oxides.* Wiley, 1998.

[188] R. E, Newnham, Properties of Materials. Oxford University Press, 2004.

[189] High-performance non-oxidic ceramics I, II (M. Jansen, ed.). *Struct. Bonding* **101**, **102** (2002).

[190] R. L. Carlin, *Magnetochemistry.* Springer, 1986.

[191] A. F. Orchard, *Magnetochemistry.* Oxford University Press, 2003.

[192] J. Parr, *Magnetochemistry.* Oxford University Press, 1999.

[193] D. W. Bruce, D. O'Hare, *Inorganic Materials.* Wiley-VCH, 1992.

[194] W. Bronger, Ternary sulfides: a model case of the relation between crystal structure and magnetism. *Angew. Chem. Int. Ed.* **20** (1981) 52.

Chapter 20

[195] G. A. Ozin, A. C. Arsenault, *Nanochemistry.* RSC Publishing, 2005.

[196] C. N. R. Rao, A. Müller, A. K. Cheetham (eds.), *The Chemistry of Nanomaterials.* Wiley-VCH, 2004.

[197] G. Schmid (ed.), *Nanoparticles.* Wiley-VCH, 2004.

[198] H. Dai, Carbon nanotubes: synthesis, integration, and properties. *Acc. Chem. Res.* **35** (2002) 1035.

[199] G. Patzke, F. Krumeich, R. Nesper, Oxidic nanotubes and nanorods – anisotropic modules for a future nanotechnology. *Angew. Chem. Int. Ed.* **41** (2002) 2462.

[200] J. Hu, T. W. Odom, C. M. Lieber, Chemistry and physics in one dimension: synthesis, and properties of nanowires and nanotubes. *Acc. Chem. Res.* **32** (1999) 435.

[201] G. J. de A. Soler-Illia, C. Sánchez, B. Lebeau, J. Patari, Chemical strategies to design textured materials: from microporous and mesoporous oxides to nanonetworks and hierarchical structures. *Chem. Rev.* **102** (2002) 4093.

[202] C. Burda, X. Chen, R. Narayanan, M. A. El-Sayed, Chemistry and properties of nanocrystals of different shapes. *Chem. Rev.* **105** (2005) 1025.

[203] C. N. R. Rao, Inorganic nanotubes. *J. Chem. Soc. Dalton* **2003**, 1.

[204] B. L. Cushing, V. L. Koesnichenko, C. J. O'Connor, Recent andvances in the liquid-phase syntheses of inorganic nanoparticles. *Chem. Rev.* **104** (2004) 3093.

[205] E. A. Turner, Y. Huang, J. F. Corrigan, Synthetic routes to the encapsulation of II-VI semiconductores in mesoporous hosts. *Eur. J. Inorg. Chem.* **2005**, 4465.

[206] R. Tenne, Advances in the synthesis of inorganic nanotubes and fullerene-like nanoparticles. *Angew. Chem. Int. Ed.* **42** (2003) 5124.

[207] H. Cölfen, M. Antonietti, Mesocrystals: inorganic superstructures by highly parallel crystallization and controlled alignment. *Angew. Chem. Int. Ed.* **44** (2005) 5576.

[208] F. Hoffmann, M. Cornelius, J. Morell, M. Fröba, Silica-based mesoporous organic–inorganic hybrid materials on the basis of silicates. *Angew. Chem. Int. Ed.* **45** (2006) 3216.

Answers to the Problems

2.1 (a) 5.18; (b) 5.91; (c) 12.53.
2.2 (a) $Fe^{o}Ti^{o}O_3^{[2n,2n]}$ or $Fe^{[6o]}Ti^{[6o]}O_3^{[2n,2n]}$;
(b) $Cd^{o}Cl_2^{[3n]}$; (c) $Mo^{[6p]}S_2^{[3n]}$; (d) $Cu_2^{[2l]}O^{t}$;
(e) $Pt^{[4l]}S^{[4t]}$ or $Pt^{s}S^{t}$; (f) $Mg^{[16FK]}Cu_2^{i}$;
(g) $Al_2^{o}Mg_3^{do}Si_3^{o}O_{12}$; (h) $U^{[6p3c]}Cl_3^{[3n]}$.
2.3 CaC_2, *I*; K_2PtCl_6, *F*; cristobalite, *F*; $CuAu_3$, *P*; K_2NiF_4, *I*; perovskite, *P*.
2.4 CsCl, 1; ZnS, 4; TiO_2, 2; $ThSi_2$, 4; ReO_3, 1; α-$ZnCl_2$, 4.
2.5 271.4 pm.
2.6 I(1)–I(2) 272.1 pm; I(2)–I(3) 250.0 pm; angle 178.4°.
2.7 210.2 and 213.2 pm; angle 101.8°.
2.8 W=O 177.5 pm; W···O 216.0 pm; W–Br 244.4 pm; angle O=W–Br 97.2°; the coordination polyhedron is a distorted octahedron (*cf.* Fig. 19.5, p. 230).
2.9 Zr–O(1), 205.1, 216.3 and 205.7 pm; Zr–O(2), 218.9, 222.0, 228.5 and 215.1 pm; c.n. 7.

3.1 H_2O, $mm2$; $HCCl_3$, $3m$; BF_3, $\bar{6}2m$; XeF_4, $4/m\,2/m\,2/m$ (for short $4/mmm$); $ClSF_5$, $4mm$; SF_6, $4/m\,\bar{3}\,2/m$ (short $m\bar{3}m$); *cis*-$SbF_4Cl_2^-$, $mm2$; *trans*-N_2F_2, $2/m$; $B(OH)_3$, $\bar{6}$; $Co(NO_2)_6^{3-}$, $2/m\bar{3}$ (short $m\bar{3}$).
3.2 Si_4^{6-}, $mm2$; As_4S_4, $\bar{4}2m$; P_4S_3, $3m$; Sn_5^{2-}, $\bar{6}2m$; As_4^{6-}, 1; As_4^{4-}, $4/m\,2/m\,2/m$ (short $4/mmm$); P_6^{6-}, $\bar{3}\,2/m\,1$ (short $\bar{3}m$); As_7^{3-}, $3m$; P_{11}^{3-}, 3; Sn_9^{2-}, $4mm$; Bi_8^{2+}, $\bar{8}2m$.
3.3 Linked tetrahedra, $2/m\,2/m\,2/m$ (short mmm) and $mm2$; linked octahedra, $4/m\,2/m\,2/m$ (short $4/mmm$), $\bar{8}2m$ and 2.
3.4 Referred to the direction of the chain: translation, 2_1 axis, and one mirror plane through each O atom (the mirror planes are perpendicular to the direction of reference); referred to the normal on the plane of the paper: mirror plane and one 2 axis through each Hg atom; referred to the direction vertical in the plane of the paper: glide plane and one 2 axis through each O atom; one inversion center in every Hg atom. If we define a coordinate system **a**,**b**,**c** with **a** perpendicular to the plane of the paper and **c** = translation vector, the Hermann–Mauguin symbol is $P(2/m\,2/c)2_1/m$; the parentheses designate the directions for which there is no translation symmetry. That is a rod group.
3.5 Hexagonal M_xWO_3, $P6/m\,2/m\,2/m$ (this is an idealized symmetry; actually the octahedra are slightly tilted and the real space group is $P6_322$); tetragonal M_xWO_3, $P4/m\,2_1/b\,2/m$ (short $P4/mbm$); CaC_2, $I4/m\,2/m\,2/m$ (short $P4/mmm$); CaB_6, $P4/m\,\bar{3}\,2/m$ (short $Pm\bar{3}m$).
3.6 (a) $2/m\,2/m\,2/m$ (short mmm), orthorhombic; (b) $4/mmm$, tetragonal; (c) $\bar{3}\,2/m$ (short $\bar{3}m$), trigonal; (d) $2/m$, monoclinic; (e) $6/m$, hexagonal; (f) 622, hexagonal; (g) 222, orthorhombic; (h) $mm2$, orthorhombic; (i) $4/m\,\bar{3}\,2/m$ (short $m\bar{3}m$), cubic.
3.7 Ti, $2a$, mmm; O, $4f$, $m2m$. There are 2 Ti and 4 O atoms in the unit cell (Z = 2)
3.8 One layer of tiles, $\overline{10}\,2m$; two layers, $\bar{5}\,2/m$ (short $\bar{5}m$).

4.1 β-Cristobalite could be converted to α and β-quartz.
4.2 At 1000 °C recrystallization will be faster.
4.3 BeF_2.
4.4 First order (hysteresis observed).
4.5 The coordination numbers of the atoms increase from 6 to 8 in the NaCl → CsCl type conversion; therefore, it is a reconstructive phase transition which can only be a first-order transition.
4.6 At −10°C, ice will melt at appproximately 100 MPa and refreeze at approximately 450 MPa, forming modification V. This will transform to ice VI at ~600 MPa, then to ice VIII at ~2.2 GPa and to ice VII at ~18 GPa.
4.7 Water will freeze at 40 °C at appproximately 1.2 GPa forming modification VI. This will transform to ice VII at 2 GPa.
4.8 $H_2O{\cdot}HF$ will crystallize, then, in addition, H_2O will freeze at −72°C.
4.9 At approximately 0.5 GPa β-quartz will transform directly to β-cristobalite.

5.1 $-\frac{8}{1}+\frac{6\sqrt{3}}{2}+\frac{12\sqrt{3}}{2\sqrt{2}}-\frac{24\sqrt{3}}{\sqrt{10}}$.
5.2 (a) 687 $kJ\,mol^{-1}$; (b) 2965 $kJ\,mol^{-1}$; (c) 3517 $kJ\,mol^{-1}$.

6.1 F···F in SiF_4 253 pm, van der Waals distance 294 pm; Cl···Cl in $SiCl_4$ 330 pm, van der Waals distance 350 pm; I···I in SiI_4 397 pm, van der Waals distance 396 pm; in SiF_4 and $SiCl_4$ the halogen atoms are squeezed.
6.2 WF_6 193, WCl_6 241, PCl_6^- 219, PBr_6^- 234, SbF_6^- 193, MnO_4^{2-} 166 pm; TiO_2 201, ReO_3 195, EuO 257, $CdCl_2$ 276 pm.

7.1 (a) Rutile; (b) rutile; (c) neither (GeO_2 is actually polymorphic, adopting the rutile and the quartz structures); (d) anti-CaF_2.
7.2 Mg^{2+} c.n. 8, Al^{3+} c.n. 6, Si^{4+} c.n. 4 (interchange of the c.n. of Mg^{2+} and Si^{4+} would also fulfill PAULING's rule, but c.n. 8 is rather improbable for Si^{4+}).
7.3 Since all cations have the same charge (+3), the electrostatic valence rule is of no help. The larger Y^{3+} ions will take the sites with c.n. 8.
7.4 No.
7.5 N is coordinated to Ag.
7.6 $s(Rb^+)=\frac{1}{10}$; $s(V^{4+})=\frac{4}{5}$ $s(V^{5+})=\frac{5}{4}$;
$p_1=1.20$; $p_2=2.25$; $p_3=2.70$; $p_4=1.55$;

Inorganic Structural Chemistry, Second Edition Ulrich Müller

$\bar{p}(V^{4+}) = 2.04$; $\bar{p}(V^{5+}) = 2.19$; Expected bond lengths: V^{4+}–O(1) 159 pm, V^{4+}–O(2) 197 pm, V^{5+}–O(2) 173 pm, V^{5+}–O(3) 180 pm, V^{5+}–O(4) 162 pm.

8.1 Linear: $BeCl_2$, Cl_3^-; angular: O_3^- (radical), S_3^{2-}; trigonal planar: BF_3; trigonal pyramidal PF_3, $TeCl_3^+$; T shaped: BrF_3, XeF_3^+; tetrahedral: $GeBr_4$, $AsCl_4^+$, $TiBr_4$, O_3BrF; square planar: ICl_4^-; trigonal bipyramidal with a missing equatorial vertex: SbF_4^-, $O_2ClF_2^-$ (F axial); trigonal bipyramid: $SbCl_5$, $SnCl_5^-$, O_2ClF_3 and O_3XeF_2 (O equatorial); square pyramidal: TeF_5^-; octahedral: $ClSF_5$.

8.2

Ta_2I_{10} like Nb_2Cl_{10} (*cf.* p. 66).

8.3 Trigonal bipyramid, CH_2 group in equatorial position perpendicular to equatorial plane. Derive it from an octahedron with bent S=C bonds.

8.4 (a) $SF_2 < SCl_2 < S_3^{2-} < S_3^- \approx OF_2$; (b) $H_3CNH_2 < [(H_3C)_2NH_2]^+$; (c) $PCl_2F_3 < PCl_3F_2$ (= 180°).

8.5 Bond lengths Al–Cl(terminal) < Al–Cl(bridge); angles Cl(bridge)–Al–Cl(bridge) ≈ 95°
< Cl(bridge)–Al–Cl(terminal) ≈ 110°
< Cl(terminal)–Al–Cl(terminal) ≈ 120°.

8.6 $SnCl_3^-$; PF_6^-; $SnCl_6^{2-}$.

8.7 $BiBr_5^{2-}$ and TeI_6^{2-}.

9.1 $[Cr(OH_2)_6]^{2+}$, $[Mn(OH_2)_6]^{3+}$, $[Cu(NH_3)_6]^{2+}$.

9.2 $CrCl_4^-$ and $NiBr_4^{2-}$, elongated tetrahedra; $CuBr_4^{2-}$, flattened tetrahedron; $FeCl_4^{2-}$ could be slightly distorted.

9.3 Tetrahedral: $Co(CO)_4^-$, $Ni(PF_3)_4$, $Cu(OH)_4^{2-}$ (distorted); square: $PtCl_2(NH_3)_2$, $Pt(NH_3)_4^{2+}$, Au_2Cl_6.

9.4 $PtCl_2(NH_2)_2(NO_2)_2$, point group 1;
$[Co(H_2N(CH_2)_2NH_2)_3]^{3+}$, point group 3 2;
$[Rh(SO_2(NH)_2)_2(H_2O)_2]^-$, point group 2;
in all three cases no inversion axes are present (including m and $\bar{1}$).

9.5 (a) 2; (b) 1; (c) 2; (d) 2; (e) 1.

10.1 The band will broaden and the DOS will decrease.

10.2 It would look like the right part of Fig. 10.7.

10.3 The s band, the p_y band and the p_z band will shift to lower energy values at Γ and X', and to higher values at X and M; the p_x band will shift to higher values at Γ and X', and to lower values at X and M.

12.1 Shorter, BeO, BN; equal, BeS, BP, AlN; longer, AlP.

12.2 Longer bonds.

12.3 Under pressure AgI could adopt the NaCl structure (it actually does).

12.4 3.

12.5 Hg_2C should have the Cu_2O structure.

13.1 (a) Simple ionic; (b) polyanionic; (c) polyanionic; (d) polyanionic; (e) polycationic; (f) polyanionic; (g) polycationic; (h) simple ionic.

13.2 (a), (b), (d).

13.3

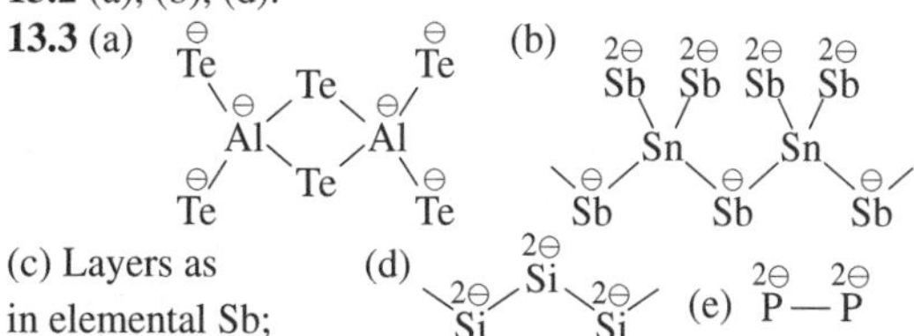

13.4 (a) Wade (26 e^-); (b) electron precise (84 e^-);
(c) 3*c*2*e* (56 e^-); (d) electron precise (72 e^-);
(e) Wade (86 e^-).

14.1 (a) *hhccc* or 41; (b) *hhhc* or 211.

14.2 (a) *ABACBC*; (b) *ABCACABCBCAB*;
(c) *ABCBABACAB*.

15.1 (a) No (because of different structures); (b) yes; (c) no; (d) no; (e) no; (f) yes; (g) yes; (h) no.

15.2

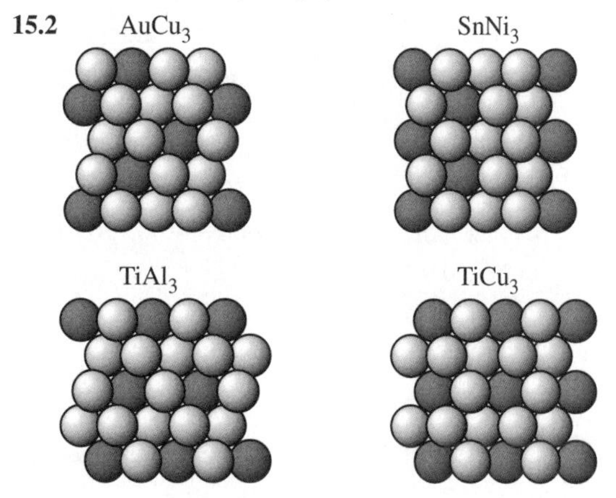

15.3 (a) CaF_2; (b) MgAgAs; (c) $MgCu_2Al$.

15.4 In both compounds each of the elements occupies two of the four different positions, but with different multiplicities: 3Cu:2Cu:2Zn:6Zn and 3Cu:2Al:2Al:6Cu.

15.5 Yes.

16.1 WO_3.

16.2 MX_4; this is the structure of a form of $ReCl_4$.

16.3 MX_4.
16.4 MoI_3 and TaS_3^{2-}.
16.5 Cristobalite.

17.1 MX_2.
17.2 Face-sharing octahedra occur only in hexagonal closest-packing.
17.3 TiN, NaCl type; FeP, MnP type; FeSb, NiAs type; CoS, NiP type; CoSb, NiAs type.
17.4 In $CaBr_2$ and RhF_3 there is a three-dimensional linking of the octahedra; CdI_2 and BiI_3 consist of layers that can mutually be displaced.
17.5 One-seventh.
17.6 MgV_2O_4, normal; VMg_2O_4, inverse; $NiGa_2O_4$, inverse; $ZnCr_2S_4$, normal; $NiFe_2O_4$, inverse.

18.1 (a) 3_1 and 3_2. (b) 3_2. (c) 2_1 axes cannot be retained upon doubling of *b*; after enlargement by an odd number 2_1 axes remain possible; therefore, *klassengleiche* subgroups after triplication are possible.
18.2 (a) *Klassengleiche*; (b) *translationengleiche*; (c) *klassengleiche*; (d) isomorphic; (e) *translationengleiche*; (f) *translationengleiche*; (g) *translationengleiche*.
18.3

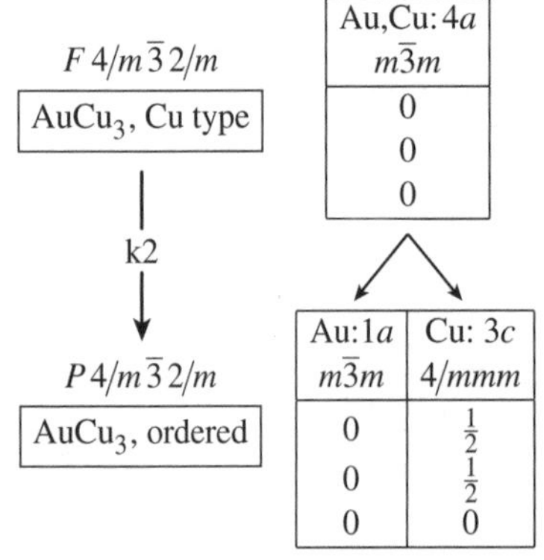

$AuCu_3$ will not form twins (k2 relation).

18.4

$P\,4/m\,\bar{3}\,2/m$ — perovskite

Ti:1*a*	Ca:1*b*	O: 3*d*
$m\bar{3}m$	$m\bar{3}m$	4/*mmm*
0	$\frac{1}{2}$	0
0	$\frac{1}{2}$	0
0	$\frac{1}{2}$	$\frac{1}{2}$

k2; 2**a**, 2**b**, 2**c**; $\frac{1}{2}x, \frac{1}{2}y, \frac{1}{2}z$

$F\,4/m\,\bar{3}\,2/m$ — elpasolite

Na:4*a*	Al: 4*b*	K:8*c*	F:24*e*
$m\bar{3}m$	$m\bar{3}m$	$\bar{4}3m$	4*mm*
0	$\frac{1}{2}$	$\frac{1}{4}$	0
0	$\frac{1}{2}$	$\frac{1}{4}$	0
0	$\frac{1}{2}$	$\frac{1}{4}$	0.25

18.5

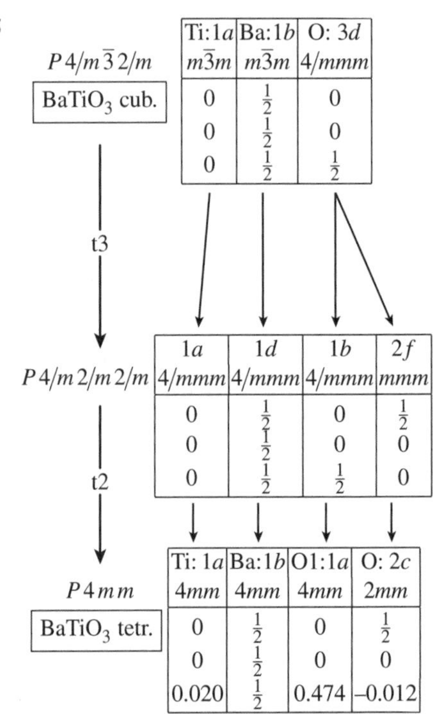

One can expect the formation of six kinds of domains, *i.e.* twins of triplet twins, due to the t3 and t2 subgroup relations.
18.6 β-Sn should form twinned crystals with domains in three orientations.
18.7 Twins with two kinds of domains will be formed (Fig. 19.4, p. 230).

Index